# THE EVERLASTING
# UNIVERSE

# THE EVERLASTING UNIVERSE

## READINGS ON THE ECOLOGICAL REVOLUTION

EDITED BY

**LORNE J. FORSTNER**
*University of Michigan*

**JOHN H. TODD**
*Woods Hole Oceanographic Institution*

D. C. HEATH AND COMPANY
Lexington, Massachusetts

Illustrations by Robert and Marilyn Dustin

Published simultaneously in Canada.

Printed in the United States of America.

Library of Congress Catalog Card Number: 78-145695

# PREFACE

The present concern for the state of the environment extends beyond the immediate question of survival to involve man once again in that ancient but always intriguing discussion of his relation to nature. Thus by means of this collection we have attempted to place a crisis in the context of time—past, present, and future. The focus of the book is the ecological eleventh hour; however, in order to present it in the wisest way possible and to keep the door to survival open with honest, constructive thought, we have attempted to remain free from the somewhat hysterical or melodramatic rhetoric that has recently engulfed the subject. This in no way implies that we are not dreadfully alarmed by the present condition of our environment. For us the anonymous line from *Newsweek*—"Death is nature's way of telling us to slow down"—has lost what comic value it might have had originally. Like so many other people in the world, we are, in all candor, damned frightened. But at the same time, we recognize that what has been referred to as the rhetoric of revolution has spread to include other areas where it is all too easy to become indignant and often irrational. Recently a reviewer of a protest drama wrote: "During the evening they spew rage at everything from air pollution to Miss America—very tired stuff, which has become easier to sleep through, than to sit through." Indignation, rage, indiscrete rhetoric of any variety eventually is reduced in the public mind to the level of futile gesture. We cannot afford the waste that comes with rash exhortation, public cant, or weak writing of any description. It goes without saying too that we cannot afford to sleep through the drama we now entitle Ecological Disaster.

Ordinarily, one might feel that a rhetoric text on a single theme would be of limited value for classroom use. However, as the reader will realize, especially as he examines the opening section which attempts a definition of ecology and man's role in nature, this particular subject organically radiates through both time and space and in the pattern of the eco-system itself emanates from and returns to a central core—the principle of life. Such topics as man's paradoxical relation to nature, the limitations and possibilities of rational thought, architecture, evolution (biological, ethical, social), political action, educational reform, mythology, consciousness, the structure of the family, the use of drugs, changes in economic theories, and the role of the female in society are integral parts of this single theme. As already mentioned, the question is much larger than the biological issue of survival.

What makes the environmental crisis so exciting (or would make it exciting if it were not so terrifying) is that it involves us in a way of thinking which critically examines so many of our assumptions about life and the way we live it. Above all, the ecologist is interested in overhauling our thought processes, which he hopes will lead to a new set of perspectives offering greater promise for happiness. It is this understanding of the problem that has dictated the organization of this collection of works by scholars and artists. The opening section of this text is an attempt at defining what is meant by ecological thought. The second section provides the historical background for the present crisis by reviewing man's past understanding of this relation to nature. In the third section an attempt is made to select a representative sampling of the various issues involved. Since it is not possible to present all the features of this complex subject, we have tried to emphasize in this section the importance of good writing for any cause. Everything in the first three sections serves as a prologue to "Toward a Future," the concluding section of this anthology. At no time in history has man had the capability to control his future as he now has. He has tremendous advantages in terms of information and the tools to use that information. All that remains in doubt is his willingness to sacrifice certain things in order to bring about change.

Finally, a word should be given to the way in which this collection was compiled. Professor Todd is an ecologist with a wide background in the natural sciences. He has done work in the Arctic, studied under Marston Bates at the University of Michigan, taught biology at San Diego State University, and is presently engaged in research at the Woods Hole Oceanographic Institution. He has also been actively engaged in an experiment in living according to the best ecological principles. It has been his job to select articles from the field of science. Professor Forstner, on the other hand, has been responsible for those selections that come from related fields and for judging the material in terms of its use in writing classes. One goal has been kept in mind throughout the evolution of this text: to provide the largest possible variety in prose styles. At all times we have tried to act in accordance with the wisdom found in J. Robert Oppenheimer's remark that "it is above all style through which power defers to reason."

Lorne J. Forstner
John H. Todd

# CONTENTS

# III THE CRISIS

# IV  TOWARD A FUTURE

# THE EVERLASTING
# UNIVERSE

# 1 ECOLOGY:
# A DEFINITION

# INTRODUCTION

From the time when man first questioned the construction of his universe, he endorsed, in theory at least, the notion of organic form. One finds, for example, in a book like the Bible or in a poem such as Dante's *The Divine Comedy* the concept represented in terms of a woman's body or an exfoliating rose. The Romantic poets of the early nineteenth century were so impressed with the image of an organically constructed world that one of them, the poet-philosopher Samuel Taylor Coleridge, could excitedly speculate:

> And what if all of animated nature
> Be but organic Harps diversely fram'd,
> That tremble into thought, as o'er them sweeps
> Plastic and vast, one intellectual breeze,
> At once the Soul of each, and God of all?

The great discoveries in the field of optics lent emphasis to the idea and in a sense confirmed that that which was previously imagined was, in fact, true. The telescope revealed the concentric and unfolding nature of the universe while the microscope explored phenomenal reality and discovered worlds within worlds.

Curiously and perhaps tragically, while endorsing the concept, man in his active life has never really subscribed to it. In other words, man recognized the truth of the organically constituted universe, but he has not solved the problem of how to behave in it. When it came to practical matters, a world organized along more simple patterns was found to be preferable because of its obvious convenience. The difficulty of living in an environment in which everything and everybody is dependent upon something else for its own being has, it appears, been found too great.

Now, however, the sad light of our own cowardice and recklessness has revealed the depths of our folly, and we are becoming aware of the organic composition of the world we share. The study of ecology specifically begins with the notion of the "eco-system," and one of ecology's major goals for the immediate future is to impress upon man the importance of what the poet knew so long ago—that the world is a web of interdependent forces involving us in a mutual assumption of responsibility. Either we apprehend the world as an eco-system and assume the responsibilities it entails or we disappear. The message is too clearly spelled out for us to avoid it any longer.

The essays in this first section are intended to illustrate the basic position of the ecologist in detail. As you read the various offerings, you will note that the subject falls into two main categories. There are those essays that infer that the solution to the environmental dilemma is scientific and technological, and there are those that appeal to our ways of thinking about man's position and role in the universe. Each in its own way involves what W. H. Auden prayed for in his poem *Petition*—"a change of heart."

# THE HISTORICAL ROOTS OF OUR ECOLOGIC CRISIS

*Lynn White, Jr.*

*There is a grave danger in believing in our own creations. Orig-
inally man fashioned the Christian faith to provide himself with a guide
for living well, but in the passing of time he has come to accept that
story as depicting certain truths about himself. The result, of course, is
arrogance. In fact, argues Lynn White, Jr., Christianity can be seen as
a major cause of the present environmental crisis. Be that as it may, our
visions, our legends, and our conceptions of ourselves act as vital factors
in the eco-system in which we live, and they must be questioned with
the intensity that greets a new drug's appearance on the market. Professor White's article is excellent in its own right. In addition it provides
the reader with a model for investigation into those other questionable
assumptions we harbor in the recesses of our minds.*

A conversation with Aldous Huxley not infrequently put one at the receiving end of an unforgettable monologue. About a year before his lamented
death he was discoursing on a favorite topic: Man's unnatural treatment of
nature and its sad results. To illustrate his point he told how, during the
previous summer, he had returned to a little valley in England where he had
spent many happy months as a child. Once it had been composed of delightful
grassy glades; now it was becoming overgrown with unsightly brush because
the rabbits that formerly kept such growth under control had largely succumbed
to a disease, myxomatosis, that was deliberately introduced by the local farmers to reduce the rabbits' destruction of crops. Being something of a Philistine,
I could be silent no longer, even in the interests of great rhetoric. I interrupted

SOURCE: Lynn White, Jr., "The Historical Roots of Our Ecologic Crisis," *Science*, vol. 155,
10 March 1967, pp. 1203–1207. © 1967 by the American Association for the Advancement
of Science.

to point out that the rabbit itself had been brought as a domestic animal to England in 1176, presumably to improve the protein diet of the peasantry.

All forms of life modify their contexts. The most spectacular and benign instance is doubtless the coral polyp. By serving its own ends, it has created a vast undersea world favorable to thousands of other kinds of animals and plants. Ever since man became a numerous species he has affected his environment notably. The hypothesis that his fire-drive method of hunting created the world's great grasslands and helped to exterminate the monster mammals of the Pleistocene from much of the globe is plausible, if not proved. For 6 millennia at least, the banks of the lower Nile have been a human artifact rather than the swampy African jungle which nature, apart from man, would have made it. The Aswan Dam, flooding 5000 square miles, is only the latest stage in a long process. In many regions terracing or irrigation, overgrazing, the cutting of forests by Romans to build ships to fight Carthaginians or by Crusaders to solve the logistics problems of their expeditions, have profoundly changed some ecologies. Observation that the French landscape falls into two basic types, the open fields of the north and the *bocage* of the south and west, inspired Marc Bloch to undertake his classic study of medieval agricultural methods. Quite unintentionally, changes in human ways often affect nonhuman nature. It has been noted, for example, that the advent of the automobile eliminated huge flocks of sparrows that once fed on the horse manure littering every street.

The history of ecologic change is still so rudimentary that we know little about what really happened, or what the results were. The extinction of the European aurochs as late as 1627 would seem to have been a simple case of overenthusiastic hunting. On more intricate matters it often is impossible to find solid information. For a thousand years or more the Frisians and Hollanders have been pushing back the North Sea, and the process is culminating in our own time in the reclamation of the Zuider Zee. What, if any, species of animals, birds, fish, shore life, or plants have died out in the process? In their epic combat with Neptune have the Netherlanders overlooked ecological values in such a way that the quality of human life in the Netherlands has suffered? I cannot discover that the questions have ever been asked, much less answered.

People, then, have often been a dynamic element in their own environment, but in the present state of historical scholarship we usually do not know exactly when, where, or with what effects man-induced changes came. As we enter the last third of the twentieth century, however, concern for the problem of ecologic backlash is mounting feverishly. Natural science, conceived as the effort to understand the nature of things, had flourished in several eras and among several peoples. Similarly there had been an age-old accumulation of technological skills, sometimes growing rapidly, sometimes slowly. But it was not until about four generations ago that Western Europe and North America arranged a marriage between science and technology, a union of the theoretical and the empirical approaches to our natural environment. The emergence in widespread practice of the Baconian creed that scientific knowledge means technological power over nature can scarcely be dated before about 1850, save in the chemical industries, where it was anticipated in the eighteenth century. Its acceptance as a normal pattern of action may mark the greatest event in human history since the invention of agriculture, and perhaps in nonhuman terrestrial history as well.

Almost at once the new situation forced the crystallization of the novel concept of ecology; indeed, the word *ecology* first appeared in the English language in 1873. Today, less than a century later, the impact of our race upon the environment has so increased in force that it has changed in essence. When the first cannons were fired, in the early fourteenth century, they affected ecology by sending workers scrambling to the forests and mountains for more potash, sulfur, iron ore, and charcoal, with some resulting erosion and deforestation. Hydrogen bombs are of a different order: a war fought with them might alter the genetics of all life on this planet. By 1285 London had a smog problem arising from the burning of soft coal, but our present combustion of fossil fuels threatens to change the chemistry of the globe's atmosphere as a whole, with consequences which we are only beginning to guess. With the population explosion, the carcinoma of planless urbanism, the new geological deposits of sewage and garbage, surely no creature older than man has ever managed to foul its nest in such short order.

There are many calls to action, but specific proposals, however worthy as individual items, seem too partial, palliative, negative: ban the bomb, tear down the billboards, give the Hindus contraceptives and tell them to eat their sacred cows. The simplest solution to any suspect change is, of course, to stop it, or, better yet, to revert to a romanticized past: make those ugly gasoline stations look like Anne Hathaway's cottage or (in the Far West) like ghost-town saloons. The "wilderness area" mentality invariably advocates deep-freezing an ecology, whether San Gimignano or the High Sierra, as it was before the first Kleenex was dropped. But neither atavism nor prettification will cope with the ecologic crisis of our time.

What shall we do? No one yet knows. Unless we think about fundamentals, our specific measures may produce new backlashes more serious than those they are designed to remedy.

As a beginning we should try to clarify our thinking by looking, in some historical depth, at the presuppositions that underlie modern technology and science. Science was traditionally aristocratic, speculative, intellectual in intent; technology was lower-class, empirical, action-oriented. The quite sudden fusion of these two, towards the middle of the nineteenth century, is surely related to the slightly prior and contemporary democratic revolutions which, by reducing social barriers, tended to assert a functional unity of brain and hand. Our ecologic crisis is the product of an emerging, entirely novel, democratic culture. The issue is whether a democratized world can survive its own implications. Presumably we cannot unless we rethink our axioms.

## The Western Traditions of Technology and Science

One thing is so certain that it seems stupid to verbalize it: both modern technology and modern science are distinctly *Occidental.* Our technology has absorbed elements from all over the world, notably from China; yet everywhere today, whether in Japan or in Nigeria, successful technology is Western. Our science is the heir to all the sciences of the past, especially perhaps to the work of the great Islamic scientists of the Middle Ages, who so often outdid the ancient Greeks in skill and perspicacity: al-Rāzī in medicine, for example; or ibn-al-Haytham in optics; or Omar Khāyyám in mathematics. Indeed, not a few works of such geniuses seem to have vanished in the original Arabic and to survive only in medieval Latin translations that helped to lay the foundations for later Western developments. Today, around the globe, all significant science is Western in style and method, whatever the pigmentation or language of the scientists.

A second pair of facts is less well recognized because they result from quite recent historical scholarship. The leadership of the West, both in technology and in science, is far older than the so-called Scientific Revolution of the seventeenth century or the so-called Industrial Revolution of the eighteenth century. These terms are in fact outmoded and obscure the true nature of what they try to describe—significant stages in two long and separate developments. By A.D. 1000 at the latest—and perhaps, feebly, as much as 200 years earlier —the West began to apply water power to industrial processes other than milling grain. This was followed in the late twelfth century by the harnessing of wind power. From simple beginnings, but with remarkable consistency of style, the West rapidly expanded its skills in the development of power machinery, labor-saving devices, and automation. Those who doubt should contemplate that most monumental achievement in the history of automation: the weight-driven mechanical clock, which appeared in two forms in the early fourteenth century. Not in craftsmanship but in basic technological capacity, the Latin West of the later Middle Ages far outstripped its elaborate, sophisticated, and esthetically magnificent sister cultures, Byzantium and Islam. In 1444 a great Greek ecclesiastic, Bessarion, who had gone to Italy, wrote a letter to a prince in Greece. He is amazed by the superiority of Western ships, arms, textiles, glass. But above all he is astonished by the spectacle of waterwheels sawing timbers and pumping the bellows of blast furnaces. Clearly, he had seen nothing of the sort in the Near East.

By the end of the fifteenth century the technological superiority of Europe was such that its small, mutually hostile nations could spill out over all the rest of the world, conquering, looting, and colonizing. The symbol of this technological superiority is the fact that Portugal, one of the weakest states of the Occident, was able to become, and to remain for a century, mistress of the East Indies. And we must remember that the technology of Vasco da Gama and Albuquerque was built by pure empiricism, drawing remarkably little support or inspiration from science.

In the present-day vernacular understanding, modern science is supposed to have begun in 1543, when both Copernicus and Vesalius published their great works. It is no derogation of their accomplishments, however, to point out that such structures as the *Fabrica* and the *De revolutionibus* do not appear overnight. The distinctive Western tradition of science, in fact, began in the late eleventh century with a massive movement of translation of Arabic and Greek scientific works into Latin. A few notable books—Theophrastus, for example— escaped the West's avid new appetite for science, but within less than 200 years effectively the entire corpus of Greek and Muslim science was available in Latin, and was being eagerly read and criticized in the new European universities. Out of criticism arose new observation, speculation, and increasing distrust of ancient authorities. By the late thirteenth century Europe had seized global leadership from the faltering hands of Islam. It would be as absurd to deny the profound originality of Newton, Galileo, or Copernicus as to deny that of the fourteenth century scholastic scientists like Buridan or Oresme on whose work they built. Before the eleventh century, science scarcely existed in the Latin West, even in Roman times. From the eleventh century onward, the scientific sector of Occidental culture has increased in a steady crescendo.

Since both our technological and our scientific movements got their start, acquired their character, and achieved world dominance in the Middle Ages, it would seem that we cannot understand their nature or their present impact upon ecology without examining fundamental medieval assumptions and developments.

## Medieval View of Man and Nature

Until recently, agriculture has been the chief occupation even in "advanced" societies; hence, any change in methods of tillage has much importance. Early plows, drawn by two oxen, did not normally turn the sod but merely scratched it. Thus, cross-plowing was needed and fields tended to be squarish. In the fairly light soils and semi-arid climates of the Near East and Mediterranean, this worked well. But such a plow was inappropriate to the wet climate and often sticky soils of northern Europe. By the latter part of the seventh century after Christ, however, following obscure beginnings, certain northern peasants were using an entirely new kind of plow, equipped with a vertical knife to cut the line of the furrow, a horizontal share to slice under the sod, and a moldboard to turn it over. The friction of this plow with the soil was so great that it normally required not two but eight oxen. It attacked the land with such violence that cross-plowing was not needed, and fields tended to be shaped in long strips.

In the days of the scratch-plow, fields were distributed generally in units capable of supporting a single family. Subsistence farming was the presupposition. But no peasant owned eight oxen: to use the new and more efficient plow, peasants pooled their oxen to form large plow-teams, originally receiving (it would appear) plowed strips in proportion to their contribution. Thus, distribution of land was based no longer on the needs of a family but, rather, on the capacity of a power machine to till the earth. Man's relation to the soil was profoundly changed. Formerly man had been part of nature; now he was the exploiter of nature. Nowhere else in the world did farmers develop any analogous agricultural implement. Is it coincidence that modern technology, with its ruthlessness toward nature, has so largely been produced by descendants of these peasants of northern Europe?

This same exploitive attitude appears slightly before A.D. 830 in Western illustrated calendars. In older calendars the months were shown as passive personifications. The new Frankish calendars, which set the style for the Middle Ages, are very different: they show men coercing the world around them—plowing, harvesting, chopping trees, butchering pigs. Man and nature are two things, and man is master.

These novelties seem to be in harmony with larger intellectual patterns. What people do about their ecology depends on what they think about themselves in relation to things around them. Human ecology is deeply conditioned by beliefs about our nature and destiny—that is, by religion. To Western eyes this is very evident in, say, India or Ceylon. It is equally true of ourselves and of our medieval ancestors.

The victory of Christianity over paganism was the greatest psychic revolution in the history of our culture. It has become fashionable today to say that, for better or worse, we live in "the post-Christian age." Certainly the forms of our thinking and language have largely ceased to be Christian, but to my eye the substance often remains amazingly akin to that of the past. Our daily habits of action, for example, are dominated by an implicit faith in perpetual progress which was unknown either to Greco-Roman antiquity or to the Orient. It is rooted in, and is indefensible apart from, Judeo-Christian teleology. The fact that Communists share it merely helps to show what can be demonstrated on many other grounds: that Marxism, like Islam, is a Judeo-Christian heresy. We continue today to live, as we have lived for about 1700 years, very largely in a context of Christian axioms.

What did Christianity tell people about their relations with the environment?

While many of the world's mythologies provide stories of creation, Greco-Roman mythology was singularly incoherent in this respect. Like Aristotle, the intellectuals of the ancient West denied that the visible world had had a beginning. Indeed, the idea of a beginning was impossible in the framework of their cyclical notion of time. In sharp contrast, Christianity inherited from Judaism not only a concept of time as nonrepetitive and linear but also a striking story of creation. By gradual stages a loving and all-powerful God had created light and darkness, the heavenly bodies, the earth and all its plants, animals, birds, and fishes. Finally, God had created Adam and, as an afterthought, Eve to keep man from being lonely. Man named all the animals, thus establishing his dominance over them. God planned all of this explicitly for man's benefit and rule: no item in the physical creation had any purpose save to serve man's purposes. And, although man's body is made of clay, he is not simply part of nature: he is made in God's image.

Especially in its Western form, Christianity is the most anthropocentric religion the world has seen. As early as the second century both Tertullian and Saint Irenaeus of Lyons were insisting that when God shaped Adam he was foreshadowing the image of the Incarnate Christ, the Second Adam. Man shares, in great measure, God's transcendence of nature. Christianity, in absolute contrast to ancient paganism and Asia's religions (except, perhaps, Zoroastrianism), not only established a dualism of man and nature but also insisted that it is God's will that man exploit nature for his proper ends.

At the level of the common people this worked out in an interesting way. In Antiquity every tree, every spring, every stream, every hill had its own *genius loci*, its guardian spirit. These spirits were accessible to men, but were very unlike men; centaurs, fauns, and mermaids show their ambivalence. Before one cut a tree, mined a mountain, or dammed a brook, it was important to placate the spirit in charge of that particular situation, and to keep it placated. By destroying pagan animism, Christianity made it possible to exploit nature in a mood of indifference to the feelings of natural objects.

It is often said that for animism the Church substituted the cult of saints. True; but the cult of saints is functionally quite different from animism. The saint is not *in* natural objects; he may have special shrines, but his citizenship is in heaven. Moreover, a saint is entirely a man; he can be approached in human terms. In addition to saints, Christianity of course also had angels and demons inherited from Judaism and perhaps, at one remove, from Zoroastrianism. But these were all as mobile as the saints themselves. The spirits *in* natural objects, which formerly had protected nature from man, evaporated. Man's effective monopoly on spirit in this world was confirmed, and the old inhibitions to the exploitation of nature crumbled.

When one speaks in such sweeping terms, a note of caution is in order. Christianity is a complex faith, and its consequences differ in differing contexts. What I have said may well apply to the medieval West, where in fact technology made spectacular advances. But the Greek East, a highly civilized realm of equal Christian devotion, seems to have produced no marked technological innovation after the late seventh century, when Greek fire was invented. The key to the contrast may perhaps be found in a difference in the tonality of piety and thought which students of comparative theology find between the Greek and the Latin Churches. The Greeks believed that sin was intellectual blindness, and that salvation was found in illumination, orthodoxy —that is, clear thinking. The Latins, on the other hand, felt that sin was moral evil, and that salvation was to be found in right conduct. Eastern theology has been intellectualist. Western theology has been voluntarist. The Greek saint contemplates; the Western saint acts. The implications of Christianity for the conquest of nature would emerge more easily in the Western atmosphere.

The Christian dogma of creation, which is found in the first clause of all the Creeds, has another meaning for our comprehension of today's ecologic crisis. By revelation, God had given man the Bible, the Book of Scripture. But since God had made nature, nature also must reveal the divine mentality. The religious study of nature for the better understanding of God was known as natural theology. In the early Church, and always in the Greek East, nature was conceived primarily as a symbolic system through which God speaks to men: the ant is a sermon to sluggards; rising flames are the symbol of the soul's aspiration. This view of nature was essentially artistic rather than scientific. While Byzantium preserved and copied great numbers of ancient Greek scientific texts, science as we conceive it could scarcely flourish in such an ambience.

However, in the Latin West by the early thirteenth century natural theology was following a very different bent. It was ceasing to be the decoding of the physical symbols of God's communication with man and was becoming the effort to understand God's mind by discovering how his creation operates. The rainbow was no longer simply a symbol of hope first sent to Noah after the Deluge: Robert Grosseteste, Friar Roger Bacon, and Theodoric of Freiberg produced startlingly sophisticated work on the optics of the rainbow, but they did it as a venture in religious understanding. From the thirteenth century onward, up to and including Leibnitz and Newton, every major scientist, in effect, explained his motivations in religious terms. Indeed, if Galileo had not been so expert an amateur theologian he would have got into far less trouble: the professionals resented his intrusion. And Newton seems to have regarded himself more as a theologian than as a scientist. It was not until the late eighteenth century that the hypothesis of God became unnecessary to many scientists.

It is often hard for the historian to judge, when men explain why they are doing what they want to do, whether they are offering real reasons or merely culturally acceptable reasons. The consistency with which scientists during the long formative centuries of Western science said that the task and the reward of the scientist was "to think God's thoughts after him" leads one to believe that this was their real motivation. If so, then modern Western science was cast in a matrix of Christian theology. The dynamism of religious devotion, shaped by the Judeo-Christian dogma of creation, gave it impetus.

## An Alternative Christian View

We would seem to be headed toward conclusions unpalatable to many Christians. Since both *science* and *technology* are blessed words in our contemporary vocabulary, some may be happy at the notions, first, that, viewed historically, modern science is an extrapolation of natural theology and, second, that modern technology is at least partly to be explained as an Occidental, voluntarist realization of the Christian dogma of man's transcendence of, and rightful mastery over, nature. But, as we now recognize, somewhat over a century ago science and technology—hitherto quite separate activities —joined to give mankind powers which, to judge by many of the ecologic effects, are out of control. If so, Christianity bears a huge burden of guilt.

I personally doubt that disastrous ecologic backlash can be avoided simply by applying to our problems more science and more technology. Our science and technology have grown out of Christian attitudes toward man's relation to nature which are almost universally held not only by Christians and neo-Christians but also by those who fondly regard themselves as post-Christians. Despite Copernicus, all the cosmos rotates around our little globe. Despite Darwin, we are *not*, in our hearts, part of the natural process. We are superior to nature, contemptuous of it, willing to use it for our slightest whim. The newly elected Governor of California, like myself a churchman but less troubled than I, spoke for the Christian tradition when he said (as is alleged), "when you've seen one redwood tree, you've seen them all." To a Christian a tree can be no more than a physical fact. The whole concept of the sacred grove is alien to Christianity and to the ethos of the West. For nearly two millennia Christian missionaries have been chopping down sacred groves, which are idolatrous because they assume spirit in nature.

What we do about ecology depends on our ideas of the man-nature relationship. More science and more technology are not going to get us out of the present ecologic crisis until we find a new religion, or rethink our old one. The beatniks, who are the basic revolutionaries of our time, show a sound instinct in their affinity for Zen Buddhism, which conceives of the man-nature relationship as very nearly the mirror image of the Christian view. Zen, however, is as deeply conditioned by Asian history as Christianity is by the experience of the West, and I am dubious of its viability among us.

Possibly we should ponder the greatest radical in Christian history since Christ: Saint Francis of Assisi. The prime miracle of Saint Francis is the fact that he did not end at the stake, as many of his left-wing followers did. He was so clearly heretical that a General of the Franciscan Order, Saint Bonaventura, a great and perceptive Christian, tried to suppress the early accounts of Franciscanism. The key to an understanding of Francis is his belief in the virtue of humility—not merely for the individual but for man as a species. Francis tried to depose man from his monarchy over creation and set up a democracy of all God's creatures. With him the ant is no longer simply a homily for the lazy, flames a sign of the thrust of the soul toward union with God; now they are Brother Ant and Sister Fire, praising the Creator in their own ways as Brother Man does in his.

Later commentators have said that Francis preached to the birds as a rebuke to men who would not listen. The records do not read so: he urged the little birds to praise God, and in spiritual ecstasy they flapped their wings and chirped rejoicing. Legends of saints, especially the Irish saints, had long told of their dealings with animals but always, I believe, to show their human dominance over creatures. With Francis it is different. The land around Gubbio in the Apennines was being ravaged by a fierce wolf. Saint Francis, says the legend, talked to the wolf and persuaded him of the error of his ways. The wolf repented, died in the odor of sanctity, and was buried in consecrated ground.

What Sir Steven Ruciman calls "the Franciscan doctrine of the animal soul" was quickly stamped out. Quite possibly it was in part inspired, consciously or unconsciously, by the belief in reincarnation held by the Cathar heretics who at that time teemed in Italy and southern France, and who presumably had got it originally from India. It is significant that at just the same moment, about 1200, traces of metempsychosis are found also in Western Judaism, in the Provençal *Cabbala*. But Francis held neither to transmigration of souls nor to pantheism. His view of nature and of man rested on a unique sort of pan-psychism of all things animate and inanimate, designed for the glorification of their transcendent Creator, who, in the ultimate gesture of cosmic humility, assumed flesh, lay helpless in a manger, and hung dying on a scaffold.

I am not suggesting that many contemporary Americans who are concerned about our ecologic crisis will be either able or willing to counsel with wolves or exhort birds. However, the present increasing disruption of the global environment is the product of a dynamic technology and science which were originating in the Western medieval world against which Saint Francis was rebelling in so original a way. Their growth cannot be understood historically apart from distinctive attitudes toward nature which are deeply grounded in Christian dogma. The fact that most people do not think of these attitudes as Christian is irrelevant. No new set of basic values has been accepted in our society to displace those of Christianity. Hence we shall continue to have a worsening ecologic crisis until we reject the Christian axiom that nature has no reason for existence save to serve man.

The greatest spiritual revolutionary in Western history, Saint Francis, proposed what he thought was an alternative Christian view of nature and man's relation to it: he tried to substitute the idea of the equality of all creatures, including man, for the idea of man's limitless rule of creation. He failed. Both our present science and our present technology are so tinctured with orthodox Christian arrogance toward nature that no solution for our ecologic crisis can be expected from them alone. Since the roots of our trouble are so largely religious, the remedy must also be essentially religious, whether we call it that or not. We must rethink and refeel our nature and destiny. The profoundly religious, but heretical, sense of the primitive Franciscans for the spiritual autonomy of all parts of nature may point a direction. I propose Francis as a patron saint for ecologists.

# THE MARGINAL WORLD

## Rachel Carson

*It can be said of the late Rachel Carson (1907–1964) that she was the first writer-biologist to penetrate the illusion of prosperity in our age and to suggest how our world is, in reality, busily engaged in its own demise. While her efforts were not always welcomed by industry, her writing, with its persuasive yet sympathetic tone, has prevailed. The following selection from* The Edge of the Sea *(1955) is a remarkable example of her unique style. Initially promising to be a descriptive sketch, her piece grows by implication to encompass the drama of evolution in all its profundity and natural beauty.*

The edge of the sea is a strange and beautiful place. All through the long history of Earth it has been an area of unrest where waves have broken heavily against the land, where the tides have pressed foward over the continents, receded, and then returned. For no two successive days is the shoreline precisely the same. Not only do the tides advance and retreat in their eternal rhythms, but the level of the sea itself is never at rest. It rises or falls as the glaciers melt or grow, as the floor of the deep ocean basins shifts under its increasing load of sediments, or as the earth's crust along the continental margins warps up or down in adjustment to strain and tension. Today a little more land may belong to the sea, tomorrow a little less. Always the edge of the sea remains an elusive and indefinable boundary.

The shore has a dual nature, changing with the swing of the tides, belonging now to the land, now to the sea. On the ebb tide it knows the harsh extremes of the land world, being exposed to heat and cold, to wind, to rain

SOURCE: Rachel Carson, *The Edge of the Sea* (Boston: Houghton Mifflin Co., 1955). Copyright © 1955 by Rachel L. Carson. Reprinted by permission of the publisher, Houghton Mifflin Co.

and drying sun. On the flood tide it is a water world, returning briefly to the relative stability of the open sea.

Only the most hardy and adaptable can survive in a region so mutable, yet the area between the tide lines is crowded with plants and animals. In this difficult world of the shore, life displays its enormous toughness and vitality by occupying almost every conceivable niche. Visibly, it carpets the intertidal rocks; or half hidden, it descends into fissures and crevices, or hides under boulders, or lurks in the wet gloom of sea caves. Invisibly, where the casual observer would say there is no life, it lies deep in the sand, in burrows and tubes and passageways. It tunnels into solid rock and bores into peat and clay. It encrusts weeds or drifting spars or the hard, chitinous shell of a lobster. It exists minutely, as the film of bacteria that spreads over a rock surface or a wharf piling; as spheres of protozoa, small as pinpricks, sparkling at the surface of the sea; and as Lilliputian beings swimming through dark pools that lie between the grains of sand.

The shore is an ancient world, for as long as there has been an earth and sea there has been this place of the meeting of land and water. Yet it is a world that keeps alive the sense of continuing creation and of the relentless drive of life. Each time that I enter it, I gain some new awareness of its beauty and its deeper meanings, sensing that intricate fabric of life by which one creature is linked with another, and each with its surroundings.

In my thoughts of the shore, one place stands apart for its revelation of exquisite beauty. It is a pool hidden within a cave that one can visit only rarely and briefly when the lowest of the year's low tides fall below it, and perhaps from that very fact it acquires some of its special beauty. Choosing such a tide, I hoped for a glimpse of the pool. The ebb was to fall early in the morning. I knew that if the wind held from the northwest and no interfering swell ran in from a distant storm the level of the sea should drop below the entrance to the pool. There had been sudden ominous showers in the night, with rain like handfuls of gravel flung on the roof. When I looked out into the early morning the sky was full of a gray dawn light but the sun had not yet risen. Water and air were pallid. Across the bay the moon was a luminous disc in the western sky, suspended above the dim line of distant shore—the full August moon, drawing the tide to the low, low levels of the threshold of the alien sea world. As I watched, a gull flew by, above the spruces. Its breast was rosy with the light of the unrisen sun. The day was, after all, to be fair.

Later, as I stood above the tide near the entrance to the pool, the promise of that rosy light was sustained. From the base of the steep wall of rock on which I stood, a moss-covered ledge jutted seaward into deep water. In the surge at the rim of the ledge the dark fronds of oarweeds swayed, smooth and gleaming as leather. The projecting ledge was the path to the small hidden cave and its pool. Occasionally a swell, stronger than the rest, rolled smoothly over the rim and broke in foam against the cliff. But the intervals between such swells were long enough to admit me to the ledge and long enough for a glimpse of that fairy pool, so seldom and so briefly exposed.

And so I knelt on the wet carpet of sea moss and looked back into the dark cavern that held the pool in a shallow basin. The floor of the cave was only a few inches below the roof, and a mirror had been created in which all that grew on the ceiling was reflected in the still water below.

Under water that was clear as glass the pool was carpeted with green sponge. Gray patches of sea squirts glistened on the ceiling and colonies of soft coral were a pale apricot color. In the moment when I looked into the cave a little elfin starfish hung down, suspended by the merest thread, perhaps by only a single tube foot. It reached down to touch its own reflection, so perfectly delineated that there might have been, not one starfish, but two. The beauty of the reflected images and of the limpid pool itself was the poignant beauty of things that are ephemeral, existing only until the sea should return to fill the little cave.

Whenever I go down into this magical zone of the lower water of the spring tides, I look for the most delicately beautiful of all the shore's inhabitants —flowers that are not plant but animal, blooming on the threshold of the deeper sea. In that fairy cave I was not disappointed. Hanging from its roof were the pendent flowers of the hydroid Tubularia, pale pink, fringed and delicate as the wind flower. Here were creatures so exquisitely fashioned that they seemed unreal, their beauty too fragile to exist in a world of crushing force. Yet every detail was functionally useful, every stalk and hydranth and petal-like tentacle fashioned for dealing with the realities of existence. I knew that they were merely waiting, in that moment of the tide's ebbing, for the return of the sea. Then in the rush of water, in the surge of surf and the pressure of the incoming tide, the delicate flower heads would stir with life. They would sway on their slender stalks, and their long tentacles would sweep the returning water, finding in it all that they needed for life.

And so in that enchanted place on the threshold of the sea the realities that possessed my mind were far from those of the land world I had left an hour before. In a different way the same sense of remoteness and of a world apart came to me in a twilight hour on a great beach on the coast of Georgia. I had come down after sunset and walked far out over sands that lay wet and gleaming, to the very edge of the retreating sea. Looking back across that immense flat, crossed by winding, water-filled gullies and here and there holding shallow pools left by the tide, I was filled with awareness that this intertidal area, although abandoned briefly and rhythmically by the sea, is always reclaimed by the rising tide. There at the edge of low water the beach with its reminders of the land seemed far away. The only sounds were those of the wind and the sea and the birds. There was one sound of wind moving over water, and another of water sliding over the sand and tumbling down the faces of its own wave forms. The flats were astir with birds, and the voice of the willet rang insistently. One of them stood at the edge of the water and gave its loud, urgent cry; an answer came from far up the beach and the two birds flew to join each other.

The flats took on a mysterious quality as dusk approached and the last evening light was reflected from the scattered pools and creeks. Then birds became only dark shadows, with no color discernible. Sanderlings scurried across the beach like little ghosts, and here and there the darker forms of the willets stood out. Often I could come very close to them before they would start up in alarm—the sanderlings running, the willets flying up, crying. Black skimmers flew along the ocean's edge silhouetted against the dull, metallic gleam, or they went flitting above the sand like large, dimly seen moths. Sometimes they "skimmed" the winding creeks of tidal water, where little spreading surface ripples marked the presence of small fish.

The shore at night is a different world, in which the very darkness that hides the distractions of daylight brings into sharper focus the elemental realities. Once, exploring the night beach, I surprised a small ghost crab in the searching beam of my torch. He was lying in a pit he had dug just above the surf, as though watching the sea and waiting. The blackness of the night possessed water, air, and beach. It was the darkness of an older world, before Man. There was no sound but the all-enveloping, primeval sounds of wind blowing over water and sand, and of waves crashing on the beach. There was no other visible life—just one small crab near the sea. I have seen hundreds of ghost crabs in other settings, but suddenly I was filled with the odd sensation that for the first time I knew the creature in its own world—that I understood, as never before, the essence of its being. In that moment time was suspended; the world to which I belonged did not exist and I might have been an onlooker from outer space. The little crab alone with the sea became a symbol that stood for life itself—for the delicate, destructible, yet incredibly vital force that somehow holds its place amid the harsh realities of the inorganic world.

The sense of creation comes with memories of a southern coast, where the sea and the mangroves, working together, are building a wilderness of thousands of small islands off the southwestern coast of Florida, separated from each other by a tortuous pattern of bays, lagoons, and narrow waterways. I remember a winter day when the sky was blue and drenched with sunlight; though there was no wind one was conscious of flowing air like cold clear crystal. I had landed on the surf-washed tip of one of those islands, and then worked my way around to the sheltered bay side. There I found the tide far out, exposing the broad mud flat of a cove bordered by the mangroves with their twisted branches, their glossy leaves, and their long prop roots reaching down, grasping and holding the mud, building the land out a little more, then again a little more.

The mud flats were strewn with the shells of that small, exquisitely colored mollusk, the rose tellin, looking like scattered petals of pink roses. There must have been a colony nearby, living buried just under the surface of the mud. At first the only creature visible was a small heron in gray and rusty plumage— a reddish egret that waded across the flat with the stealthy, hesitant movements of its kind. But other land creatures had been there, for a line of fresh tracks wound in and out among the mangrove roots, marking the path of a raccoon feeding on the oysters that gripped the supporting roots with projections from their shells. Soon I found the tracks of a shore bird, probably a sanderling, and followed them a little; then they turned toward the water and were lost, for the tide had erased them and made them as though they had never been.

Looking out over the cove I felt a strong sense of the interchangeability of land and sea in this marginal world of the shore, and of the links between the life of the two. There was also an awareness of the past and of the continuing flow of time, obliterating much that had gone before, as the sea had that morning washed away the tracks of the bird.

The sequence and meaning of the drift of time were quietly summarized in the existence of hundreds of small snails—the mangrove periwinkles— browsing on the branches and roots of the trees. Once their ancestors had been sea dwellers, bound to the salt waters by every tie of their life processes. Little by little over the thousands and millions of years the ties had been broken, the snails had adjusted themselves to life out of water, and now today they were living many feet above the tide to which they only occasionally returned. And perhaps, who could say how many ages hence, there would be in their descendants not even this gesture of remembrance for the sea.

The spiral shells of other snails—these quite minute—left winding tracks on the mud as they moved about in search of food. They were horn shells, and when I saw them I had a nostalgic moment when I wished I might see what Audubon saw, a century and more ago. For such little horn shells were the food of the flamingo, once so numerous on this coast, and when I half closed my eyes I could almost imagine a flock of these magnificent flame birds feeding in that cove, filling it with their color. It was a mere yesterday in the life of the earth that they were there; in nature, time and space are relative matters, perhaps most truly perceived subjectively in occasional flashes of insight, sparked by such a magical hour and place.

There is a common thread that links these scenes and memories—the spectacle of life in all its varied manifestations as it has appeared, evolved, and sometimes died out. Underlying the beauty of the spectacle there is meaning and significance. It is the elusiveness of that meaning that haunts us, that sends us again and again into the natural world where the key to the riddle is hidden. It sends us back to the edge of the sea, where the drama of life played its first scene on earth and perhaps even its prelude; where the forces of evolution are at work today, as they have been since the appearance of what we know as life; and where the spectacle of living creatures faced by the cosmic realities of their world is crystal clear.

# ECOLOGY AND MAN—A VIEWPOINT

*Paul Shepard*

*If it is anything, ecology is an instinctual grasping of the subtle complexity of the world mosaic. It is not a science in search of technical answers, it is not a branch of philosophic thought based on some unique comprehension of what man is, but it is an attractively inclusive attitude which we are all capable of sharing if we show a willingness to clear our minds of some of their most misleading beliefs. Paul Shepard makes these and other distinctions in what will be recognized as a remarkably coherent and insightful essay. One is immediately impressed by the author's easy fusion of learning and style. In its way, the essay is a singular model for the very point it attempts to make about man's behavior in this world.*

Ecology is sometimes characterized as the study of a natural "web of life." It would follow that man is somewhere in the web or that he in fact manipulates its strands, exemplifying what Thomas Huxley called "man's place in nature." But the image of a web is too meager and simple for the reality. A web is flat and finished and has the mortal frailty of the individual spider. Although elastic, it has insufficient depth. However solid to the touch of the spider, for us it fails to denote the *eikos*—the habitation—and to suggest the enduring integration of the primitive Greek domicile with its sacred hearth, bonding the earth to all aspects of society.

Ecology deals with organisms in an environment and with the processes that link organism and place. But ecology as such cannot be studied, only organisms, earth, air, and sea can be studied. It is not a discipline: there is

SOURCE: Paul Shepard and Daniel McKinley, *The Subversive Science* (Boston: Houghton Mifflin Co., 1969). Copyright © 1969 by Paul Shepard and Daniel McKinley. Reprinted by permission of the publisher, Houghton Mifflin Co.

no body of thought and technique which frames an ecology of man.[1] It must be therefore a scope or a way of seeing. Such a *perspective* on the human situation is very old and has been part of philosophy and art for thousands of years. It badly needs attention and revival.

Man is in the world and his ecology is the nature of that *inness*. He is in the world as in a room, and in transience, as in the belly of a tiger or in love. What does he do there in nature? What does nature do there *in him*? What is the nature of the transaction? Biology tells us that the transaction is always circular, always a mutual feedback. Human ecology cannot be limited strictly to biological concepts, but it cannot ignore them. It cannot even transcend them. It emerges from biological reality and grows from the fact of interconnection as a general principle of life. It must take a long view of human life and nature as they form a mesh or pattern going beyond historical time and beyond the conceptual bounds of other humane studies. As a natural history of what it means to be human, ecology might proceed the same way one would define a stomach, for example, by attention to its nervous and circulatory connections as well as its entrance, exit, and muscular walls.

Many educated people today believe that only what is unique to the individual is important or creative, and turn away from talk of populations and species as they would from talk of the masses. I once knew a director of a wealthy conservation foundation who had misgivings about the approach of ecology to urgent environmental problems in America because its concepts of communities and systems seemed to discount the individual. Communities to him suggested only followers, gray masses without the tradition of the individual. He looked instead—or in reaction—to the profit motive and capitalistic formulas, in terms of efficiency, investment, and production. It seemed to me that he had missed a singular opportunity. He had shied from the very aspect of the world now beginning to interest industry, business, and technology as the biological basis of their—and our—affluence, and which his foundation could have shown to be the ultimate basis of all economics.

Individual man *has* his particular integrity, to be sure. Oak trees, even mountains, have selves or integrities too (a poor word for my meaning, but it will have to do). To our knowledge, those other forms are not troubled by seeing themselves in more than one way, as man is. In one aspect the self is an arrangement of organs, feelings, and thoughts—a "me"—surrounded by a hard body boundary: skin, clothes, and insular habits. This idea needs no defense. It is conferred on us by the whole history of our civilization. Its virtue is verified by our affluence. The alternative is a self as a center of organization, constantly drawing on and influencing the surroundings, whose skin and behavior are soft zones contacting the world instead of excluding it.

[1] There is a branch of sociology called Human Ecology, but it is mostly about urban geography.

Both views are real and their reciprocity significant. We need them both to have a healthy social and human maturity.

The second view—that of relatedness of the self—has been given short shrift. Attitudes toward ourselves do not change easily. The conventional image of a man, like that of the heraldic lion, is iconographic; its outlines are stylized to fit the fixed curves of our vision. We are hidden from ourselves by habits of perception. Because we learn to talk at the same time we learn to think, our language, for example, encourages us to see ourselves—or a plant or animal—as an isolated sack, a thing, a contained self. Ecological thinking, on the other hand, requires a kind of vision across boundaries. The epidermis of the skin is ecologically like a pond surface or a forest soil, not a shell so much as a delicate interpenetration. It reveals the self ennobled and extended rather than threatened as part of the landscape and the ecosystem, because the beauty and complexity of nature are continuous with ourselves.

And so ecology as applied to man faces the task of renewing a balanced view where now there is man-centeredness, even pathology of isolation and fear. It implies that we must find room in "our" world for all plants and animals, even for their otherness and their opposition. It further implies exploration and openness across an inner boundary—an ego boundary—and appreciative understanding of the animal in ourselves which our heritage of Platonism, Christian morbidity, duality, and mechanism have long held repellent and degrading. The older countercurrents—relics of pagan myth, the universal application of Christian compassion, philosophical naturalism, nature romanticism and pantheism—have been swept away, leaving only odd bits of wreckage. Now we find ourselves in a deteriorating environment which breeds aggressiveness and hostility toward ourselves and our world.

How simple our relationship to nature would be if we only had to choose between protecting our natural home and destroying it. Most of our efforts to provide for the natural in our philosophy have failed—run aground on their own determination to work out a peace at arm's length. Our harsh reaction against the peaceable kingdom of sentimental romanticism was evoked partly by the tone of its dulcet façade but also by the disillusion to which it led. Natural dependence and contingency suggests togetherness and emotional surrender to mass behavior and other lowest common denominators. The environmentalists matching culture and geography provoke outrage for their oversimple theories of cause and effect, against the sciences which sponsor them and even against a natural world in which the theories may or may not be true. Our historical disappointment in the nature of nature has created a cold climate for ecologists who assert once again that we are limited and obligated. Somehow they must manage in spite of the chill to reach the centers of humanism and technology, to convey there a sense of our place in a universal vascular system without depriving us of our self-esteem and confidence.

Their message is not, after all, all bad news. Our natural affiliations define and illumine freedom instead of denying it. They demonstrate it better than any dialectic. Being more enduring than we individuals, ecological patterns—spatial distributions, symbioses, the streams of energy and matter and communication—create among individuals the tensions and polarities so different from dichotomy and separateness. The responses, or what theologians call "the sensibilities" of creatures (including ourselves) to such arrangements grow in part from a healthy union of the two kinds of self already mentioned, one emphasizing integrity, the other relatedness. But it goes beyond that to something better known to twelfth century Europeans or Paleolithic hunters than to ourselves. If nature is not a prison and earth a shoddy waystation, we must find the faith and force to affirm its metabolism as our own—or rather, our own as part of it. To do so means nothing less than a shift in our whole frame of reference and our attitude toward life itself, a wider perception of the landscape as a creative, harmonious being where relationships of things are as real as the things. Without losing our sense of a great human destiny and without intellectual surrender, we must affirm that the world is a being, a part of our own body.[2]

Such a being may be called an ecosystem or simply a forest or landscape. Its members are engaged in a kind of choreography of materials and energy and information, the creation of order and organization. (Analogy to corporate organization here is misleading, for the distinction between social [one species] and ecological [many species] is fundamental.) The pond is an example. Its ecology includes all events: the conversion of sunlight to food and the food-chains within and around it, man drinking, bathing, fishing, plowing the slopes of the watershed, drawing a picture of it, and formulating theories about the world based on what he sees in the pond. He and all the other organisms at and in the pond act upon one another, engage the earth and atmosphere, and are linked to other ponds by a network of connections like the threads of protoplasm connecting cells in living tissues.

The elegance of such systems and delicacy of equilibrium are the outcome of a long evolution of interdependence. Even society, mind and culture are parts of that evolution. There is an essential relationship between them and the natural habitat: that is, between the emergence of higher primates and flowering plants, pollinating insects, seeds, humus, and arboreal life. It is unlikely that a man-like creature could arise by any other means than a long arboreal sojourn following and followed by a time of terrestriality. The fruit's complex construction and the mammalian brain are twin offspring of the maturing earth, impossible, even meaningless, without the deepening soil and the mutual development of savannas and their faunas in the last geological epoch.

[2] See Alan Watts, "The World Is Your Body," in *The Book on the Taboo Against Knowing Who You Are* (New York: Pantheon Books, 1966).

Internal complexity, as the mind of a primate, is an extension of natural complexity, measured by the variety of plants and animals and the variety of nerve cells—organic extensions of each other.

The exuberance of kinds as the setting in which a good mind could evolve (to deal with a complex world) was not only a past condition. Man did not arrive in the world as though disembarking from a train in the city. He continues to arrive, somewhat like the birth of art, a train in Roger Fry's definition, passing through many stations, none of which is wholly left behind. This idea of natural complexity as a counterpart to human intricacy is central to an ecology of man. The creation of order, of which man is an example, is realized also in the number of species and habitats, an abundance of landscapes lush and poor. Even deserts and tundras increase the planetary opulence. Curiously, only man and possibly a few birds can appreciate this opulence, being the world's travelers. Reduction of this variegation would, by extension then, be an amputation of man. To convert all "wastes"—all deserts, estuaries, tundras, ice-fields, marshes, steppes and moors—into cultivated fields and cities would impoverish rather than enrich life esthetically as well as ecologically. By esthetically, I do not mean that weasel term connoting the pleasure of baubles. We have diverted ourselves with litterbug campaigns and greenbelts in the name of esthetics while the fabric of our very environment is unraveling. In the name of conservation, too, such things are done, so that conservation becomes ambiguous. Nature is a fundamental "resource" to be sustained for our own well-being. But it loses in the translation into usable energy and commodities. Ecology may testify as often against our uses of the world, even against conservation techniques of control and management for sustained yield, as it does for them. Although ecology may be treated as a science, its greater and overriding wisdom is universal.

That wisdom can be approached mathematically, chemically, or it can be danced or told as a myth. It has been embodied in widely scattered economically different cultures. It is manifest, for example, among pre-Classical Greeks, in Navajo religion and social orientation, in Romantic poetry of the eighteenth and nineteenth centuries, in Chinese landscape painting of the eleventh century, in current Whiteheadian philosophy, in Zen Buddhism, in the world view of the cult of the Cretan Great Mother, in the ceremonials of Bushman hunters, and in the medieval Christian metaphysics of light. What is common among all of them is a deep sense of engagement with the landscape, with profound connections to surroundings and to natural processes central to all life.

It is difficult in our language even to describe that sense. English becomes imprecise or mystical—and therefore suspicious—as it struggles with "process" thought. Its noun and verb organization shapes a divided world of static doers separate from the doing. It belongs to an idiom of social hierarchy in which all nature is made to mimic man. The living world is perceived in that idiom as an upright ladder, a "great chain of being," an image which seems at first ecological but is basically rigid, linear, condescending, lacking humility and love of otherness.

We are all familiar from childhood with its classifications of everything on a scale from the lowest to the highest: inanimate matter/vegetative life/lower animals/higher animals/men/angels/gods. It ranks animals themselves in categories of increasing good: the vicious and lowly parasites, pathogens and predators/the filthy decay and scavenging organisms/indifferent wild or merely useless forms/good tame creatures/and virtuous beasts domesticated for human service. It shadows the great man-centered political scheme upon the world, derived from the ordered ascendency from parishioners to clerics to bishops to cardinals to popes, or in a secular form from criminals to proletarians to aldermen to mayors to senators to presidents.

And so is nature pigeonholed. The sardonic phrase, "the place of nature in man's world," offers, tongue-in-cheek, a clever footing for confronting a world made in man's image and conforming to words. It satirizes the prevailing philosophy of anti-nature and human omniscience. It is possible because of an attitude which—like ecology—has ancient roots, but whose modern form was shaped when Aquinas reconciled Aristotelian homocentrism with Judeo-Christian dogma. In a later setting of machine technology, puritanical capitalism, and an urban ethos it carves its own version of reality into the landscape like a schoolboy initialing a tree. For such a philosophy nothing in nature has inherent merit. As one professor recently put it, "The only reason anything is done on this earth is for people. Did the rivers, winds, animals, rocks, or dust ever consider my wishes or needs? Surely, we do all our acts in an earthly environment, but I have never had a tree, valley, mountain, or flower thank me for preserving it."[3] This view carries great force, epitomized in history by Bacon, Descartes, Hegel, Hobbes, and Marx.

Some other post-Renaissance thinkers are wrongly accused of undermining our assurance of natural order. The theories of the heliocentric solar system, of biological evolution, and of the unconscious mind are held to have deprived the universe of the beneficence and purpose to which man was a special heir and to have evoked feelings of separation, of antipathy toward a meaningless existence in a neutral cosmos. Modern despair, the arts of anxiety, the politics of pathological individualism and predatory socialism were not, however, the results of Copernicus, Darwin and Freud. If man was not the center of the universe, was not created by a single stroke of Providence, and is not ruled solely by rational intelligence, it does not follow therefore that nature is defective where we thought it perfect. The astronomer, biologist and psychiatrist each achieved for mankind corrections in sensibility. Each showed

[3] Clare A. Gunn in Landscape Architecture, July 1966, p. 260.

the interpenetration of human life and the universe to be richer and more mysterious than had been thought.

Darwin's theory of evolution has been crucial to ecology. Indeed, it might have helped rather than aggravated the growing sense of human alienation had its interpreters emphasized predation and competition less (and, for this reason, one is tempted to add, had Thomas Huxley, Herbert Spencer, Samuel Butler and G. B. Shaw had less to say about it). Its bases of universal kinship and common bonds of function, experience and value among organisms were obscured by preexisting ideas of animal depravity. Evolutionary theory was exploited to justify the worst in men and was misused in defense of social and economic injustice. Nor was it better used by humanitarians. They opposed the degradation of men in the service of industrial progress, the slaughter of American Indians, and child labor, because each treated men "like animals." That is to say, men were not animals, and the temper of social reform was to find good only in attributes separating men from animals. Kindness both toward and among animals was still a rare idea in the nineteenth century, so that using men as animals could mean only cruelty.

Since Thomas Huxley's day the nonanimal forces have developed a more subtle dictum to the effect that, "Man may be an animal, but he is more than an animal, too!" The *more* is really what is important. This appealing aphorism is a kind of anesthetic. The truth is that we are ignorant of what it is like or what it means to be any other kind of creature than we are. If we are unable to truly define the animal's experience of life or "being an animal" how can we isolate our animal part?

The rejection of animality is a rejection of nature as a whole. As a teacher, I see students develop in their humanities studies a proper distrust of science and technology. What concerns me is that the stigma spreads to the natural world itself. C. P. Snow's "Two Cultures," setting the sciences against the humanities, can be misunderstood as placing nature against art. The idea that the current destruction of people and environment is scientific and would be corrected by more communication with the arts neglects the hatred for this world carried by our whole culture. Yet science as it is now taught does not promote a respect for nature. Western civilization breeds no more ecology in Western science than in Western philosophy. Snow's two cultures cannot explain the antithesis that splits the world, nor is the division ideological, economic or political in the strict sense. The antidote he proposes is roughly equivalent to a liberal education, the traditional prescription for making broad and well-rounded men. Unfortunately, there is little even in the liberal education of ecology-and-man. Nature is usually synonymous with either natural resources or scenery, the great stereotypes in the minds of middle class, college-educated Americans.

One might suppose that the study of biology would mitigate the humanistic—largely literary—confusion between materialism and a concern for nature. But biology made the mistake at the end of the seventeenth century of adopting a *modus operandi* or life style from physics, in which the question why was not to be asked, only the question how. Biology succumbed to its own image as an esoteric prologue to technics and encouraged the whole society to mistrust naturalists. When scholars realized what the sciences were about it is not surprising that they threw out the babies with the bathwater: the information content and naturalistic lore with the rest of it. This is the setting in which academia and intellectual America undertook the single-minded pursuit of human uniqueness, and uncovered a great mass of pseudo distinctions such as language, tradition, culture, love, consciousness, history and awe of the supernatural. Only men were found to be capable of escape from predictability, determinism, environmental control, instincts and other mechanisms which "imprison" other life. Even biologists, such as Julian Huxley, announced that the purpose of the world was to produce man, whose social evolution excused him forever from biological evolution. Such a view incorporated three important presumptions: that nature is a power structure shaped after human political hierarchies; that man has a monopoly of immortal souls; and omnipotence will come through technology. It seems to me that all of these foster a failure of responsible behavior in what Paul Sears calls "the living landscape" except within the limits of immediate self-interest.

What ecology must communicate to the humanities—indeed, as a humanity—is that such an image of the world and the society so conceived [is] incomplete. There is overwhelming evidence of likeness, from molecular to mental, between men and animals. But the dispersal of this information is not necessarily a solution. The Two Culture idea that the problem is an information bottleneck is only partly true; advances in biochemistry, genetics, ethology, paleoanthropology, comparative physiology and psychobiology are not self-evidently unifying. They need a unifying principle not found in any of them, a wisdom in the sense that Walter B. Cannon used the word in his book *Wisdom of the Body*[4] about the community of self-regulating systems within the organism. If the ecological extension of that perspective is correct, societies and ecosystems as well as cells have a physiology, and insight into it is built into organisms, including man. What was intuitively apparent last year—whether esthetically or romantically—is a find of this year's inductive analysis. It seems apparent to me that there is an ecological instinct which probes deeper and more comprehensively than science, and which anticipates every scientific confirmation of the natural history of man.

It is not surprising, therefore, to find substantial ecological insight in art. Of course there is nothing wrong with a poem or dance which is ecologically neutral; its merit may have nothing to do with the transaction of man and nature. It is my impression, however, that students of the arts no longer feel

[4] New York: W. W. Norton, 1932.

that the subject of a work of art—what it "represents"—is without importance, as was said about forty years ago. But there are poems and dances as there are prayers and laws attending to ecology. Some are more than mere comments on it. Such creations become part of all life. Essays on nature are an element of a functional or feedback system influencing men's reactions to their environment, messages projected by men to themselves through some act of design, the manipulation of paints or written words. They are natural objects, like bird nests. The essay is as real a part of the community—in both the one-species sociological and many-species ecological senses—as are the songs of choirs or crickets. An essay is an Orphic sound, words that make knowing possible, for it was Orpheus as Adam who named and thus made intelligible all creatures.

What is the conflict of Two Cultures if it is not between science and art or between national ideologies? The distinction rather divides science and art within themselves. An example within science was the controversy over the atmospheric testing of nuclear bombs and the effect of radioactive fallout from the explosions. Opposing views were widely published and personified when Linus Pauling, a biochemist, and Edward Teller, a physicist, disagreed. Teller, one of the "fathers" of the bomb, pictured the fallout as a small factor in a worldwide struggle, the possible damage to life in tiny fractions of a percent, and even noted that evolutionary progress comes from mutations. Pauling, an expert on the hereditary material, knowing that most mutations are detrimental, argued that a large absolute number of people might be injured, as well as other life in the world's biosphere.

The humanness of ecology is that the dilemma of our emerging world ecological crises (overpopulation, environmental pollution, etc.) is at least in part a matter of values and ideas. It does not divide men as much by their trades as by the complex of personality and experience shaping their feelings toward other people and the world at large. I have mentioned the disillusion generated by the collapse of unsound nature philosophies. The anti-nature position today is often associated with the focusing of general fears and hostilities on the natural world. It can be seen in the behavior of control-obsessed engineers, corporation people selling consumption itself, academic superhumanists and media professionals fixated on political and economic crisis, neurotics working out psychic problems in the realm of power over men or nature, artistic symbol-manipulators disgusted by anything organic. It includes many normal, earnest people who are unconsciously defending themselves or their families against a vaguely threatening universe. The dangerous eruption of humanity in a deteriorating environment does not show itself as such in the daily experience of most people, but is felt as general tension and anxiety. We feel the pressure of events not as direct causes but more like omens. A kind of madness arises from the prevailing nature-conquering, nature-hating and self- and world-denial. Although in many ways most

Americans live comfortable, satiated lives, there is a nameless frustration born of an increasing nullity. The aseptic home and society are progressively cut off from direct organic sources of health and increasingly isolated from the means of altering the course of events. Success, where its price is the misuse of landscapes, the deterioration of air and water and the loss of wild things, becomes a pointless glut, experience one-sided, time on our hands an unlocalized ache.

The unrest can be exploited to perpetuate itself. One familiar prescription for our sick society and its loss of environmental equilibrium is an increase in the intangible Good Things: more Culture, more Security and more Escape from pressures and tempo. The "search for identity" is not only a social but an ecological problem having to do with a sense of place and time in the context of all life. The pain of that search can be cleverly manipulated to keep the *status* quo by urging that what we need is only improved forms and more energetic expressions of what now occupy us: engrossment with ideological struggle and military power, with productivity and consumption as public and private goals, with commerce and urban growth, with amusements, with fixation on one's navel, with those tokens of escape or success already belabored by so many idealists and social critics so ineffectually.

To come back to those Good Things: the need for culture, security and escape [is] just near enough to the truth to take us in. But the real cultural deficiency is the absence of a true *cultus* with its significant ceremony, relevant mythical cosmos, and artifacts. The real failure in security is the disappearance from our personal lives of the small human group as the functional unit of society and the web of other creatures, domestic and wild, which are part of our humanity. As for escape, the idea of simple remission and avoidance fails to provide for the value of solitude, to integrate leisure and natural encounter. Instead of these, what are foisted on the puzzled and troubled soul as Culture, Security and Escape are more art museums, more psychiatry, and more automobiles.

The ideological status of ecology is that of a resistance movement. Its Rachel Carsons and Aldo Leopolds are subversive (as Sears recently called ecology itself[5]). They challenge the public or private right to pollute the environment, to systematically destroy predatory animals, to spread chemical pesticides indiscriminately, to meddle chemically with food and water, to appropriate without hindrance space and surface for technological and military ends; they oppose the uninhibited growth of human populations, some forms of "aid" to "underdeveloped" peoples, the needless addition of radioactivity to the landscape, the extinction of species of plants and animals, the domestication of all wild places, large-scale manipulation of the atmosphere or the sea, and most other purely engineering solutions to problems of and intrusions into the organic world.

[5] Paul P. Sears, "Ecology—a subversive subject," *BioScience*, 14(7):11, July 1964.

If naturalists seem always to be *against* something it is because they feel a responsibility to share their understanding, and their opposition constitutes a defense of the natural systems to which man is committed as an organic being. Sometimes naturalists propose projects too, but the project approach is itself partly the fault, the need for projects a consequence of linear, compartmental thinking, of machine-like units to be controlled and manipulated. If the ecological crisis were merely a matter of alternative techniques, the issue would belong among the technicians and developers (where most schools and departments of conservation have put it).

Truly ecological thinking need not be incompatible with our place and time. It does have an element of humility which is foreign to our thought, which moves us to silent wonder and glad affirmation. But it offers an essential factor, like a necessary vitamin, to all our engineering and social planning, to our poetry and our understanding. There is only one ecology, not a human ecology on one hand and another for the subhuman. No one school or theory or project or agency controls it. For us it means seeing the world mosaic from the human vantage without being man-fanatic. We must use it to confront the great philosophical problems of man—transience, meaning, and limitation—without fear. Affirmation of its own organic essence will be the ultimate test of the human mind.

# THE ECOLOGICAL APPROACH
# TO THE SOCIAL SCIENCES

## F. Fraser Darling

*Fraser Darling goes further than most ecologists by placing man as well as man's natural environment under the inquiring eye of his microscope. Ordinarily, we consider the effects of nature on man and not the effects of man on nature. Darling follows the latter course and as a consequence gains insights which others miss. Legitimate problems for him are measuring the effects of marriage on crops or the effects of family life on milk production. In other words, he provides an example of the mobile mind required to confront the complexities of our age. This inventive spirit also permeates his prose, which constantly seeks effect primarily because of his ability to select unique illustrations.*

Some years ago there was a great advance in Britain in the methods of growing grass, the basic food of livestock. We learned how to grow more grass, how to lay down new pastures, to select leafy strains, and to compound seed mixtures which would give early and late grazing. Indeed, such was the thrill of power, some agricultural scientists became grass fanciers and forgot the livestock in what should have been a cow's millennium.

Nobody asked the cow.

Nevertheless, the grass fanciers were sure they were in the position to supply the best of all possible cows' worlds and could point to results in increased stocking capacity, more beef, more milk. The problem of cows' lives was solved.

But was it? The cows had a habit of searching diligently in the hedge bottoms and some were so perverse as to break out of heaven and graze the

SOURCE: F. Fraser Darling, "The Ecological Approach to the Social Sciences," *American Scientist* 39, no. 2, 1951. Reprinted by permission of the author and *American Scientist*.

roadside roughage. And lately, investigators have come upon a number of digestive disturbances and conditions which can really only be called poisoning, occurring on these improved, artificial pastures.

Within the last five years, students of animal behaviour have begun to study cows and record their observations. You can present the cow with a questionnaire, but she is inarticulate, like most human beings. Yet she has quite decided opinions and all sorts of little preferences, dislikes, and fussinesses which are important to the good life—of cows. We are learning more of how to keep cows in mental and physical health by watching them. When do they feed, when do they rest, what is it they seek in the hedge bottoms; and if they find it, how much of it do they want; what is the physical and chemical quality of the plant sought? Do cows like trees, and if so, what for—cover, browse, back-scratchers, or what? What is the structure and nature of their community life? A cow's world, you will see, is becoming a complex one, and it is quite difficult to assess scientifically the various environmental factors which influence her well-being. Her life cannot be planned from the material end with such omniscience that she can be popped down in the environment which we are assured provides the greatest good for the greatest number—of cows. She has shown us that the environment should be planned around her as a sentient organism and a personality in a social group. She had forest-roaming ancestors.

So had we.

The much greater complexity of human communities and the more baffling mental and physical sicknesses we suffer as a result of having tried to create for ourselves a grass fancier's world, are the reasons for my choosing this title for my paper. You cannot turn a highly bred dairy cow back into the forest again, and it is quite certain that we are not going "back to nature." Yet we must go on learning something of the natural history of man.

I am merely a biologist whose main interest is ecology and animal behaviour in relation to conservation, but during the last seventeen years I have been applying the methods of ecology to studying the life of the West Highland people among whom I have lived; the study was desultory for ten years or more, but has been intensive since then. The underlying principle in conservation today is to study the complexity of the habitat, the wholeness of the environment, and the relations and behaviour of the animals within it in time as well as in space; and if you can keep the habitat going, in sufficient quantity that it is not dying on the fringes, there is no difficulty in conserving any particular animal within it. The social life of the animal is now recognized as being an important part of its environment. Conservation in this sense is closely associated with the pressures between human communities and their environment and between themselves. The study of these is human ecology.

At the end of these six years the conviction has grown that the ecological approach to a study of human communities can be an illuminating one, but I would not be so bold as to say that I could now set down a sound statement of what human ecology is. Rather I have learned what a great deal we do not know, and the good idea of human ecology will need much hard thinking and careful discipline before it is good science. That is what many of us are seeking in our different ways: to make the social investigation of man into good science. Human ecology deals with the structure of animal communities which man dominates and their development through the ecological principle of succession. As Paul Sears says, "The social function of ecology is to provide a scientific basis whereby man may shape the environment and his relations to it as he expresses himself in and through his culture patterns." Perhaps in these early days of human ecology it would be better not to set it up as a science, but rather to say that human problems may be nearer solution if we tackle them ecologically.

I believe that human ecology and social science can be good science, but we should not confuse it with social service. If I may say so, the natural history of man and the emergent social sciences are not missionary endeavour. If, as scientists, we come upon an outbreak of wife-beating, the men's immortal souls and the women's suffering backs are not our primary concern, as investigators. We would seek causes for the phenomenon, and possibly find it in a hectoring foreman and the operation of peck-orders. Doormats among animals and men have a habit of being hard on their females and children.

There are different levels in what might be called the social management of man. These are exploration and fact-finding, research and the development of ideas, application, and maintenance. We should not confuse the first two of these strata with the second two.

I was once asked by a social anthropologist what human ecology was that social anthropology was not. This was a very right and proper question to which the reply should be that there is no difference. But I ventured to say that human ecology deals essentially with *process*. The value of the ecologist in society will be in his power and accuracy in elucidating causes and forecasting consequences.

## The West Highland Problem

The relatively small West Highland and Hebridean populations live close to their physical elemental environment, and to the natural resources on which they have largely depended for their existence. It seemed to me, when I began the West Highland Survey seven years ago, that the problem of the Highlands should be investigated from the biological point of view, looking on the people—without the least disrespect—as members of the indigenous fauna and social animals, and inquiring what were the factors of change in the environment, or in them, which were rendering man a slowly failing species in that environment. This was an essay in human ecology, the approach of a naturalist in conservation as contrasted with the economic attitude of mind which tends to be that of the grass fancier towards the cow. The West Highland problem cannot be described here in detail, but will serve for illustration of what I consider to be the ecological approach to the study of social behaviour.

Broadly, the Highland problem is that of a *very old and in many ways primitive human culture existing in an administratively awkward and physically refractory terrain set on the fringe of a highly industrialized urban civilization,* which itself is situated in one of the greenest, kindest lands on earth. Highlanders have been part and parcel of our national structure for only two hundred years, having until then lived a very different kind of life, in standards, laws, language, and techniques, than had the rest of Britain. Yet Highlanders are not New Hebrideans or Eskimos over whom, try as we may to the contrary, we feel some kind of mental superiority. Here is a race of people of probably greater average intelligence and intellect than the dominant group, indistinguishable from it in physical appearance. And as members of this race moved so smoothly and successfully in the dominant civilization, it was overlooked how different were the inner rhythm of life and the style of thought and tradition. The new centralized British government of that day merely extended its administrative, economic, and social regime to include the Highlands, and with some ameliorations and some encrustations this applies today.

In human ecology we can never neglect history, for we are studying process; I would say, therefore, that a cross-sectional social survey is not ecological unless it studies origins and successions, in other words, process. We must always remember the significance of political action as an environmental factor. For example, the manipulation of the Salt Tax in the last part of the eighteenth and early nineteenth centuries had profound results on the lives of Highlanders, and the transposition of the English system of poor relief had some fantastic consequences. Again, imagine the island of Islay being immune from Spirit Duty, as it was in the late eighteenth century: distillers flocked in, the bread corn of the people was deflected to whisky, the distillers were soon making money advances (at their own rates) on the barley crops of small tenants; drunkenness was rife and the people were reduced to an appalling social state. The detailed research into population movement conducted by the West Highland Survey shows that this favoured island has suffered more than any other part of the Highlands from excessive emigration.

Another historical factor at the root of the Highland problem of today, is the exploitation which the natural resources suffered in the past. The Tudor monarchs in England were already conscious that the supply of oak was dwindling, and there were prohibitions on the felling of English oak. This sent the shipbuilders northward to the Scottish forests; and a hundred years later, when the iron districts of Surrey and Sussex had lost their trees, there was a determined attack on Highland forests to provide fuel and charcoal. The iron ore was shipped up there. The ultimate disappearance of the forests followed the introduction of sheep-farming on the extensive, extractive system in the second half of the eighteenth century.

The countryside was one of steep hills, initially poor rocks, and of high rainfall. The climax vegetation which conserved fertility was broken, and there was rapid deterioration of the habitat. That is the core of the problem today: the people are living in a devastated habitat. And now we come to another important ecological factor, the age of the culture. The Gael is living where he has lived for several thousand years and is tenacious of place and culture. How different from North Wisconsin, where settlers went in to still virgin forests in the 1920's, devastated their environment in a very short time, and left! "Ghost towns" remain. The administration which furthered the movement had forgotten the podsol conditions of the soil in relation to climate. A hetero-geneous aggregation of people would not continue to inhabit a devastated terrain in the way an old culture hangs on to its place, even in decay.

Before leaving the historical aspect, we might consider briefly the effect of a change of food habits. Dr. Salaman of Cambridge has recently published his great book, *The History and Social Influence of the Potato;* it is a mine of wealth for the human ecologist. The acceptance of the potato as the staff of life allowed an immense increase in the number of mouths so long as a low standard of existence was accepted. The history of the west of Ireland and the West Highlands and islands of Scotland—both places where wheat was not grown and where the bread corn was relatively difficult to harvest—shows that the potato, coinciding with the practice of vaccination, did bring about a swarming of the population and a very marked depression in the standard of living. Potatoes and maize meal were staples of diet at the most chaotic period. Arthur Young tried to make the potato the food of the rural working class in England; Cobbett fought the potato school tooth and nail, and the English labourer stood firm by his wheaten loaf.

We may take it for granted that when a countryside begins to feed on much the same diet as its pigs, social problems are piling up ahead. And that is the right order; the change in diet precedes the social trouble. The human ecologist will never neglect the belly of the people. Professor Paul Sears has noted an interesting situation that occurred in Mexico. The government had prohibited the fermentation of a beer, pulque, from a plant called maguéy that is grown as a stiff hedge of spiky leaves. The result of the prohibition was a high incidence of diseases associated with deficiency of vitamin B, and only when the plant was ceasing to be grown was it discovered that it was one of the most efficient anti-erosion plants on the plateau.

To return to the Highlands: the destruction of the forests has meant the removal of cover, and this environmental factor is of great importance in human lives. Humanity needs cover for all sorts of things—shelter for crops and stock; cover to enable a man to do a little experimentation which he dare not try if the eyes of every household in the township are upon him; and cover for courting and love-making. It is obvious what a social problem lack of cover imposes in certain types of urban communities. In the Highlands it has imposed a set of conventions almost the exact opposite of our own. Darkness is the only cover, but this is supplemented by a build-up of psychological cover. The Tiree crofter visits the Duke of Argyll's factor on the nights of no moon, though he could just as well go in the day. A fellow and a girl in the Hebrides will ignore each other in daytime should they meet on the road, but he will be calling at her home just about the time of night when in our culture we should have taken our leave. Good manners require that he be gone before it is light.

I have mentioned the value of cover in experimentation. We tend to forget how important it is in primitive communities that people should not be different, and the initial attempts to be different are the most dangerous ones. Think how in our own lives we like to experiment in private and avoid being different in the beginning. The Anglo-Saxon races have a firm belief in the power of demonstration in changing methods of doing things. This is a fallacy. The Gael or the Mexican is wiser. It does not matter that a changed practice will reap him a bigger material reward. That is not recompense for having to that extent placed himself outside his group. If the material reward is real, he will be envied by his fellows, and that is not a good state to be in. If the reward is illusory, he will be ridiculed, and that is not good either in a society where there is no privacy.

I have seen the sudden loss of cover depress a small community psychologically, because of the sudden cessation of the opportunity to grow flowers and fruit. Nor should we neglect an animate factor such as the rabbit as a creator of deserts, and as an animal weed of poor land. I have seen a community give up all effort at gardening because of rabbits, and looking forth on a deteriorating habitat fostered psychological ills of frustration and ultimate indifference. Village halls do not correct this kind of situation. The first requirement is a coordinated scheme of habitat rehabilitation. It is in this way that the Tennessee Valley Authority has been such a splendid ecological project.

## Deterioration of the Habitat

The science of ecology deals with causes of observed biological phenomena, and it should be expected to lay bare multiple-factor causation, which is a very difficult field. But it is also concerned with consequences and ramifications. The practical value of ecology, as I have said, is the ability to forecast consequences of certain courses of action and of observable trends. The politician has to be very careful here, and I would suggest that the ecologist is as necessary a servant to the statesman as the economist. Let me take examples from the Highland problem. I have said that the destruction of the region's greatest natural wealth, its forests, was followed by the establishment of large-scale sheep-farming on the ranching, extractive system. The immediate social consequences of this were unfortunate, in that the people were pushed to the coast and suffered a forcible social break. This kind of sheep-farming meant a very heavy preponderance of sheep over cattle, and I have managed to discover in detail how this style of grazing destroys the habitat over a period of a century or so.

The soil is in general sour and peaty, and the roots of trees reaching down to the rock and possible glacial drift were an essential means of bringing mineral matter of a basic nature to the surface. First it went to the leaves, and as a proportion of the calcium-rich young leaves were eaten by caterpillars, there was a rain of their faeces onto the surface of the ground, where they were consumed by earthworms, which are so necessary to the British terrain to the production of a porous, well-mixed soil. Removal of the trees has broken the circulatory system of basic salts and destroyed the continuum. Earthworms disappear if the calcium level of their medium is not maintained; the soil becomes a tough, peaty skin and loses its absorptive as well as its nutritive qualities. Sheep graze much more selectively than cattle and tend to remove the more palatable components of the herbage, especially the ameliorative legumes. Sheep also neglect tall and toughened herbage, so that burning of the terrain is necessay when the sheep-cattle ratio is wide. This practice in itself impoverishes the variety of the herbage, helps the spread of the bracken fern, and tends to produce a biotic climax of a few dominants of poor nutritive quality. Burning on peat slopes also tends to produce an impervious surface which accelerates lateral runoff. This runoff water, being heavily charged with carbonic acid as it runs over the acid peat, itself helps in souring the land in the glens. This is a story of impoverishment of habitat by imposing a foreign land use.

Where the sheep-cattle ratio is grossly disturbed, conditions for a peasantry become desperate. I have now reached the stage in the Highlands when I can say: "Tell me the cattle-sheep ratio in an area, and I shall know the social health of the people." If the ratio is wide, 30–50 or more sheep to one cattle-beast, there is serious trouble; if it is under 10, things are not so bad. One can also correlate the cattle-sheep ratio with the age-structure of the population. Another thing that becomes evident is that it is the children who keep milk cows on the land; when the age-structure gets top-heavy like that of Assynt in 1931 (Figure 1), down goes the number of milk cows.

The descending spiral of fertility of the general habitat, as outlined above, is continued on the inbye land of the croft: when a man replaces his cattle with sheep, he finds he has no manure for his arable plot, and the yields go down so far that his capacity for winter cattle is decreased. He also finds that he must bring his ewes onto the inbye land to lamb, and there they stay, nibbling the heart out of the grass until the end of May. Such meadow land cannot be expected to yield a good crop of hay, and being relieved of grazing so late means that the hay crop is not ready to cut until a time when heavy rain is general. This means the hay will not be gotten well and its nutritive value will be poor, so that once more the ability to maintain a cattle stock and the fertility of the arable land is being assailed. It is quite definitely an ecological story, and to attempt to study social and economic problems apart from the biological background would be to blindfold oneself.

Now, where do the politician and economist conflict with the ecologist in the example just given? You may have heard that ten years ago hill sheep-farming was not paying, yet a supply of hill sheep was necessary for the stratification of crosses leading to the low-ground farms, which were paying. The economist finds many good reasons for the discrepancy, though deterioration of the habitat by the hill sheep is not one of them. He says: "We must take some of the high profit from the fat-lamb end of the chain and put it back at the fountainhead." And this has been done by giving a substantial subsidy to hill ewes. The idea may have worked well in the Southern Uplands of Scotland, but in the poor terrain of the Highlands the ewe subsidy might have been specially designed for further deterioration of the habitat and for fostering social unhealth. The politician says we cannot start differentiating between one countryside and another in a measure of this kind. All he could do was to slap a still bigger subsidy on hill cattle and another new one on calves. The economists here will admit that this is a dangerous path to follow, and I as an ecologist will say, from close observation of this particular measure, that the ultimate good it can do is negligible unless it is linked with vigorous rehabilitation of habitat, which is the basis of social health.

### Other Sources of Social Problems

Depopulation and distortion of the age-structure go together and bring a new set of social problems. People in and out of the Highlands have said often enough that industries should be established there, industries of the kind where wheels go round in an important way. But what do we find? Where such industry has been established, there has been even greater depopulation in the adjacent rural areas, yet the big problem is how to maintain dispersion. Fort William and Kinlochleven may have provided Britain with aluminum, but they have created new social problems and solved none.

The remoter areas of the Highlands need roads and better transport, and scarcely anyone can be found to question the benefit that might accrue. But again, this obvious measure of amelioration must be considered ecologically. For example, I happen to know well the townships on either side of a long sea loch, one side of which has a road and the other has not. The living conditions of the people either side are different. Those on the road buy Glasgow bread (untouched by hand) and packeted goods of all sorts, and I

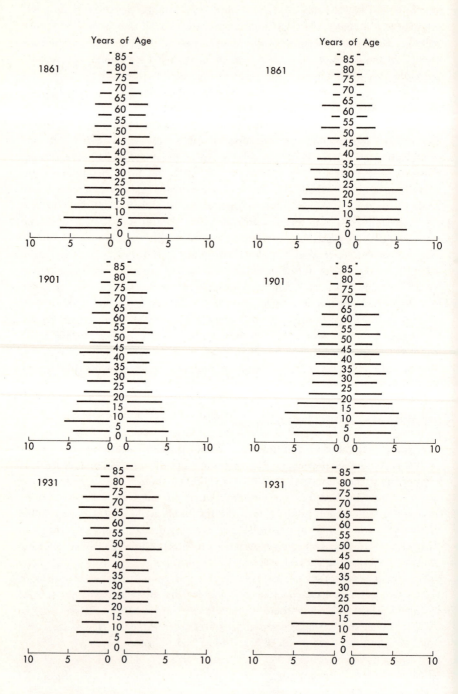

have seen tinned porridge sold from the vans. The communities are absolutely dependent on the vans, and their standard of husbandry is low. On the other side of the loch, more cows are kept; cheese and butter are made; homemade oatcakes and porridge are the cereal staple rather than bought bread; the men fish more, and the standard of husbandry is higher. So what has the road done? It has given those people the benefit of our well-known brands of this and that and a daily paper. But it has not so reorganized the habitat that the so-called higher standard of living can be paid for out of the greater amount of produce exported. Indeed, quite apart from the loss of social health and skills, these people are in a worse economic plight. On the roadless side there is still self-sufficiency, competence, and a realization that the croft must be well farmed. A road can be a benefit only if the environmental factors are closely studied and integrated. Here is seen clearly the effect on this small, old, subsistence culture, of being on the fringe of the most highly developed urban culture of its day. Had most of Britain been like the Highlands, the impact might have been less severe.

The ecologist asks that unquestioned beliefs should be questioned. Good communications is one of these; education is another. Consider, for example, the problem of educating the Reindeer Lapps in Scandinavia. How do you do it? The convenient way is to put the children in schools in the winter season when the Lapps are at the southernmost end of their pastoral migration. But if this is done the families and their reindeer are unduly immobilized, and the secret of pastoralism in poor terrain is to keep on the move. The winter range of the Reindeer Lapps is thus being overbrowsed, and as it is the amount of winter range which determines numbers of livestock, the damage to the birch and willow forests means that the high summer potential of the tundra is being less used. We can still believe in education, of course, but at least let us ponder methods of applying it, in terms of consequences on the habitat.

I have mentioned depopulation and the distortion of the age-group pyramids. The problems of human ecology arising from the phenomenon in small communities are manifold. In the first place, the old remain in power and so prevail that they can initiate an era of reaction in the life of a community, so that in a region of hard-shell Presbyterianism all gaiety for the young is frowned upon. And nowhere do the young show greater consideration for the old than in the Highlands. I know of townships where there are

Figure 1. (opposite page) Population pyramids for two localities in the West Highlands. Age classes have been reduced to percentage of total population, with males represented at the left and females at the right of each pyramid. Series at left is for Assynt, a village on the mainland, showing top-heavy age-structure in a deteriorated habitat. Series at right is for the island of South Uist, Outer Hebrides, a somewhat healthier social environment.

but few married couples now. Brothers and sisters have cared for their old folk, and now that they are gone they continue living in their parents' houses and cannot bring themselves to the considerable upset of getting married. The social urge and necessary gaiety are not there. This depression of the vivid social life of man is likely to lead to such undesirable consequences as burning of the hills in an excitement bordering on hysteria. The crass burnings of the heather are made ostensibly to further the growth of young grass; in actual fact they further the devastation of the habitat. The fires occur at Beltane, which was once the breeding season. Where the social life is in better order, burning is under control. That great American ecologist, Charles Adams, who has now turned his attention to mankind, told me recently of an almost identical phenomenon in one of the southern states, and of how the problem had solved itself with improvement in social conditions.

It is difficult to avoid the impression that religion is a considerable ecological factor, but it must always be related with other environmental characteristics. The areas of most pronounced depopulation in the Highlands have the harshest sects of Presbyterianism; but I do not want to overdo this idea or give a wrong impression, because the area of greatest congestion, Lewis, follows the same faith. What I would say is this: that in the Highlands a small, remote community with poor services would have more chance of survival if it were Catholic than if it were Free Presbyterian. This is because there is more sense of community to be found in the districts of the old, liberal—almost Columban—style of Catholicism. The culture is stronger altogether; music and folk tales have not been dimmed, and the status of women is higher. Birth rates are exactly the same. It is in these small, isolated communities, where the social pattern of humankind can scarcely be completed, that a factor which is associated with the old culture can be critical.

The human ecologist must always be on the lookout for these marginal factors, the comprehension of which may illuminate a much wider field where complexity defeats scientific investigation.

There is one more illustration that I want to give from the Island of Lewis, which, as I have said, is a congested island. The terrain is poor, but the people have been there for 4000 years or so, with various immigrant waves which have accepted the old culture and have not imposed their own. The old Celtic custom of subdivision of land, and the intense conviction that the land is theirs, have resulted in the island's being entirely held by crofters, all doing much the same things. It is a one-class society worthy of very close study. Weaving has given prosperity, and though the land is tending to be neglected because it is more profitable to weave, the people cannot effect the social revolution of relinquishing at least lip service to subsistence husbandry, and thereby achieving division of labour and social stratification. There is an intense social life from house to house among the young in Lewis, who are numerous enough to maintain a fine gaiety in the face of religious proscription, but there is little knowledge of the constructive or artistic use of leisure. Nearly everybody is a peasant except for a handful of professional people in Stornoway. Prosperity has come as money—pound notes—but in rural Lewis there is nothing much to spend money on.[1] Social evolution would seem to have stuck, and needs a catalyst. Lewis will not allow itself to evolve, and the observer cannot help comparing the tremendous social vitality maintained by the good proportion of young folk, with the stricken life of the dying communities on the mainland shore. The right hand of Lewis reaches out for all that the world can offer, but her left hand holds fast to the croft in the unenclosed township, and she is anchored in time. The fact that the crofting townships are unenclosed, precludes differentiation of husbandry and agricultural improvement.

I want to close these remarks on the natural history of society by pointing the obvious: that tradition and accumulated experience are part of man's environment, and for all the importance of the physical and biological factors I have mentioned, the ethos is still the biggest ecological factor of all on the life of the individual. Here I would digress for a moment on methods of approach in gathering data. The ecologist must distrust the questionnaire so beloved of the sociologists, because it fails to take sufficient notice of the ethos of a people. The questionnaire will not necessarily give you scientific data. In the course of the West Highland Survey we compiled a punch-card Domesday of factual data about crofting townships and it is immensely valuable, but we never asked questions on personal household matters or questions of opinion. Had we done so we should either have come up against a brick wall or, with such a sensitive and penetrating people, we should have got the answers they thought we should like. Much the best way is observation and soaking in the culture. Ability to observe closely and interpret accurately, by way of a large grasp of the organism of a society in its habitat, is the essence of human ecology. It is an integrative science as much as an analytical one, with observation as its basis.

If the psychologists could devise courses in development of the power of observation as part of the training for a research career, we should at least be able to pick out at an early stage those graduates who are fitted to study man as a social animal. After that must come the faculty to use several disciplines. Teamwork in human ecology will be essential, but still each specialist will have to have the quality of delighting in another man's work and linking his own to it; and he cannot be the traditionally remote academic type, but must be inquisitive about what humankind is doing to itself.

[1] At the moment of going to press, the export market for Harris tweed has suffered a relapse; there is depression in the weaving districts of Lewis.

# MAN'S PLACE IN NATURE

## Marston Bates

We cannot command nature except by obeying her.

—Francis Bacon, in *Novum Organum*

*Like Rachel Carson, Marston Bates saw our present situation clearly when others considered talk of the environmental crisis hysterical. He was especially aware of the dangers of unlimited population growth. Extracted from* The Forest and the Sea, *this selection, as the book title suggests, is the result of a comparative study of rain forests and coral reefs. Professor Bates sees these two extreme environments in the light of their similarities and their position in the larger eco-sphere. A practicing humanist, he is constantly aware of the ethical implications of his biological studies; and this connection is reflected in his prose, which is unique for its wise compassion. The student of writing should pay particular attention to the relation between subject matter and the quality of the speaker's voice in the following essay.*

We started . . . in the tropical forest, thinking about the sea, looking at the similarities in the way life is organized in these different circumstances. This led us to some reflections on the continuity of life in space, in the biosphere, and on its continuity in time, in evolution. The grand design of this system of life includes many different patterns—seas, reefs, lakes, rivers, forests, grasslands and deserts—and we looked at some of these. The structure of the system is everywhere similar, though everywhere complex, turning on the relations among individuals, populations and communities of organisms, and on relations with the physical environment. We have only glanced at these relations, though I hope the glance has been lingering enough to reveal some of the infinite possibilities for study and contemplation.

Then we came to man and his place in this system of life. We could have left man out, playing the ecological game of "let's pretend man doesn't exist." But this seems as unfair as the corresponding game of the economists, "let's pretend nature doesn't exist." The economy of nature and the ecology of man are inseparable and attempts to separate them are more than misleading, they are dangerous. Man's destiny is tied to nature's destiny and the arrogance of the engineering mind does not change this. Man may be a very peculiar animal, but he is still a part of the system of nature.

From the point of view of zoological classification, man is easy enough to deal with. To be sure, there are all sorts of variations among men in color of the skin, eyes and hair; in texture of hair and in its distribution over the body; in details of the face and in bodily physique. But no special study is needed to show that these varieties breed together easily and freely whenever they come in contact: that they form a single species.

The classification of the varieties does present problems. They can be lumped together into three or four main types, or they can be split into thirty or forty different races. Whatever system is used, many individuals will be found that do not fit into any category. The varieties, however, show a rough geographical pattern and it is likely that the differences arose through geographical separation, a common enough phenomenon with many kinds of animals.

This human species . . . can logically be placed in a genus and family by itself; and men, along with the great apes, Old World monkeys, New World monkeys and a few other animals, can be grouped as an order (the primates) in the general class of mammals.

From the point of view of ecology, man is less easy to deal with. Essentially he is a predatory animal, a second- or third-order consumer. But he shows a tremendous variety of food habits and in many parts of the world he is primarily a first-order consumer, living directly off vegetation. This is probably a rather recent development (in geological terms) because man's plant-eating habits depend to a large degree on processing with fire. The tubers and grains that make up the basic starches of the human diet in most parts of the world are inedible for man unless cooked. He can and does, on the other hand, eat meat of all sorts (including fish and molluscs) without cooking. His vegetable diet, without fire, would be limited to things like fruits and nuts.

It is hard to define the human habitat, because men are found everywhere on land and around the margins of the seas, except under conditions of extreme dryness or extreme cold. But this wide ecological distribution depends on cultural rather than biological traits: on the use of fire, the wearing of clothes, the construction of shelters, the management of boats. It looks as though naked, uncultured man would be a tropical or subtropical species adapted to the rain forest or to transition zones between forest and scrub or forest and grassland. But we have no specimens of uncultured man for study.

When we try to study the relations of man with the physical and biological environment, we always come up against the problem of how to deal with cultural traits. Should we consider culture as a part of man, as an essential attribute of the human species; or should we consider culture as a part of the environment in which this human species lives? This seems like a quibbling, academic sort of question, but it has worried me for a long time. And this specific and striking case has contributed to my general feeling that it is often misleading to attempt to distinguish sharply between organism and environment, whether dealing with men or mice. We are always concerned with interacting systems—which sometimes act as single systems.

I have no substitute for the idea of environment, and I wouldn't know how to get along without the word, but it is tricky. To get back to the problem of man: it seems to me that in general, psychologists tend to treat culture as a part of the environment, while anthropologists tend to regard culture as a part of man, as a part of his equipment for dealing with the environment. To put it another way, psychologists tend to regard culture as a constant, something to which all men are subjected, and they are interested in the ways in which individuals cope with this: how they learn to conform or, consciously or unconsciously, to rebel. They find that frustrations, joys, neuroses, psychoses, all sorts of human behavioral patterns, derive from the interactions between animal man and this cultural environment. At least it often seems to me that this is what they are doing—though the psychologists themselves are not particularly apt to use the word *culture*.

In contrast, anthropologists tend to take "human nature" for granted, to regard animal man as more or less the same everywhere and to explain all differences in cultural terms. They find culture to be adaptive: the Eskimo way of life fitted to the arctic and subarctic environment, the Dyak way of life to the rain forest conditions of Borneo. They are preoccupied with the description and analysis of all the different kinds of culture they can find, with the study of cultural evolution, and with the effects of cultural diffusion and contact.

In its extreme form, this point of view in anthropology ignores animal man altogether. Culture, I suppose they would have to admit, could not continue without continuing men, and men in this sense created the cultural forces. But men have long since become the helpless victims of their cultures, and developments go on inexorably, according to the laws governing cultural evolution, regardless of the will or desire or power of any individual.

From the point of view of biology, it is most convenient to treat culture sometimes as an attribute of man, sometimes as a part of the human environment. The biologist, trying to look at the human species, cannot think in terms of man and environment: he must deal always with man-culture-environment. Pygmies in the Congo, Bantu in the Congo, Belgians in the Congo, are all men, *Homo sapiens,* in a particular geographical and ecological environment. But they behave quite differently; the environment has quite different meanings for them; they, in turn, have quite different effects on the environment. The whole ecological situation is different.

The problem of man's place in nature, then, is the problem of the relations between man's developing cultures and other aspects of the biosphere. The understanding of these is greatly handicapped by the way in which we have come to organize knowledge. To be sure, man with his varying cultures and cultural traits forms a special phenomenon which requires special means of study and the accumulation of special sorts of information. But still, man has not escaped from the biosphere. He has got into a new, unprecedented kind of relationship with the biosphere; and his success in maintaining this may well depend not only on his understanding of himself, but on his understanding of this world in which he lives.

This makes the split between the social and biological sciences particularly unfortunate. *Economics* and *ecology,* as words, have the same root; but that is about all they have in common. As fields of knowledge, they are cultivated in remotely separated parts of our universities, through the use of quite different methods, by scholars who would hardly recognize anything in common. The world of the ecologists is "unspoiled nature." They tend to avoid cities, parks, fields, orchards. The real world of the economists is like Plato's, it is a world of ideas, of abstractions—money, labor, market, goods, capital. There is no room for squirrels scolding in the oak trees, no room for robins on the lawn. There is no room for people either, for that matter—people loving and hating and dreaming. People become the labor force or the market.

More and more, in all areas, we tend to separate the study of man from the study of nature. The separation is one of the basic lines of division in the way we have organized knowledge, in our pattern of specialization. The natural sciences and the social sciences exist in practically complete isolation from one another. Man's body, curiously, has been left with the natural sciences while the social sciences have taken over his mind—at a time when we are most aware of the artificiality of the body-mind separation.

Our third great division of basic knowledge, the humanities, has long since forgotten about nature. Joseph Wood Krutch can well remark: "There are many courses in 'The Nature Poets' in American colleges. But nature is usually left out of them." Surely there is some way of putting all of these things together, of achieving a more balanced view of ourselves and the rest of the natural world. The matter, I think, has some urgency.

Ours has been aptly called the age of anxiety, and this is curious. We should be able to look about us and feel a certain self-satisfaction. We have learned to develop and direct tremendous power; we can create the kind of conditions we find comfortable; we can produce large quantities of a great variety of foods; we have achieved a surprising degree of control over disease and physical pain. In almost any way we assess man's relations with his environment, he seems to be doing well when compared with the past, even though there is still obvious room for great improvement.

Yet, despite this abundance and progress, almost all attempts to look at man's future are gloomy. I can't think of any recently written image of the future that sounds very attractive, even when the author was trying hard to look for glories. The glories mostly turn out to be bigger and better gadgets, faster trips to a dismal Mars, or better adjusted husbands and wives who no longer take to drink. Usually the author looking into the future doesn't pretend to like his 1984 or his brave new world: but looking about him, this is what he sees coming.

Our anxiety about the future, when we analyze it, turns largely on three related things: the likelihood of continuing warfare, the dizzy rate of human population growth, and the exhaustion of resources. But these don't look like insoluble problems. Surely men who can manufacture a moon can learn to stop killing each other; men who can control infectious disease can learn to breed more thoughtfully than guinea pigs; men who can measure the universe can learn to act wisely in handling the materials of the universe. Why are we so pessimistic?

Chiefly, I suspect, because we have come more and more to doubt our ability to act rationally. Reason seems to be a property of individual men, not of the species or of organized groups. Somewhere we have lost the faith of the eighteenth century French philosophers in the perfectibility of man, and the rather different faith of the nineteenth century in the idea of progress.

Maybe the anthropologists are right when they say that culture acts as a thing in itself, sweeping along according to inexorable laws, no more under man's control than rodent evolution is under the control of the mice in the fields. The difference between men and mice, then, would be a matter only of awareness, of self-consciousness. We can study the laws of cultural evolution—or organic evolution—but we can't change them. We can foretell our doom but we can't forestall it.

I don't believe this, and I doubt whether the extreme culturists really believe it either. If they believed what they say, I think they wouldn't talk so much. They are like the disciples of Karl Marx who say they believe in the inexorable dialectic of history, but continually try to give history a push in the right direction. Man can't change the laws of cultural evolution or organic evolution—true enough, no doubt—but understanding the laws and acting with the laws, he can influence the consequences. He has in his hands a certain measure of control over his destiny, but this control depends on understanding, and on the spread and proper use of knowledge.

The great immediate threat, of course, is the misuse of nuclear power, the danger of catastrophic war. The long-term threat is the cancerous multiplication of the numbers of men: a new human population the size of the city of Detroit every month, year after year. The thought is dizzying. And then the thought of a nuclear blast capable of killing last month's millions in a few seconds is hardly reassuring. It looks as though, as a part of nature, we have become a disease of nature—perhaps a fatal disease. And when the host dies, so does the pathogen.

How, in the face of our power, in the face of our danger, do we develop a guiding philosophy?

No single man, no single field of knowledge, holds the answer to that. But all men and all knowledge can contribute to the answer. Insofar as man's relations with the rest of nature are concerned, I think we must make every effort to maintain diversity—that we must make this effort even though it requires constant compromise with apparent immediate needs. To look at this, it may be most convenient to sort out the arguments into those that are primarily ethical, those that are primarily esthetic, and those that are essentially utilitarian.

Albert Schweitzer remarks in his autobiography that "the great fault of all ethics hitherto has been that they believed themselves to have to deal only with the relations of man to man." This is particularly true in the Western, Christian tradition. The present material world, in the philosophy of this tradition, is unimportant, no more than a transient scene for the testing of the soul's fitness for eternity. The material universe is completely man-centered. Nature, insofar as it is noticed, is only a convenience—or a temptation—with no positive value in itself.

Animals are unimportant because they have no souls. God may notice the sparrows, but this is an example of His omniscience rather than of His preoccupation. Even Christ gave no thought to the Gadarene swine. The first arguments against bear-baiting, cockfighting and the like were not that they were liable to cause injury and pain to the animals, but that they were liable to demoralize the human character, leading to gambling, thievery and the like.

For a considerable part of humanity, however, this world has direct religious significance. Many primitive religions have various forms of nature worship, of animism and totemism. But in some of the great religions, particularly Buddhism and Hinduism, attitudes toward nature—toward animals in particular—have an ethical basis. For many millions of Hindus it is a sin to kill any animal. With the Jains, this is carried to an extreme to avoid possible injury even to the tiniest of insects.

We deplore the Hindu attitude toward cattle as uneconomical—which it certainly is—and a handicap to the development of India. In countries within the Western tradition, however, attitudes toward animals often cannot be explained on practical or rational grounds. I suspect that a visitor from Mars, observing our treatment of dogs, cats and other domestic pets, would conclude that they were sacred animals. Horses in some Western subcultures are also treated as sacred animals. The horror of eating horse meat—or dog meat—seems not too different from the Muslim horror of eating pig or the Hindu horror of eating any animal.

There have always been individuals within the Christian tradition with a love of nature, with a kind feeling toward animals. St. Francis of Assisi rightfully is their patron. In modern times this has grown into a cult of great emotional force, leading to the development of a variety of formal organizations for the prevention of cruelty to animals, for the protection of wildlife, which reaches an extreme in the anti-vivisectionist groups. This attitude is most highly developed in the industrialized regions since it goes along with economic security and relative leisure. It is a characteristic of "affluent societies." It is reassuring in the sense that kindness and tolerance and sympathy—whether for slaves, for children or for animals—seem to gain force and spread with economic development.

This kindness and sympathy for animals might well be classed as an ethical attitude. Curiously, along with the cult of kindness to animals, we have a parallel development in the same societies and circumstances of the cult of the sportsman, in which killing becomes a good in itself. As hunting ceased to be a necessity, it became a luxury for men; and hunting as play, hunting as sport, has long characterized classes of men with the leisure to indulge in it. Hunting is sometimes thought to represent a basic "instinct" in human nature, and certainly there is something elemental and primitive in the thrill of the chase. Intellectually, I have abandoned hunting as a sport since, when a boy, I watched the agonies of a raccoon I had wounded. But often enough, hunting for some worthy "scientific" purpose, I have felt my intellectual pretensions slide away and I have become lost in the purely emotional absorption of getting my game.

The sport of kings and noblemen has now become the sport of millions, of anyone with an automobile and a rifle or shotgun. It is recreation. But also a philosophy has developed whereby this killing of deer and ducks and quail is supposed to inculcate virtue. Krutch quotes the propaganda slogan of a gun company: "Go hunting with your boy and you'll never have to go hunting for him."

I get lost in the ethical issues involved in these problems. Intellectually I sympathize with the teachings of Buddha, that all life is sacred. But practically, I see no way of acting on this. There is no logical stopping place before the end reached by the people of Samuel Butler's *Erewhon*. They became vegetarian out of respect for the rights of animals. But as one of their learned men pointed out, vegetables are equally alive, and equally have rights. So the Erewhonians, to be consistent, are reduced to eating cabbages certified to have died a natural death. Monkeys, deer, cows, rats, quail, songbirds, lizards, fish, insects, molluscs, vegetables—where do you draw the line between what can be properly killed and eaten, and what not? It so happens that I don't like decayed cabbages and I do like rare roast beef—which leaves me, as usual, blundering around in a quandary.

The ethical question is difficult. We have drifted in the modern world into a position of ethical relativism which leaves us with no absolutes of good and bad, right and wrong. Things are good or right according to the context, depending on the values of the society or culture. Yet one feels that there must be some basis of right conduct, applicable to all men and all places and not depending on any particular dogma or any specific revelation. Science has undermined the dogmas and revelations; and it provides, for many working scientists, a sort of faith, a sort of humanism, that can replace the need for an articulated code of conduct. But our scientists and philosophers have so far failed to explain this in a way that reaches any very large number of people. This, it seems to me, is one of the great tasks of modern philosophy, which the philosophers, dallying in their academic groves, have shunned.

When some thinker does come forth to provide us with a rationale for conduct, he will have to consider not only the problems of man's conduct with his fellow men, but also of man's conduct toward nature. Life is a unity; the biosphere is a complex network of interrelations among all the host of living things. Man, in gaining the godlike quality of awareness, has also acquired a godlike responsibility. The questions of the nature of his relationships with the birds and the beasts, with the trees of the forests and the fish of the seas, become ethical questions: questions of what is good and right not only for man himself, but for the living world as a whole. In the words of Aldo Leopold, we need to develop an ecological conscience.

It is sometimes said that the esthetic appreciation of nature is relatively new, that the Greeks, for instance, did not admire landscapes. The matter can be argued and I don't know that anyone has made a careful study of changing attitudes, or of differences in attitude among the great civilizations. Within our own civilization, it looks as though the conscious appreciation of the beauties of nature had its roots in the so-called Romantic movement of the eighteenth century. We can see this most plainly in literature, in landscape painting and in landscape architecture. It is less clear in the other arts, though Lovejoy plausibly equates it with the love of diversity and the search for new forms that characterize Western art generally in the last two centuries.

It looks as though man's esthetic appreciation of nature increases as the development of his civilization removes him from constant and immediate contact with nature. The peasant hardly notices the grandeur of the view from his fields; the woodsman is not impressed by stately trees, nor the fisherman by the forms and colors on the reefs. In part, this is the general problem of not seeing the familiar, of not appreciating what we have until it is lost.

The reasons behind the conservation movement, from this point of view, are similar to the reasons for preserving antiquities, for maintaining museums of art or history or science. Nature is beautiful, therefore it should not be wantonly destroyed. Representative landscapes should be preserved because of their esthetic value, because of their importance in scientific study, and because of their possibilities for recreation.

I have often wished, as I saw a tropical forest being cleared, that this beautiful place could somehow be protected and preserved for the future to enjoy. The idea, to the people involved in the clearing, seems absurd. The forest is an enemy, to be fought and destroyed; beauty lies in the fields and orchards that will replace it. This was the attitude of our ancestors who in the end effectively cleared the great deciduous forest that once covered the eastern United States, leaving only accidental and incidental traces. How we would love now to have a fair sample of that great forest! But the idea of deliberately saving a part of the wilderness they were conquering never occurred to the pioneers. Nor does it occur to pioneers now in parts of the world where pioneering is still possible.

There must be some way in which one nation can profit by the experience of another nation; some way of saving examples of the landscapes and wildlife that have not yet been devastated by the onrush of industrial civilization. In Africa there is a danger that the national parks will be regarded as toys of colonial administrations, and fade with the fading of those administrations. And the colonial powers, even with the experience of loss in their homelands, are not always too careful about the preservation and maintenance of samples of the natural world under their care.

In tropical America we have the effect of the Spanish tradition. The Romantic movement never crossed the Pyrenees. Spanish thought and art remain essentially man-centered. Some of my Spanish friends have suggested that the relative failure of science to develop in that tradition may be a consequence of this indifference, on the part of most of the people, to the world of nature. The correction for this might be deliberate attempts to foster nature study in the school systems. Whatever the cause, the conservation movement has not made great headway in the parts of the world dominated by Spanish culture.

In the United States, we have a National Park system, and various sorts of reservations and wildlife refuges under national, state and private auspices. This is largely the consequence of the dedicated efforts of a few people, and we are still far from the point where we can sit back and congratulate ourselves. Conservation interests fall under different branches of government and efforts to form a coherent and unified national policy have not been very successful; we still have no Department of Conservation with cabinet rank. The struggle for financial support is always hard. And there is a constant, eroding pressure from conflicting private and governmental interests.

Ugliness—by any esthetic standard—remains the predominant characteristic of development, of urbanization, of industrialization. We talk about regional planning, diversification, working with the landscape—and we build vast stretches of the new suburbia. The ideas so forcefully developed by Patrick Geddes, Lewis Mumford and others like them, fall on deaf ears. We need an ecological conscience. We also need to develop ecological appreciation. The Romantic movement, despite its two hundred year history, has not yet reached our city councils or our highway engineers.

Practical considerations are—and perhaps ought to be—overwhelmingly important in governing man's relations with the rest of nature. Utility, at first thought, requires man to concentrate selfishly and arrogantly on his own immediate needs and convenience, to regard nature purely as a subject for exploitation. A little further thought, however, shows the fallacy of this. The danger of complete man-centeredness in relation to nature is like the danger of immediate and thoughtless selfishness everywhere: the momentary gain results in ultimate loss and defeat. "Enlightened self-interest" requires some consideration for the other fellow, for the other nation, for the other point of view; some giving with the taking. This applies with particular force to relations between man and the rest of nature.

The trend of human modification of the biological community is toward simplification. The object of agriculture is to grow pure stands of crops, single species of plants that can be eaten directly by man; or single crops that provide food for animals that can be eaten. The shorter the food chain, the more efficient the conversion of solar energy into human food. The logical end result of this process, sometimes foreseen by science fiction writers, would be the removal of all competing forms of life—with the planet left inhabited by man alone, growing his food in the form of algal soup cultivated in vast tanks. Perhaps ultimately the algae could be dispensed with, and there would be only man, living through chemical manipulations.

Efficient, perhaps; dismal, certainly; and also dangerous. A general principle is gradually emerging from ecological study to the effect that the more complex the biological community, the more stable. The intricate checks and balances among the different populations in a forest or a sea look inefficient and hampering from the point of view of any particular population, but they insure the stability and continuity of the system as a whole and thus, however indirectly, contribute to the survival of particular populations.

Just as health in a nation is, in the long run, promoted by a diversified economy, so is the health of the biosphere promoted by a diversified ecology. The single crop system is always in precarious equilibrium. It is created by man and it has to be maintained by man, ever alert with chemicals and machinery, with no other protection against the hazards of some new development in the wounded natural system. It is man working against nature: an artificial system with the uncertainties of artifacts. Epidemic catastrophe becomes an ever present threat.

This is one of the dangers inherent in man's mad spree of population growth—he is being forced into an even more arbitrary, more artificial, more precarious relation with the resources of the planet. The other great danger is related. With teeming numbers, an ever tighter system of control becomes necessary. Complex organization, totalitarian government, becomes inevitable; the individual man becomes a worker ant, a sterile robot. This surely is not our inevitable destiny.

I am not advocating a return to the neolithic. Obviously we have to have the most efficient systems possible for agriculture and resource use. But long run efficiency would seem to require certain compromises with nature—hedgerows and woodlots along with orchards and fields, the development of a variegated landscape, leaving some leeway for the checks and balances and diversity of the system of nature.

Ethical, esthetic and utilitarian reasons thus all support the attempt to conserve the diversity of nature. It is morally the right thing to do; it will provide, for future generations, a richer and more satisfying experience than would otherwise be possible; and it provides a much needed insurance against ecological catastrophe. "Unless one merely thinks man was intended to be an all-conquering and sterilizing power in the world," Charles Elton has remarked, "there must be some general basis for understanding what it is best to do. This means looking for some wise principle of co-existence between man and nature, even if it be a modified kind of man and a modified kind of nature. This is what I understand by *conservation*."

In defying nature, in destroying nature, in building an arrogantly selfish, man-centered, artificial world, I do not see how man can gain peace or freedom or joy. I have faith in man's future, faith in the possibilities latent in the human experiment: but it is faith in man as a part of nature, working with the forces that govern the forests and the seas; faith in man sharing life, not destroying it.

# II NATURE: BACKGROUND FOR THE CRISIS

BE FRUITFUL
AND MULTIPLY
AND REPLENISH THE
EARTH, AND SUBDUE IT;
AND HAVE DOMINION
OVER THE FISH OF THE SEA, AND
OVER THE FOWL OF THE AIR, AND
OVER EVERY LIVING THING
THAT MOVETH UPON
THE EARTH,
AND GOD SAID, BEHOLD,
I HAVE GIVEN YOU
EVERY HERB BEARING SEED,
WHICH IS UPON THE FACE
OF ALL THE EARTH,

AND EVERY TREE, IN THE
WHICH IS THE FRUIT OF
A TREE YEILDING
SEED; TO YOU IT SHALL BE
FOR MEAT. AND TO EVERY
BEAST OF THE EARTH AND TO
EVERY FOWL OF THE AIR.
AND TO EVERYTHING THAT
CREEPETH UPON
THE EARTH
WHEREIN
THERE
IS
LIFE
I HAVE
GIVEN
EVERY
GREEN
HERB
FOR MEAT
AND SO
IT WAS.

# INTRODUCTION

Our present understanding of nature is extremely curious. We seldom pay any attention to it except on Saturdays when we slice and hack at it with finely honed electric swords emasculating the tenacious yew or simply intimidating an old and disinterested rose bush. Work has long been the American Adam's curse; we apparently cannot justify two days of rest per week when the Lord Himself required but one. Yet, lash a tent to the same man's back, give him four hundred horses to drag his one hundred and fifty horsepowered outboard to the side of some unsuspecting lake or river's edge, and his response to nature will match Pascal's enthusiasm as he described it more than three hundred years ago.

> All this visible world is but an imperceptible point in the ample bosom of nature. No idea approaches it. In vain we extend our conceptions beyond imaginable spaces: we bring forth but atoms in comparison with the reality of things. It is an infinite sphere, of which the centre is everywhere, the circumference nowhere. In fine, it is the greatest discernible character of the omnipotence of God that our imagination loses itself in this thought.

We, of course, might express the emotion differently. Pascal was not given to "wow" or "gee"; and regrettably, he did not live long enough to know of the exclamation "blows your mind," or he certainly would have used it.

Fortunately, there are more than fifty meanings for the word *nature*. The attitudes just described need not be taken as typical of everyone in America. On the more serious side (and do not think that the man with his hundreds of mechanical horses is not a serious matter), twentieth-century man's view of nature comes from his great interest in history and biology. It is a complex view in that it pictures nature as representing both permanence and continual change. Nature, says modern man, is repetitive; it extends our understanding of ourselves back through history and onward into the future. As Mary McCarthy has written recently, nature "gives us the awareness of being an instant reverberant in time, as distinct as the echoing sound of our football in a silent forest or the plash of a stone dropped into a pool." That is to say, though we are transient beings, nature lends us the feeling of eternity. At the same time, since Darwin we have become aware of the mutability of nature, her infinite capacity for change. This notion has been so influential on the modern psyche that many think of doomsday as an evolutionary event and not the kind of happening imagined in the Bible. D. H. Lawrence, for example, recognized the connection between Darwin's theory and social and industrial progress. For him, when man gave in to the machine, he came to the final stage of his own evolution; what remains, had we the wit to recognize it, is nothing more than death.

Be that as it may, man has not always held the present understandings of nature. The Greeks in the time of Aristotle, for instance, pictured a world full of motion and regulated by mind or soul. Mind permeated everything, lending meaning and order to existence. Man belonged to a world that was both alive and intelligent. Even vegetable life participated according to its capacity in the physical and psychological harmony of the universe. For Aristotle fire rises and rocks fall because it is their nature and desire to do so. Similarly, other phenomena of a more complex order behaved according to the natural desires of the individual objects involved. It was a world of psychical and intellectual kinship that is fascinating to contemplate and one that, in the minds of ecologists at least, is making a comeback.

During the sixteenth and seventeenth centuries, a major shift in thought occurred. The Copernican revolution revised the Greek view by robbing nature of its life and its intelligence, and by reducing the act of living at one with natural phenomena to a science of observation. The genius formerly owned by nature was now accounted for by certain laws which were found to operate the world more or less as a sophisticated machine operated. The machine, in turn, was believed to be run by a brilliant but remote mind. Eventually, the analogy of the world as clock and God as Clockmaker became very popular (see William Paley, *Natural Theology*, pp. 116–117).

There were efforts made to adjust this view, but they were for the most part futile. In time, man himself participated in the split between the thing and the intelligence. René Descartes, the French mathematician and philosopher, divided man into body and soul—body going the way of natural objects, mind joining God—and despite the efforts of great minds like Spinoza, Berkeley, Coleridge, Goethe, and so on, no one was capable of uniting God and nature, mind and body, once again. Thus, nature has come to be something that exists apart from man, and the problem it presents to man is one of relation. For the most part, we have seen this relation as a matter of dependency. Nature exists in the sense that it is not an illusion, but it is a mindless thing requiring all the guidance we can provide. Once we had assumed this much, it was a short ego trip to the assumption that nature is whatever we want her to be. Where the Greek, working by analogy with natural objects, saw himself as a microcosm and nature as the macrocosm, the Renaissance man and those who followed, working with the analogy of the machine, have arrived at a radically different view.

There are obviously few advantages to the present environmental dilemma, but one of them may very well be that we will be forced into re-examining our relationship with the natural world. This section provides a start in that direction. If we are to open the inquiry that has obsessed man's mind throughout history, we best begin by renewing what has been thought and said in the past. Read the following prose pieces carefully and entertain the ideas they contain with a naive intelligence—one that is not clogged with the prejudice which vanity has created.

# WHERE I LIVED, AND WHAT I LIVED FOR

*Henry David Thoreau*

*If, as Joseph Wood Krutch claims, Henry David Thoreau (1817–1862) is read more today than other great prose poets from the nineteenth century, there is little indication that he has become easier to live with. Like his Concord neighbors, we still consider him an impertinent critic of the American way of life. In all likelihood our reaction to his audacious manifesto—simplify—can be explained by the frustration caused by our inability to conquer complexity. Yet Thoreau is more than a nagging conscience; he is a man in possession of the knowledge of beauty. In this brilliantly developed selection from Walden (1892), it is the beauty of his vision that is communicated. Perhaps once we have appreciated the aesthetic profundity of his position, we can attend his practical advice.*

Every morning was a cheerful invitation to make my life of equal simplicity, and I may say innocence, with Nature herself. I have been as sincere a worshipper of Aurora as the Greeks. I got up early and bathed in the pond; that was a religious exercise, and one of the best things which I did. They say that characters were engraven on the bathing tub of king Tching-thang to this effect: "Renew thyself completely each day; do it again, and again, and forever again." I can understand that. Morning brings back the heroic ages. I was as much affected by the faint hum of a mosquito making its invisible and unimaginable tour through my apartment at earliest dawn, when I was sitting with door and windows open, as I could be by any trumpet that ever sang of fame. It was Homer's requiem; itself an Iliad and Odyssey in the air, singing its own wrath and wanderings. There was something cosmical about it; a

SOURCE: Henry D. Thoreau, *Walden*, The Riverside Aldine Series, 2 vols. (Boston and New York, 1892), vol. I.

standing advertisement, till forbidden, of the everlasting vigor and fertility of the world. The morning, which is the most memorable season of the day, is the awakening hour. Then there is least somnolence in us; and for an hour, at least, some part of us awakes which slumbers all the rest of the day and night. Little is to be expected of that day, if it can be called a day, to which we are not awakened by our Genius, but by the mechanical nudgings of some servitor, are not awakened by our own newly acquired force and aspirations from within, accompanied by the undulations of celestial music, instead of factory bells, and a fragrance filling the air—to a higher life than we fell asleep from; and thus the darkness bear its fruit, and prove itself to be good, no less than the light. That man who does not believe that each day contains an earlier, more sacred, and auroral hour than he has yet profaned, has despaired of life, and is pursuing a descending and darkening way. After a partial cessation of his sensuous life, the soul of man, or its organs rather, are reinvigorated each day, and his Genius tries again what noble life it can make. All memorable events, I should say, transpire in morning time and in a morning atmosphere. The Vedas say, "All intelligences awake with the morning." Poetry and art, and the fairest and most memorable of the actions of men, date from such an hour. All poets and heroes, like Memnon, are the children of Aurora, and emit their music at sunrise. To him whose elastic and vigorous thought keeps pace with the sun, the day is a perpetual morning. It matters not what the clocks say or the attitudes and labors of men. Morning is when I am awake and there is a dawn in me. Moral reform is the effort to throw off sleep. Why is it that men give so poor an account of their day if they have not been slumbering? They are not such poor calculators. If they had not been overcome with drowsiness they would have performed something. The millions are awake enough for physical labor; but only one in a million is awake enough for effective intellectual exertion, only one in a hundred millions to a poetic or divine life. To be awake is to be alive. I have never yet met a man who was quite awake. How could I have looked him in the face?

We must learn to reawaken and keep ourselves awake, not by mechanical aids, but by an infinite expectation of the dawn, which does not forsake us in our soundest sleep. I know of no more encouraging fact than the unquestionable ability of man to elevate his life by a conscious endeavor. It is something to be able to paint a particular picture, or to carve a statue, and so to make a few objects beautiful; but it is far more glorious to carve and paint the very atmosphere and medium through which we look, which morally we can do. To affect the quality of the day, that is the highest of arts. Every man is tasked to make his life, even in its details, worthy of the contemplation of his most elevated and critical hour. If we refused, or rather used up, such paltry information as we get, the oracles would distinctly inform us how this might be done.

I went to the woods because I wished to live deliberately, to front only the essential facts of life, and see if I could not learn what it had to teach, and not, when I came to die, discover that I had not lived. I did not wish to live what was not life, living is so dear; nor did I wish to practise resignation, unless it was quite necessary. I wanted to live deep and suck out all the marrow of life, to live so sturdily and Spartan-like as to put to rout all that was not life, to cut a broad swath and shave close, to drive life into a corner, and reduce it to its lowest terms, and, if it proved to be mean, why then to get the whole and genuine meanness of it, and publish its meanness to the world; or if it were sublime, to know it by experience, and be able to give a true account of it in my next excursion. For most men, it appears to me, are in a strange uncertainty about it, whether it is of the devil or of God, and have *somewhat hastily* concluded that it is the chief end of man here to "glorify God and enjoy him forever."

Still we live meanly, like ants; though the fable tells us that we were long ago changed into men; like pygmies we fight with cranes; it is error upon error, and clout upon clout, and our best virtue has for its occasion a superfluous and evitable wretchedness. Our life is frittered away by detail. An honest man has hardly need to count more than his ten fingers, or in extreme cases he may add his ten toes, and lump the rest. Simplicity, simplicity, simplicity! I say, let your affairs be as two or three, and not a hundred or a thousand; instead of a million count half a dozen, and keep your accounts on your thumb nail. In the midst of this chopping sea of civilized life, such are the clouds and storms and quick-sands and thousand-and-one items to be allowed for, that a man has to live, if he would not founder and go to the bottom and not make his port at all, by dead reckoning, and he must be a great calculator indeed who succeeds. Simplify, simplify. Instead of three meals a day, if it be necessary eat but one; instead of a hundred dishes, five; and reduce other things in proportion. Our life is like a German Confederacy, made up of petty states, with its boundary forever fluctuating, so that even a German cannot tell you how it is bounded at any moment. The nation itself, with all its so-called internal improvements, which, by the way, are all external and superficial, is just such an unwieldly and overgrown establishment, cluttered with furniture and tripped up by its own traps, ruined by luxury and heedless expense, by want of calculation and a worthy aim, as the million households in the land; and the only cure for it as for them is in a rigid economy, a stern and more than Spartan simplicity of life and elevation of purpose. It lives too fast. Men think that it is essential that the *Nation* have commerce, and export ice, and talk through a telegraph, and ride thirty miles an hour, without a doubt, whether *they* do or not; but whether we should live like baboons or like men, is a little uncertain. If we do not get out sleepers, and forge rails, and devote days and nights to the work, but go to tinkering upon our *lives* to improve *them,* who will build railroads? And if rail-roads are not built, how shall we get to heaven in season? But if we stay at home and mind our business, who will want railroads? We do not ride on the railroad; it rides upon us. Did you ever think what those sleepers are that underlie the railroad? Each one is a man, an Irishman, or a Yankee man.

The rails are laid on them, and they are covered with sand, and the cars run smoothly over them. They are sound sleepers, I assure you. And every few years a new lot is laid down and run over; so that, if some have the pleasure of riding on a rail, others have the misfortune to be ridden upon. And when they run over a man that is walking in his sleep, a supernumerary sleeper in the wrong position, and wake him up, they suddenly stop the cars, and make a hue and cry about it, as if this were an exception. I am glad to know that it takes a gang of men for every five miles to keep the sleepers down and level in their beds as it is, for this is a sign that they may sometime get up again.

Why should we live with such hurry and waste of life? We are determined to be starved before we are hungry. Men say that a stitch in time saves nine, and so they take a thousand stitches to-day to save nine to-morrow. As for *work*, we haven't any of any consequence. We have the Saint Vitus' dance, and cannot possibly keep our heads still. If I should only give a few pulls at the parish bell-rope, as for a fire, that is, without setting the bell, there is hardly a man on his farm in the outskirts of Concord notwithstanding that press of engagements which was his excuse so many times this morning, nor a boy, nor a woman, I might almost say, but would forsake all and follow that sound, not mainly to save property from the flames, but, if we will confess the truth, much more to see it burn, since burn it must, and we, be it known, did not set it on fire,—or to see it put out, and have a hand in it, if that is done as handsomely; yes, even if it were the parish church itself. Hardly a man takes a half hour's nap after dinner, but when he wakes he holds up his head and asks, "What's the news?" as if the rest of mankind had stood his sentinels. Some give directions to be waked every half hour, doubtless for no other purpose; and then, to pay for it, they tell what they have dreamed. After a night's sleep the news is as indispensable as the breakfast. "Pray tell me anything new that has happened to a man anywhere on this globe,"—and he reads it over his coffee and rolls, that a man has had his eyes gouged out this morning on the Wachito River; never dreaming the while that he lives in the dark unfathomed mammoth cave of this world, and has but the rudiment of an eye himself.

For my part, I could easily do without the post-office. I think that there are very few important communications made through it. To speak critically, I never received more than one or two letters in my life—I wrote this some years ago—that were worth the postage. The penny-post is, commonly, an institution through which you seriously offer a man that penny for his thought which is so often safely offered in jest. And I am sure that I never read any memorable news in a newspaper. If we read of one man robbed, or murdered, or killed by accident, or one house burned, or one vessel wrecked, or one steamboat blown up, or one cow run over on the Western Railroad, or one mad dog killed, or one lot of grasshoppers in the winter,—we never need read of another. One is enough. If you are acquainted with the principle, what do you care for a myriad instances and applications? To a philosopher all *news*, as it is called, is gossip, and they who edit and read it are old women over their tea. Yet not a few are greedy after this gossip. There was such a rush, as I hear, the other day at one of the offices to learn the foreign news by the last arrival, that several large squares of plate glass belonging to the establishment were broken by the pressure,—news which I seriously think a ready wit might write a twelvemonth or twelve years beforehand with sufficient accuracy. As for Spain, for instance, if you know how to throw in Don Carlos and the Infanta, and Don Pedro and Seville and Granada, from time to time in the right proportions,—they may have changed the names a little since I saw the papers,—and serve up a bull-fight when other entertainments fail, it will be true to the letter, and give us as good an idea of the exact state or ruin of things in Spain as the most succinct and lucid reports under this head in the newspapers: and as for England, almost the last significant scrap of news from that quarter was the revolution of 1649; and if you have learned the history of her crops for an average year, you never need attend to that thing again, unless your speculations are of a merely pecuniary character. If one may judge who rarely looks into the newspapers, nothing new does ever happen in foreign parts, a French revolution not excepted.

What news! how much more important to know what that is which was never old! "Kieouhe-yu (great dignitary of the state of Wei) sent a man to Khoung-tseu to know his news. Khoung-tseu caused the messenger to be seated near him, and questioned him in these terms: What is your master doing? The messenger answered with respect: My master desires to diminish the number of his faults, but he cannot come to the end of them. The messenger being gone, the philosopher remarked: What a worthy messenger! What a worthy messenger!" The preacher, instead of vexing the ears of drowsy farmers on their day of rest at the end of the week,—for Sunday is the fit conclusion of an ill-spent week, and not the fresh and brave beginning of a new one,—with this one other draggletail of a sermon, should shout with thundering voice,—"Pause! Avast! Why so seeming fast, but deadly slow?"

Shams and delusions are esteemed for soundest truths, while reality is fabulous. If men would steadily observe realities only, and not allow themselves to be deluded, life, to compare it with such things as we know, would be like a fairy tale and the Arabian Nights' Entertainments. If we respected only what is inevitable and has a right to be, music and poetry would resound along the streets. When we are unhurried and wise, we perceive that only great and worthy things have any permanent and absolute existence,—that petty fears and petty pleasures are but the shadow of the reality. This is always exhilarating and sublime. By closing the eyes and slumbering, and consenting to be deceived by shows, men establish and confirm their daily life of routine and habit everywhere, which still is built on purely illusory foundations. Children, who play life, discern its true law and relations more clearly than men, who fail to live it worthily, but who think that they are wiser by experience, that is, by failure. I have read in a Hindoo book that "There was a king's son, who, being expelled in infancy from his native city, was brought up by a forester, and, growing up to maturity in that state, imagined himself to belong to the barbarous race with which he lived. One of his father's ministers having discovered him, revealed to him what he was, and the misconception of his character was removed, and he knew himself to be a prince. So soul," continues the Hindoo philosopher, "from the circumstances in which it is placed, mistakes its own character, until the truth is revealed to it by some holy teacher, and then it knows itself to be *Brahme*." I perceive that we inhabitants of New England live this mean life that we do because our vision does not penetrate the surface of things. We think that that *is* which *appears* to be. If a man should walk through this town and see only the reality, where, think you, would the "Mill-dam" go to? If he should give us an account of the realities he beheld there, we should not recognize the place in his description. Look at a meeting-house, or a court-house, or a jail, or a shop, or a dwelling-house, and say what that thing really is before a true gaze, and they would all go to pieces in your account of them. Men esteem truth remote, in the outskirts of the system, behind the farthest star, before Adam and after the last man. In eternity there is indeed something true and sublime. But all these times and places and occasions are now and here. God Himself culminates in the present moment, and will never be more divine in the lapse of all the ages. And we are enabled to apprehend at all what is sublime and noble only by the perpetual instilling and drenching of the reality that surrounds us. The universe constantly and obediently answers to our conceptions; whether we travel fast or slow, the track is laid for us. Let us spend our lives in conceiving then. The poet or the artist never yet had so fair and noble a design but some of his posterity at least could accomplish it.

Let us spend one day as deliberately as Nature, and not be thrown off the track by every nutshell and mosquito's wing that falls on the rails. Let us rise early and fast, or break fast, gently and without perturbation; let company come and let company go, let the bells ring and the children cry,—determined to make a day of it. Why should we knock under and go with the stream? Let us not be upset and overwhelmed in that terrible rapid and whirlpool called a dinner, situated in the meridian shallows. Weather this danger and you are safe, for the rest of the way is down hill. With unrelaxed nerves, with morning vigor, sail by it, looking another way, tied to the mast like Ulysses. If the engine whistles, let it whistle till it is hoarse for its pains. If the bell rings, why should we run? We will consider what kind of music they are like. Let us settle ourselves, and work and wedge our feet downward through the mud and slush of opinion, and prejudice, and tradition, and delusion and appearance, that alluvion which covers the globe, through Paris and London, through New York and Boston and Concord, through church and state, through poetry and philosophy and religion, till we come to a hard bottom and rocks in place, which we can call *reality,* and say, This is, and no mistake; and then begin, having a *point d'appui,* below freshet and frost and fire, a place where you might found a wall or a state, or set a lamppost safely, or perhaps a gauge, not a Nilometer, but a Realometer, that future ages might know how deep a freshet of shams and appearances had gathered from time to time. If you stand right fronting and face to face to a fact, you will see the sun glimmer on both its surfaces, as if it were a cimeter, and feel its sweet edge dividing you through the heart and marrow, and so you will happily conclude your mortal career. Be it life or death, we crave only reality. If we are really dying, let us hear the rattle in our throats and feel cold in the extremities; if we are alive, let us go about our business.

Time is but the stream I go a-fishing in. I drink at it; but while I drink I see the sandy bottom and detect how shallow it is. Its thin current slides away, but eternity remains. I would drink deeper; fish in the sky, whose bottom is pebbly with stars. I cannot count one. I know not the first letter of the alphabet, I have always been regretting that I was not as wise as the day I was born. The intellect is a cleaver; it discerns and rifts its way into the secret of things. I do not wish to be any more busy with my hands than is necessary. My head is hands and feet. I feel all my best faculties concentrated in it. My instinct tells me that my head is an organ for burrowing, as some creatures use their snout and forepaws, and with it I would mine and burrow my way through these hills. I think that the richest vein is somewhere hereabouts; so by the divining rod and thin rising vapors I judge; and here I will begin to mine.

# from NATURE

## Ralph Waldo Emerson

Nature is but an image or imitation of wisdom, the last thing of the soul; nature being a thing which doth only do, but not know.

—Plotinus (Motto of 1836)

A subtle chain of countless rings
The next unto the farthest brings;
The eye reads omens where it goes,
And speaks all languages the rose;
And, striving to be man, the worm
Mounts through all the spires of form.

(Motto of 1849)

*The thought of Ralph Waldo Emerson (1803–1882) begins with the conception of unity. Nothing, as far as he was concerned, is ultimately unanswerable. Existence has an order, which will withstand all inquiries man would make. Of course, we presently do not see the world clearly but, says Emerson, every man is a microcosm of the universe; and if we care to understand the unity of the whole, we must begin with attempting to understand ourselves. Furthermore, since we share ourselves with nature, the inquiry must begin with an interrogation of that "great apparition." The selection that follows marks the beginning of that great inquiry. Do not be unduly concerned about Emerson's language. Like most Romantics, the influence of the East can be felt in his writing. One's response should not be Western or discursive at the cost of missing the importance of what Emerson suggests. Do not, for example, look for or*

SOURCE: Ralph Waldo Emerson, *Nature, Addresses, and Lectures*, vol. I, *Emerson's Complete Works*, Riverside Edition (Boston: Houghton Mifflin Co. and Cambridge: The Riverside Press, 1886).

*demand a rigorous logic from Emerson. As he himself has written, "inaccuracy is not material."*

## Introduction

Our age is retrospective. It builds the sepulchres of the fathers. It writes biographies, histories, and criticism. The foregoing generations beheld God and nature face to face; we, through their eyes. Why should not we also enjoy an original relation to the universe? Why should not we have a poetry and philosophy of insight and not of tradition, and a religion by revelation to us, and not the history of theirs? Embosomed for a season in nature, whose floods of life stream around and through us, and invite us, by the powers they supply, to action proportioned to nature, why should we grope among the dry bones of the past, or put the living generation into masquerade out of its faded wardrobe? The sun shines today also. There is more wool and flax in the fields. There are new lands, new men, new thoughts. Let us demand our own works and laws and worship.

Undoubtedly we have no questions to ask which are unanswerable. We must trust the perfection of the creation so far as to believe that whatever curiosity the order of things has awakened in our minds, the order of things can satisfy. Every man's condition is a solution in hieroglyphic to those inquiries he would put. He acts it as life, before he apprehends it as truth. In like manner, nature is already, in its forms and tendencies, describing its own design. Let us interrogate the great apparition that shines so peacefully around us. Let us inquire, to what end is nature?

All science has one aim, namely, to find a theory of nature. We have theories of races and of functions, but scarcely yet a remote approach to an idea of creation. We are now so far from the road to truth, that religious teachers dispute and hate each other, and speculative men are esteemed unsound and frivolous. But to a sound judgment, the most abstract truth is the most practical. Whenever a true theory appears, it will be its own evidence. Its test is, that it will explain all phenomena. Now many are thought not only unexplained but inexplicable; as language, sleep, madness, dreams, beasts, sex.

Philosophically considered, the universe is composed of Nature and the Soul. Strictly speaking, therefore, all that is separate from us, all which Philosophy distinguishes as the NOT ME, that is, both nature and art, all other men and my own body, must be ranked under this name, NATURE. In enumerating the values of nature and casting up their sum, I shall use the word in both senses;—in its common and in its philosophical import. In inquiries so general as our present one, the inaccuracy is not material; no confusion of thought will occur. *Nature,* in the common sense, refers to essences unchanged by man; space, the air, the river, the leaf. *Art* is applied to the mixture of his will with the same things, as in a house, a canal, a statue, a picture. But his operations taken together are so insignificant, a little chipping, baking, patching, and washing, that in an impression so grand as that of the world on the human mind, they do not vary the result.

## I. Nature

To go into solitude, a man needs to retire as much from his chamber as from society. I am not solitary whilst I read and write, though nobody is with me. But if a man would be alone, let him look at the stars. The rays that come from those heavenly worlds will separate between him and what he touches. One might think the atmosphere was made transparent with this design, to give man, in the heavenly bodies, the perpetual presence of the sublime. Seen in the streets of cities, how great they are! If the stars should appear one night in a thousand years, how would men believe and adore; and preserve for many generations the remembrance of the city of God which had been shown! But every night come out these envoys of beauty, and light the universe with their admonishing smile.

The stars awaken a certain reverence, because though always present, they are inaccessible; but all natural objects make a kindred impression, when the mind is open to their influence. Nature never wears a mean appearance. Neither does the wisest man extort her secret, and lose his curiosity by finding out all her perfection. Nature never became a toy to a wise spirit. The flowers, the animals, the mountains, reflected the wisdom of his best hour, as much as they had delighted the simplicity of his childhood.

When we speak of nature in this manner, we have a distinct but most poetical sense in the mind. We mean the integrity of impression made by manifold natural objects. It is this which distinguishes the stick of timber of the wood-cutter from the tree of the poet. The charming landscape which I saw this morning is indubitably made up of some twenty or thirty farms. Miller owns this field, Locke that, and Manning the woodland beyond. But none of them owns the landscape. There is a property in the horizon which no man has but he whose eye can integrate all the parts, that is, the poet. This is the best part of these men's farms, yet to this their warranty-deeds give no title.

To speak truly, few adult persons can see nature. Most persons do not see the sun. At least they have a very superficial seeing. The sun illuminates only the eye of the man, but shines into the eye and the heart of the child. The lover of nature is he whose inward and outward senses are still truly adjusted to each other; who has retained the spirit of infancy even into the era of manhood. His intercourse with heaven and earth becomes part of his daily food. In the presence of nature a wild delight runs through the man, in spite of real sorrows. Nature says,—he is my creature, and maugre all his impertinent griefs, he shall be glad with me. Not the sun or the summer alone, but every hour and season yields its tribute of delight; for every hour and change corresponds to and authorizes a different state of the mind, from breathless noon to grimmest midnight. Nature is a setting that fits equally well a comic or a mourning piece. In good health, the air is a cordial of incredible virtue. Crossing a bare common, in snow puddles, at twilight, under a clouded sky, without having in my thoughts any occurrence of special good fortune, I have enjoyed a perfect exhilaration. I am glad to the brink of fear. In the woods, too, a man casts off his years, as the snake his slough, and at what period soever of life is always a child. In the woods is perpetual youth. Within these plantations of God, a decorum and sanctity reign, a perennial festival is dressed, and the guest sees not how he should tire of them in a thousand years. In the woods, we return to reason and faith. There I feel that nothing can befall me in life,—no disgrace, no calamity (leaving me my eyes), which nature cannot repair. Standing on the bare ground,—my head bathed by the blithe air and uplifted into infinite space,—all mean egotism vanishes. I become a transparent eyeball; I am nothing; I see all; the currents of the Universal Being circulate through me; I am part or parcel of God. The name of the nearest friend sounds then foreign and accidental: to be brothers, to be acquaintances, master or servant, is then a trifle and a disturbance. I am the lover of uncontained and immortal beauty. In the wilderness, I find something more dear and connate than in streets or villages. In the tranquil landscape, and especially in the distant line of the horizon, man beholds somewhat as beautiful as his own nature.

The greatest delight which the fields and woods minister is the suggestion of an occult relation between man and the vegetable. I am not alone and unacknowledged. They nod to me, and I to them. The waving of the boughs in the storm is new to me and old. It takes me by surprise, and yet is not unknown. Its effect is like that of a higher thought or a better emotion coming over me, when I deemed I was thinking justly or doing right.

Yet it is certain that the power to produce this delight does not reside in nature, but in man, or in a harmony of both. It is necessary to use these pleasures with great temperance. For nature is not always tricked in holiday attire, but the same scene which yesterday breathed perfume and glittered as for the frolic of the nymphs is overspread with melancholy today. Nature always wears the colors of the spirit. To a man laboring under calamity, the heat of his own fire hath sadness in it. Then there is a kind of contempt of the landscape felt by him who has just lost by death a dear friend. The sky is less grand as it shuts down over less worth in the population.

*     *     *

## III. Beauty

A nobler want of man is served by nature, namely, the love of Beauty. The ancient Greeks called the world Κόσμος,[1] beauty. Such is the constitution of all things, or such the plastic power of the human eye, that the primary forms, as the sky, the mountain, the tree, the animal, give us a delight *in and for themselves*; a pleasure arising from outline, color, motion, and grouping. This seems partly owing to the eye itself. The eye is the best of artists. By the mutual action of its structure and of the laws of light, perspective is produced, which integrates every mass of objects, of what character soever, into a well colored and shaded globe, so that where the particular objects are mean and unaffecting, the landscape which they compose is round and symmetrical. And as the eye is the best composer, so light is the first of painters. There is no object so foul that intense light will not make beautiful. And the stimulus it affords to the sense, and a sort of infinitude which it hath, like space and time, make all matter gay. Even the corpse has its own beauty. But besides this general grace diffused over nature, almost all the individual forms are agreeable to the eye, as is proved by our endless imitations of some of them, as the acorn, the grape, the pine-cone, the wheat-ear, the egg, the wings and forms of most birds, the lion's claw, the serpent, the butterfly, seashells, flames, clouds, buds, leaves, and the forms of many trees, as the palm.

For better consideration, we may distribute the aspects of Beauty in a threefold manner.

1. First, the simple perception of natural forms is a delight. The influence of the forms and actions in nature is so needful to man that, in its lowest functions, it seems to lie on the confines of commodity and beauty. To the body and mind which have been cramped by noxious work or company, nature is medicinal and restores their tone. The tradesman, the attorney comes out of the din and craft of the street and sees the sky and the woods, and is a man again. In their eternal calm, he finds himself. The health of the eye seems to demand a horizon. We are never tired, so long as we can see far enough.

But in other hours, Nature satisfies by its loveliness, and without any mixture of corporeal benefit. I see the spectacle of morning from the hilltop over against my house, from daybreak to sunrise, with emotions which an angel might share. The long slender bars of cloud float like fishes in the sea of crimson light. From the earth as a shore, I look out into that silent sea. I seem to partake its rapid transformations; the active enchantment reaches my dust, and I dilate and conspire with the morning wind. How does Nature deify us with a few and cheap elements! Give me health and a day, and I will make the pomp of emperors ridiculous. The dawn is my Assyria; the sunset and moonrise my Paphos, and unimaginable realms of faerie; broad noon shall be

[1] Kosmos, order.

my England of the senses and the understanding; the night shall be my Germany of mystic philosophy and dreams.

Not less excellent, except for our less susceptibilty in the afternoon, was the charm, last evening, of a January sunset. The western clouds divided and subdivided themselves into pink flakes modulated with tints of unspeakable softness, and the air had so much life and sweetness that it was a pain to come within doors. What was it that nature would say? Was there no meaning in the live repose of the valley behind the mill, and which Homer or Shakespeare could not re-form for me in words? The leafless trees become spires of flame in the sunset, with the blue east for their background, and the stars of the dead calices of flowers, and every withered stem and stubble rimed with frost, contribute something to the mute music.

The inhabitants of cities suppose that the country landscape is pleasant only half the year. I please myself with the graces of the winter scenery, and believe that we are as much touched by it as by the genial influences of summer. To the attentive eye, each moment of the year has its own beauty, and in the same field, it beholds, every hour, a picture which was never seen before, and which shall never be seen again. The heavens change every moment, and reflect their glory or gloom on the plains beneath. The state of the crop in the surrounding farms alters the expression of the earth from week to week. The succession of native plants in the pastures and roadsides, which makes the silent clock by which time tells the summer hours, will make even the divisions of the day sensible to a keen observer. The tribes of birds and insects, like the plants punctual to their time, follow each other, and the year has room for all. By watercourses, the variety is greater. In July, the blue pontederia or pickerel-weed blooms in large beds in the shallow parts of our pleasant river, and swarms with yellow butterflies in continual motion. Art cannot rival this pomp of purple and gold. Indeed the river is a perpetual gala, and boasts each month a new ornament.

But this beauty of Nature which is seen and felt as beauty, is the least part. The shows of day, the dewy morning, the rainbow, mountains, orchards in blossom, stars, moonlight, shadows in still water, and the like, if too eagerly hunted, become shows merely, and mock us with their unreality. Go out of the house to see the moon, and 'tis mere tinsel; it will not please as when its light shines upon your necessary journey. The beauty that shimmers in the yellow afternoons of October, who ever could clutch it? Go forth to find it, and it is gone; 'tis only a mirage as you look from the windows of diligence.

2. The presence of a higher, namely, of the spiritual element is essential to its perfection. The high and divine beauty which can be loved without effeminacy, is that which is found in combination with the human will. Beauty is the mark God sets upon virtue. Every natural action is graceful. Every heroic act is also decent, and causes the place and the bystanders to shine. We are taught by great actions that the universe is the property of every indi-

vidual in it. Every rational creature has all nature for his dowry and estate. It is his, if he will. He may divest himself of it; he may creep into a corner, and abdicate his kingdom, as most men do, but he is entitled to the world by his constitution. In proportion to the energy of his thought and will, he takes up the world into himself. "All those things for which men plough, build, or sail, obey virtue," said Sallust. "The winds and waves," said Gibbon, "are always on the side of the ablest navigators." So are the sun and moon and all the stars of heaven. When a noble act is done—perchance in a scene of great natural beauty; when Leonidas and his three hundred martyrs consume one day in dying, and the sun and moon come each and look at them once in the steep defile of Thermopylae; when Arnold Winkelried, in the high Alps, under the shadow of the avalanche, gathers in his side a sheaf of Austrian spears to break the line for his comrades; are not these heroes entitled to add the beauty of the scene to the beauty of the deed? When the bark of Columbus nears the shore of America;—before it the beach lined with savages, fleeing out of all their huts of cane; the sea behind; and the purple mountains of the Indian Archipelago around, can we separate the man from the living picture? Does not the New World clothe his form with her palm-groves and savannahs as fit drapery? Ever does natural beauty steal in like air, and envelope great actions. When Sir Harry Vane was dragged up the Tower-hill, sitting on a sled, to suffer death as the champion of the English laws, one of the multitude cried out to him, "You never sate on so glorious a seat!" Charles II, to intimidate the citizens of London, caused the patriot Lord Russell to be drawn in an open coach through the principal streets of the city on his way to the scaffold. "But," his biographer says, "the multitude imagined they saw liberty and virtue sitting by his side." In private places, among sordid objects, an act of truth or heroism seems at once to draw to itself the sky as its temple, the sun as its candle. Nature stretches out her arms to embrace man, only let his thoughts be of equal greatness. Willingly does she follow his steps with the rose and the violet, and bend her lines of grandeur and grace to the decoration of her darling child. Only let his thoughts be of equal scope, and the frame will suit the picture. A virtuous man is in unison with her works, and makes the central figure of the visible sphere. Homer, Pindar, Socrates, Phocion, associate themselves fitly in our memory with the geography and climate of Greece. The visible heavens and earth sympathize with Jesus. And in common life whosoever has seen a person of powerful character and happy genius, will have remarked how easily he took all things along with him,—the persons, the opinions, and the day, and nature become ancillary to a man.

3. There is still another aspect under which the beauty of the world may be viewed, namely, as it becomes an object of the intellect. Beside the relation of things to virtue, they have a relation to thought. The intellect searches out the absolute order of things as they stand in the mind of God, and without the colors of affection. The intellectual and the active powers seem to succeed

each other, and the exclusive activity of the one generates the exclusive activity of the other. There is something unfriendly in each to the other, but they are like the alternate periods of feeding and working in animals; each prepares and will be followed by the other. Therefore does beauty, which, in relation to actions, as we have seen, comes unsought, and comes because it is unsought, remain for the apprehension and pursuit of the intellect; and then again, in its turn, of the active power. Nothing divine dies. All good is eternally reproductive. The beauty of nature re-forms itself in the mind, and not for barren contemplation, but for new creation.

All men are in some degree impressed by the face of the world; some men even to delight. This love of beauty is Taste. Others have the same love in such excess, that, not content with admiring, they seek to embody it in new forms. The creation of beauty is Art.

The production of a work of art throws a light upon the mystery of humanity. A work of art is an abstract or epitome of the world. It is the result or expression of nature, in miniature. For although the works of nature are innumerable and all different, the result or the expression of them all is similar and single. Nature is a sea of forms radically alike and even unique. A leaf, a sunbeam, a landscape, the ocean, make an analogous impression on the mind. What is common to them all,—that perfectness and harmony, is beauty. The standard of beauty is the entire circuit of natural forms,—the totality of nature; which the Italians expressed by defining beauty "il più nell' uno."[2] Nothing is quite beautiful alone; nothing but is beautiful in the whole. A single object is only so far beautiful as it suggests this universal grace. The poet, the painter, the sculptor, the musician, the architect, seek each to concentrate this radiance of the world on one point, and each in his several work to satisfy the love of beauty which stimulates him to produce. Thus is Art a nature passed through the alembic of man. Thus in art does Nature work through the will of a man filled with the beauty of her first works.

The world thus exists to the soul to satisfy the desire of beauty. This element I call an ultimate end. No reason can be asked or given why the soul seeks beauty. Beauty, in its largest and profoundest sense, is one expression for the universe. God is the all-fair. Truth, and goodness, and beauty, are but different faces of the same All. But beauty in nature is not ultimate. It is the herald of inward and eternal beauty, and is not alone a solid and satisfactory good. It must stand as a part, and not as yet the last or highest expression of the final cause of Nature.

*   *   *

[2] The many in one.

## V. Discipline

In view of the significance of nature, we arrive at once at a new fact, that nature is a discipline. This use of the world includes the preceding uses, as parts of itself.

Space, time, society, labor, climate, food, locomotion, the animals, the mechanical forces, give us sincerest lessons, day by day, whose meaning is unlimited. They educate both the Understanding and the Reason. Every property of matter is a school for the understanding,—its solidity or resistance, its inertia, its extension, its figure, its divisibility. The understanding adds, divides, combines, measures, and finds nutriment and room for its activity in this worthy scene. Meantime, Reason transfers all these lessons into its own world of thought, by perceiving the analogy that marries Matter and Mind.

1. Nature is a discipline of the understanding in intellectual truths. Our dealing with sensible objects is a constant exercise in the necessary lessons of difference, of likeness, of order, of being and seeming, of progressive arrangement; of ascent from particular to general; of combination to one end of manifold forces. Proportioned to the importance of the organ to be formed, is the extreme care with which its tuition is provided,—a care pretermitted in no single case. What tedious training, day after day, year after year, never ending, to form the common sense; what continual reproduction of annoyances, inconveniences, dilemmas; what rejoicing over us of little men; what disputing of prices, what reckonings of interest,—and all to form the Hand of the mind;—to instruct us that "good thoughts are no better than good dreams, unless they be executed!"

The same good office is performed by Property and its filial systems of debt and credit. Debt, grinding debt, whose iron face the widow, the orphan, and the sons of genius fear and hate;—debt, which consumes so much time, which so cripples and disheartens a great spirit with cares that seem so base, is a preceptor whose lessons cannot be foregone, and is needed most by those who suffer from it most. Moreover, property, which has been well compared to snow,—"if it fall level today, it will be blown into drifts tomorrow,"—is the surface action of internal machinery, like the index on the face of a clock. Whilst now it is the gymnastics of the understanding, it is hiving, in the foresight of the spirit, experience in profounder laws.

The whole character and fortune of the individual are affected by the least inequalities in the culture of the understanding; for example, in the perception of differences. Therefore is Space, and therefore Time, that man may know that things are not huddled and lumped, but sundered and individual. A bell and a plough have each their use, and neither can do the office of the other. Water is good to drink, coal to burn, wool to wear; but wool cannot be drunk, nor water spun, nor coal eaten. The wise man shows his wisdom in separation, in gradation, and his scale of creatures and of merits is as wide as nature. The foolish have no range in their scale, but suppose every man is as every other man. What is not good they call the worst, and what is not hateful, they call the best.

In like manner, what good heed Nature forms in us! She pardons no mistakes. Her yea is yea, and her nay, nay.

The first steps in Agriculture, Astronomy, Zoology (those first steps which the farmer, the hunter, and the sailor take), teach that Nature's dice are always loaded; that in her heaps and rubbish are concealed sure and useful results.

How calmly and genially the mind apprehends one after another the laws of physics! What noble emotions dilate the mortal as he enters into the councils of the creation, and feels by knowledge the privilege to BE! His insight refines him. The beauty of nature shines in his own breast. Man is greater that he can see this, and the universe less, because Time and Space relations vanish as laws are known.

Here again we are impressed and even daunted by the immense Universe to be explored. "What we know is a point to what we do not know." Open any recent journal of science, and weigh the problems suggested concerning Light, Heat, Electricity, Magnetism, Physiology, Geology, and judge whether the interest of natural science is likely to be soon exhausted.

Passing by many particulars of the discipline of nature, we must not omit to specify two.

The exercise of the Will, or the lesson of power, is taught in every event. From the child's successive possession of his several senses up to the hour when he saith, "Thy will be done!" he is learning the secret that he can reduce under his will not only particular events but great classes, nay, the whole series of events, and so conform all facts to his character. Nature is thoroughly mediate. It is made to serve. It receives the dominion of man as meekly as the ass on which the Saviour rode. It offers all its kingdoms to man as the raw material which he may mold into what is useful. Man is never weary of working it up. He forges the subtle and delicate air into wise and melodious words, and gives them wing as angels of persuasion and command. One after another his victorious thought comes up with and reduces all things, until the world becomes at last only a realized will,—the double of the man.

2. Sensible objects conform to the premonitions of Reason and reflect the conscience. All things are moral; and in their boundless changes have an unceasing reference to spiritual nature. Therefore is nature glorious with form, color, and motion; that every globe in the remotest heaven, every chemical change from the rudest crystal up to the laws of life, every change of vegetation from the first principle of growth in the eye of a leaf, to the tropical forest and antediluvian coal-mine, every animal function from the sponge up to Hercules, shall hint or thunder to man the laws of right and wrong, and

echo the Ten Commandments. Therefore is Nature ever the ally of Religion: lends all her pomp and riches to the religious sentiment. Prophet and priest, David, Isaiah, Jesus, have drawn deeply from this source. This ethical character so penetrates the bone and marrow of nature, as to seem the end for which it was made. Whatever private purpose is answered by any member or part, this is its public and universal function, and is never omitted. Nothing in nature is exhausted in its first use. When a thing has served an end to the uttermost, it is wholly new for an ulterior service. In God, every end is converted into a new means. Thus the use of commodity, regarded by itself, is mean and squalid. But it is to the mind an education in the doctrine of Use, namely, that a thing is good only so far as it serves; that a conspiring of parts and efforts to the production of an end is essential to any being. The first and gross manifestation of this truth is our inevitable and hated training in values and wants, in corn and meat.

It has already been illustrated, that every natural process is a version of a moral sentence. The moral law lies at the center of nature and radiates to the circumference. It is the pith and marrow of every substance, every relation, and every process. All things with which we deal, preach to us. What is a farm but a mute gospel? The chaff and the wheat, weeds and plants, blight, rain, insects, sun,—it is a sacred emblem from the first furrow of spring to the last stack which the snow of winter overtakes in the fields. But the sailor, the shepherd, the miner, the merchant, in their several resorts, have each an experience precisely parallel, and leading to the same conclusion: because all organizations are radically alike. Nor can it be doubted that this moral sentiment which thus scents the air, grows in the grain, and impregnates the waters of the world, is caught by man and sinks into his soul. The moral influence of nature upon every individual is that amount of truth which it illustrates to him. Who can estimate this? Who can guess how much firmness the sea-beaten rock has taught the fisherman? how much tranquillity has been reflected to man from the azure sky, over whose unspotted deeps the winds forevermore drive flocks of stormy clouds, and leave no wrinkle or stain? how much industry and providence and affection we have caught from the pantomime of brutes? What a searching preacher of self-command is the varying phenomenon of Health!

Herein is especially apprehended the unity of Nature,—the unity in variety,—which meets us everywhere. All the endless variety of things make an identical impression. Xenophanes complained in his old age, that, look where he would, all things hastened back to Unity. He was weary of seeing the same entity in the tedious variety of forms. The fable of Proteus has a cordial truth. A leaf, a drop, a crystal, a moment of time, is related to the whole, and partakes of the perfection of the whole. Each particle is a microcosm, and faithfully renders the likeness of the world.

Not only resemblances exist in things whose analogy is obvious, as when we detect the type of the human hand in the flipper of the fossil saurus, but also in objects wherein there is great superficial unlikeness. Thus architecture is called "frozen music," by De Staël and Goethe. Vitruvius thought an architect should be a musician. "A Gothic church," said Coleridge, "is a petrified religion." Michael Angelo maintained, that, to an architect, a knowledge of anatomy is essential. In Haydn's oratorios, the notes present to the imagination not only motions, as of the snake, the stag, and the elephant, but colors also; as the green grass. The law of harmonic sounds reappears in the harmonic colors. The granite is differenced in its laws only by the more or less of heat from the river that wears it away. The river, as it flows, resembles the air that flows over it; the air resembles the light which traverses it with more subtile currents; the light resembles the heat which rides with it through Space. Each creature is only a modification of the other; the likeness in them is more than the difference, and their radical law is one and the same. A rule of one art, or a law of one organization, holds true throughout nature. So intimate is this Unity, that, it is easily seen, it lies under the undermost garment of Nature, and betrays its source in Universal Spirit. For it pervades Thought also. Every universal truth which we express in words, implies or supposes every other truth. *Omne verum vero consonat.*[3] It is like a great circle on a sphere, comprising all possible circles; which, however, may be drawn and comprise it in like manner. Every such truth is the absolute Ens[4] seen from one side. But it has innumerable sides.

The central Unity is still more conspicuous in actions. Words are finite organs of the infinite mind. They cannot cover the dimensions of what is in truth. They break, chop, and impoverish it. An action is the perfection and publication of thought. A right action seems to fill the eye, and to be related to all nature. "The wise man, in doing one thing, does all; or, in the one thing he does rightly, he sees the likeness of all which is done rightly."

Words and actions are not the attributes of brute nature. They introduce us to the human form, of which all other organizations appear to be degradations. When this appears among so many that surround it, the spirit prefers it to all others. It says, "From such as this have I drawn joy and knowledge; in such as this have I found and beheld myself; I will speak to it; it can speak again; it can yield me thought already formed and alive." In fact, the eye,— the mind,—is always accompanied by these forms, male and female; and these are incomparably the richest informations of the power and order that lie at the heart of things. Unfortunately every one of them bears the marks as of some injury; is marred and superficially defective. Nevertheless, far different

[3] All truth accords with truth.
[4] Being.

from the deaf and dumb nature around them, these all rest like fountain-pipes on the unfathomed sea of thought and virtue whereto they alone, of all organizations, are the entrances.

It were a pleasant inquiry to follow into detail their ministry to our education, but where would it stop? We are associated in adolescent and adult life with some friends, who, like skies and waters, are coextensive with our idea; who, answering each to a certain affection of the soul, satisfy our desire on that side; whom we lack power to put at such focal distance from us, that we can mend or even analyze them. We cannot choose but love them. When much intercourse with a friend has supplied us with a standard of excellence, and has increased our respect for the resources of God who thus sends a real person to outgo our ideal; when he has, moreover, become an object of thought, and, whilst his character retains all its unconscious effect, is converted in the mind into solid and sweet wisdom,—it is a sign to us that his office is closing, and he is commonly withdrawn from our sight in a short time.

*     *     *

## VIII. Prospects

In inquiries respecting the laws of the world and the frame of things, the highest reason is always the truest. That which seems faintly possible, it is so refined, is often faint and dim because it is deepest seated in the mind among the eternal verities. Empirical science is apt to cloud the sight, and by the very knowledge of functions and processes to bereave the student of the manly contemplation of the whole. The savant becomes unpoetic. But the best read naturalist who lends an entire and devout attention to truth, will see that there remains much to learn of his relation to the world, and that it is not to be learned by any addition or subtraction or other comparison of known quantities, but is arrived at by untaught sallies of the spirit, by a continual self-recovery, and by entire humility. He will perceive that there are far more excellent qualities in the student than preciseness and infallibility; that a guess is more often fruitful than an indisputable affirmation, and that a dream may let us deeper into the secret of nature than a hundred concerted experiments.

For the problems to be solved are precisely those which the physiologist and the naturalist omit to state. It is not so pertinent to man to know all the individuals of the animal kingdom, as it is to know whence and whereto is this tyrannizing unity in his constitution, which evermore separates and classifies things, endeavoring to reduce the most diverse to one form. When I behold a rich landscape, it is less to my purpose to recite correctly the order and superposition of the strata, than to know why all thought of multitude is lost in a tranquil sense of unity. I cannot greatly honor minuteness in details, so long as there is no hint to explain the relation between things and thoughts; no ray upon the *metaphysics* of conchology, of botany, of the arts, to show the relation of the forms of flowers, shells, animals, architecture, to the mind, and build science upon ideas. In a cabinet of natural history, we become sensible of a certain occult recognition and sympathy in regard to the most unwieldy and eccentric forms of beast, fish, and insect. The American who has been confined, in his own country, to the sight of buildings designed after foreign models, is surprised on entering York Minster or St. Peter's at Rome, by the feeling that these structures are imitations also,—faint copies of an invisible archetype. Nor has science sufficient humanity, so long as the naturalist overlooks that wonderful congruity which subsists between man and the world; of which he is lord, not because he is the most subtile inhabitant, but because he is its head and heart, and finds something of himself in every great and small thing, in every mountain stratum, in every new law of color, fact of astronomy, or atmospheric influence which observation or analysis lays open. A perception of this mystery inspires the muse of George Herbert, the beautiful psalmist of the seventeenth century. The following lines are part of his little poem on Man.

Man is all symmetry,
Full of proportions, one limb to another,
And all to all the world besides.
Each part may call the farthest, brother;
For head with foot hath private amity,
And both with moons and tides.

Nothing hath got so far
But man hath caught and kept it as his prey;
His eyes dismount the highest star:
He is in little all the sphere.
Herbs gladly cure our flesh, because that they
Find their acquaintance there.

For us, the winds do blow,
The earth doth rest, heaven move, and fountains flow;
Nothing we see, but means our good,
As our delight, or as our treasure;
The whole is either our cupboard of food,
Or cabinet of pleasure.

The stars have us to bed:
Night draws the curtain; which the sun withdraws.
Music and light attend our head.
All things unto our flesh are kind,
In their descent and being; to our mind,
In their ascent and cause.

More servants wait on man
Than he'll take notice of. In every path,
  He treads down that which doth befriend him
  When sickness makes him pale and wan.
Oh mighty love! Man is one world, and hath
  Another to attend him.

The perception of this class of truths makes the attraction which draws men to science, but the end is lost sight of in attention to the means. In view of this half-sight of science, we accept the sentence of Plato, that "poetry comes nearer to vital truth than history." Every surmise and vatication of the mind is entitled to a certain respect, and we learn to prefer imperfect theories, and sentences which contain glimpses of truth, to digested systems which have no one valuable suggestion. A wise writer will feel that the ends of study and composition are best answered by announcing undiscovered regions of thought, and so communicating, through hope, new activity to the torpid spirit.

I shall therefore conclude this essay with some traditions of man and nature, which a certain poet sang to me; and which, as they have always been in the world, and perhaps reappear to every bard, may be both history and prophecy.

The foundations of man are not in matter, but in spirit. But the element of spirit is eternity. To it, therefore, the longest series of events, the oldest chronologies are young and recent. In the cycle of the universal man, from whom the known individuals proceed, centuries are points, and all history is but the epoch of one degradation.

We distrust and deny inwardly our sympathy with nature. We own and disown our relation to it, by turns. We are like Nebuchadnezzar, dethroned, bereft of reason, and eating grass like an ox. But who can set limits to the remedial force of spirit?

A man is a god in ruins. When men are innocent, life shall be longer, and shall pass into the immortal as gently as we awake from dreams. Now, the world would be insane and rabid, if these disorganizations should last for hundreds of years. It is kept in check by death and infancy. Infancy is the perpetual Messiah, which comes into the arms of fallen men, and pleads with them to return to paradise.

Man is the dwarf of himself. Once he was permeated and dissolved by spirit. He filled nature with his overflowing currents. Out from him sprang the sun and moon; from man the sun, from woman the moon. The laws of his mind, the periods of his actions externized themselves into day and night, into the year and the seasons. But, having made for himself this huge shell, his waters retired; he no longer fills the veins and veinlets; he is shrunk to a drop. He sees that the structure still fits him, but fits him colossally. Say, rather, once it fitted him, now it corresponds to him from far and on high. He adores timidly his own work. Now is man the follower of the sun, and woman the follower of the moon. Yet sometimes he starts in his slumber, and wonders at himself and his house, and muses strangely at the resemblance betwixt him and it. He perceives that if his law is still paramount, if still he have elemental power, if his word is sterling yet in nature, it is not conscious power, it is not inferior but superior to his will. It is instinct.

Thus my Orphic poet sang.

At present, man applies to nature but half his force. He works on the world with his understanding alone. He lives in it and masters it by a penny-wisdom; and he that works most in it is but a half-man, and whilst his arms are strong and his digestion good, his mind is imbruted, and he is a selfish savage. His relation to nature, his power over it, is through the understanding, as by manure; the economic use of fire, wind, water, and the mariner's needle; steam, coal, chemical agriculture; the repairs of the human body by the dentist and the surgeon. This is such a resumption of power as if a banished king should buy his territories inch by inch, instead of vaulting at once into his throne. Meantime, in the thick darkness, there are not wanting gleams of a better light,—occasional examples of the action of man upon nature with his entire force,—with reason as well as understanding. Such examples are, the traditions of miracles in the earliest antiquity of all nations; the history of Jesus Christ; the achievements of a principle, as in religious and political revolutions, and in the abolition of the slave-trade; the miracles of enthusiasm, as those reported of Swedenborg, Hohenlohe, and the Shakers; many obscure and yet contested facts, now arranged under the name of Animal Magnetism; prayer; eloquence; self-healing; and the wisdom of children. These are examples of Reason's momentary grasp of the scepter; the exertions of a power which exists not in time or space, but an instantaneous in-streaming causing power. The difference between the actual and the ideal force of man is happily figured by the schoolmen, in saying, that the knowledge of man is an evening knowledge, *vespertina cognitio,* but that of God is a morning knowledge, *matutina cognitio.*

The problem of restoring to the world original and eternal beauty is solved by the redemption of the soul. The ruin or the blank that we see when we look at nature, is in our own eye. The axis of vision is not coincident with the axis of things, and so they appear not transparent but opaque. The reason why the world lacks unity, and lies broken and in heaps, is because man is disunited with himself. He cannot be a naturalist until he satisfies all the demands of the spirit. Love is as much its demand as perception. Indeed, neither can be perfect without the other. In the uttermost meaning of the words, thought is devout, and devotion is thought. Deep calls unto deep. But in actual life, the marriage is not celebrated. There are innocent men who worship God after the tradition of their fathers, but their sense of duty has not yet extended to the use of all their faculties. And there are patient naturalists, but they freeze their subject under the wintry light of the understanding. Is not prayer also a study of truth,—a sally of the soul into the unfound infinite? No man ever prayed heartily without learning something. But when a faithful thinker, resolute to detach every object from personal relations and see it in the light of thought, shall, at the same time, kindle science with the fire of the holiest affections, then will God go forth anew into the creation.

It will not need, when the mind is prepared for study, to search for objects. The invariable mark of wisdom is to see the miraculous in the common. What is a day? What is a year? What is summer? What is a woman? What is a child? What is sleep? To our blindness, these things seem unaffecting. We make fables to hide the baldness of the fact and conform it, as we say, to the higher law of the mind. But when the fact is seen under the light of an idea, the gaudy fable fades and shrivels. We behold the real higher law. To the wise, therefore, a fact is true poetry, and the most beautiful of fables. These wonders are brought to our own door. You also are a man. Man and woman and their social life, poverty, labor, sleep, fear, fortune, are known to you. Learn that none of these things is superficial, but that each phenomenon has its roots in the faculties and affections of the mind. Whilst the abstract question occupies your intellect, nature brings it in the concrete to be solved by your hands. It were a wise inquiry for the closet, to compare, point by point, especially at remarkable crises in life, our daily history with the rise and progress of ideas in the mind.

So shall we come to look at the world with new eyes. It shall answer the endless inquiry of the intellect,—What is truth? and of the affections,—What is good? by yielding itself passive to the educated Will. Then shall come to pass what my poet said:

> Nature is not fixed but fluid. Spirit alters, molds, makes it. The immobility or bruteness of nature is the absence of spirit; to pure spirit it is fluid, it is volatile, it is obedient. Every spirit builds itself a house, and beyond its house a world, and beyond its world a heaven. Know then that the world exists for you. For you is the phenomenon perfect. What we are, that only can we see. All that Adam had, all that Caesar could, you have and can do. Adam called his house, heaven and earth; Caesar called his house, Rome; you perhaps call yours, a cobbler's trade; a hundred acres of ploughed land; or a scholar's garret. Yet line for line and point for point your dominion is as great as theirs, though without fine names. Build therefore your own world. As fast as you conform your life to the pure idea in your mind, that will unfold its great proportions. A correspondent revolution in things will attend the influx of the spirit. So fast will disagreeable appearances, swine, spiders, snakes, pests, madhouses, prisons, enemies, vanish; they are temporary and shall be no more seen. The sordor and filths of nature, the sun shall dry up and the wind exhale. As when the summer comes from the south the snow-banks melt and the face of the earth becomes green before it, so shall the advancing spirit create its ornaments along its path, and carry with it the beauty it visits and the song which

enchants it; it shall draw beautiful faces, warm hearts, wise discourse, and heroic acts, around its way, until evil is no more seen. The kingdom of man over nature, which cometh not with observation,—a dominion such as now is beyond his dream of God,—he shall enter without more wonder than the blind man feels who is gradually restored to perfect sight.

# THE MOUNTAIN SPRING

*John Steinbeck*

*For many the reputation of John Steinbeck (1902–1969) has been based on what may be called his literature of protest dating back to the social turmoil of the 1930's.* In Dubious Battle *(1936),* Of Mice and Men *(1937), and* The Grapes of Wrath *(1939) are typical examples of that period of his history as a writer. Lately, however, these novels have been replaced in the public mind by works that are more lyrical, more relaxed, and more descriptive in character.* Travels with Charlie in Search of America *(1962), for instance, is an account of his tour of forty states accompanied by his poodle. Of course, no writer changes. He grows by adding, not discarding, experience. In the following excerpt from* The Pearl *(1945), we find a description that is filled with at least the implications of a design intended to guide our behavior.*

I.

High in the gray stone mountains, under a frowning peak, a little spring bubbled out of a rupture in the stone. It was fed by shade-preserved snow in the summer, and now and then it died completely and bare rocks and dry algae were on its bottom. But nearly always it gushed out, cold and clean and lovely. In the times when the quick rains fell, it might become a freshet and send its column of white water crashing down the mountain cleft, but nearly always it was a lean little spring. It bubbled out into a pool and then fell a hundred feet to another pool, and this one, overflowing, dropped again, so that it continued, down and down, until it came to the rubble of the upland,

and there it disappeared altogether. There wasn't much left of it then anyway, for every time it fell over an escarpment the thirsty air drank it, and it splashed from the pools to the dry vegetation. The animals from miles around came to drink from the little pools, and the wild sheep and the deer, the pumas and raccoons, and the mice—all came to drink. And the birds which spent the day in the brushland came at night to the little pools that were like steps in the mountain cleft. Beside this tiny stream, wherever enough earth collected for a root-hold, colonies of plants grew, wild grape and little palms, maidenhair fern, hibiscus, and tall pampas grass with feathery rods raised about the spike leaves. And in the pool lived frogs and water-skaters, and waterworms crawled on the bottom of the pool. Everything that loved water came to these few shallow places. The cats took their prey there, and strewed feathers and lapped water through their bloody teeth. The little pools were places of life because of the water, and places of killing because of the water, too.

## II.

The lowest step, where the stream collected before it tumbled down a hundred feet and disappeared into the rubbly desert, was a little platform of stone and sand. Only a pencil of water fell into the pool, but it was enough to keep the pool full and to keep the ferns green in the underhang of the cliff, and wild grape climbed the stone mountain and all manner of little plants found comfort there. The freshets had made a small sandy beach through which the pool flowed, and bright green watercress grew in the damp sand. The beach was cut and scarred and padded by the feet of animals that had come to drink and to hunt.

# SPRING RAIN

*John Updike*

*There is a large segment of the American reading public that has not recovered from Time Magazine's discovery that John Updike is over thirty years of age (actually he is nearly forty). His highly impressionistic writing has for them been synonymous with youthful awareness. Of course, it matters little that Updike has written some nine books of short and long fiction and a good number of children's stories, and has been a regular contributor to The New Yorker; his style remains refreshingly youthful. This sample of his art, for instance, is typically vivid. As you read it, note how what is apparently description can, with one subtle stroke, involve the reader in a surprisingly large drama.*

As the sky is pushed farther and farther away by the stiff-arms of this and that new steel frame, we sometimes wonder if what is reaching us is really weather at all. Whenever we have looked down at the street this spring, the perpetually raincoated figures have appeared to be marching, jerkily fore-shortened and steadfastly down-staring, under a kind of sooty fluorescence bearing little relation to the expansive and variable light of outdoors. The other day, as if at the repeated invitation of all those raincoats, it *did* rain, and we ventured outdoors ourself; that is to say, we made our way down several corridors and shafts and into a broader corridor called Forty-fourth Street, whose ceiling, if one bothered to look, consisted of that vaguely tonic, vaporish semi-opacity old-fashionedly termed the Firmament. On this day, the Firmament, which showed as a little, ragged strip wedged between the upper edges of the buildings, seemed in a heavy temper. Water was being silently inserted in the slots between the building tops, and a snappy little secondary rain was dripping from marquees, overhead signs, fire escapes, and ledges. On the street itself, whose asphalt had emerged from the blanket of winter as creased and bumpy as a slept-on sheet, the water was conducting itself an extravagantly complicated debate of ripple and counter-ripple, flow and anti-flow. It looked black but not dirty, and we thought, in that decisive syntactical way we reserve for such occasions, how all water is in passage from purity to purity. Puddles, gutters, sewers are incidental disguises: the casual avatars of perpetually reincarnated cloud droplets; momentary embarrassments, having nothing to do with the ineluctable poise of $H_2O$. Throw her on the street, mix her with candy wrappers, splash her with taxi wheels, she remains a virgin and a lady.

The breeze caught its breath, the rain slackened, and the crowds that had been clustered in entranceways and under overhangs shattered and scattered like drying pods. We went over to Fifth Avenue; the buildings there, steeped in humidity, seemed to be a kind of print of their own images, a slightly too inky impression of an etching entitled "Fifth Avenue, Manhattan, c. 1962."

No matter how long we live among rectangular stones, we still listen, in the pauses of a rain, for the sound of birds chirping as they shake themselves. No birds chirped, but the cars and buses squawked in deeper, openly humorous voices, and a trash can and a mailbox broke into conversation. CAST YOUR BALLOT HERE FOR A CLEANER NEW YORK, the trash can said, and MAIL EARLY IN THE DAY IT'S THE BETTER WAY, the mailbox beside it quickly responded. Both seemed to be rejoicing in the knowledge of their own inner snugness—of all the paper, folded or crumpled, addressed or discarded, that they had kept dry through the shower.

The façades of the buildings darkened in tint, the lights within windows seemed not merely to burn but to blaze, and abruptly the rain was upon us again. In the instant before it fell, the air felt full of soft circular motions and a silent cry of "Hurry!" Pedestrians hustled for shelter. The search converted Fifth Avenue into a romantic and primitive setting for adventure. Pelted, we gained the cave of Finchley's Tudor arcade, with its patio-red floor and plastic orange tree and California sports jackets. The next instant, we ran on into the green glade of the Olivetti entrance, with its typewriter-tipped stalagmite. Finally, we lodged in the narrow but deep shelter of Brentano's leafy *allée* of best-sellers, and from there we observed how the rain, a gusty downpour now, had the effect of exquisitely pressing the city down into itself. Everything— taxi roofs, umbrellas, cellophane-skinned hats, even squinting eyebrows—conveyed a sharp impression of shelter. Just as in a Miro painting the ovals and ellipses and lima beans of color sail across the canvas, so the city seemed a mobile conglomerate of dabs of dryness swimming through a fabric of wet. The rain intensified yet one more notch; the Fred F. French Building developed a positively livid stain along its bricks and the scene seemed squeezed so tight that it yielded the essence of granite, the very idea of a city. In a younger century, we might have wept for joy.

And when the rain stopped at last, a supernaturally well-staged effect was produced in the north. Owing to the arrangement of the slabs of Rockefeller Center, the low, westward-moving sun had laid an exclusive shaft of light upon the face of St. Patrick's Cathedral. Like two elegant conical bottles, the steeples were brimful of a mildly creamy glow. We hastened toward the omen, but by the time we reached the site the sunshine had faded. Yet, looking up through the skeleton globe upheld here by the grimacing Atlas, we saw beyond the metal framework what was, patchy blue and scudding gray, indisputably sky.

# NATURE AS ORGANISM

## Alfred North Whitehead

*It has been said that the twentieth century is an age of philosophy professors and not an age of philosophers. Alfred North Whitehead (1861–1947) was both. He was educated at Cambridge and taught philosophy at Harvard (1924–1936). He also wrote a number of important philosophical and mathematical works, including* **Science and the Modern World** *(1925),* **Religion in the Making** *(1926),* **Symbolism** *(1927), and* **Process and Reality** *(1929). The following excerpt treats a problem that has been at the center of Western thought for hundreds of years.*

So far as concerns English literature we find, as might be anticipated, the most interesting criticism of the thoughts of science among the leaders of the romantic reaction which accompanied and succeeded the epoch of the French Revolution. In English literature, the deepest thinkers of this school were Coleridge, Wordsworth, and Shelley. Keats is an example of literature untouched by science. We may neglect Coleridge's attempt at an explicit philosophical formulation. It was influential in his own generation; but in these lectures it is my object only to mention those elements of the thought of the past which stand for all time. Even with this limitation, only a selection is possible. For our purposes Coleridge is only important by his influence on Wordsworth. Thus Wordsworth and Shelley remain.

Wordsworth was passionately absorbed in nature. It has been said of Spinoza, that he was drunk with God. It is equally true that Wordsworth was drunk with nature. But he was a thoughtful, well-read man, with philosophical

SOURCE: Reprinted with permission of The Macmillan Company from *Science and the Modern World* by Alfred North Whitehead. Copyright 1925 by The Macmillan Company, renewed 1953 by Evelyn Whitehead.

interests, and sane even to the point of prosiness. In addition, he was a genius. He weakens his evidence by his dislike of science. We all remember his scorn of the poor man whom he somewhat hastily accuses of peeping and botanising on his mother's grave. Passage after passage could be quoted from him, expressing his repulsion. In this respect, his characteristic thought can be summed up in his phrase, "We murder to dissect."

In this latter passage, he discloses the intellectual basis of his criticism of science. He alleges against science its absorption in abstractions. His consistent theme is that the important facts of nature elude the scientific method. It is important therefore to ask, what Wordsworth found in nature that failed to receive expression in science. I ask this question in the interest of science itself; for one main position in these lectures is a protest against the idea that the abstractions of science are irreformable and unalterable. Now it is emphatically not the case that Wordsworth hands over inorganic matter to the mercy of science, and concentrates on the faith that in the living organism there is some element that sciences cannot analyse. Of course he recognises, what no one doubts, that in some sense living things are different from lifeless things. But that is not his main point. It is the brooding presence of the hills which haunts him. His theme is nature *insolido,* that is to say, he dwells on that mysterious presence of surrounding things, which imposes itself on any separate element that we set up as an individual for its own sake. He always grasps the whole of nature as involved in the tonality of the particular instance. That is why he laughs with the daffodils, and finds in the primrose thoughts "too deep for tears."

Wordsworth's greatest poem is, by far, the first book of *The Prelude.* It is pervaded by this sense of the haunting presences of nature. A series of magnificent passages, too long for quotation, express this idea. Of course, Wordsworth is a poet writing a poem, and is not concerned with dry philosophical statements. But it would hardly be possible to express more clearly a feeling for nature, as exhibiting entwined prehensive unities, each suffused with modal presences of others:

Ye Presences of Nature in the sky
And on the earth! Ye visions of the hills!
And Souls of lonely places! can I think
A vulgar hope was yours when ye employed
Such ministry, when ye through many a year
Haunting me thus among my boyish sports,
On caves and trees, upon the woods and hills,
Impressed upon all forms the characters
Of danger or desire; and thus did make
The surface of the universal earth,
With triumph and delight, with hope and fear,
Work like a sea? . . .

In thus citing Wordsworth, the point which I wish to make is that we forget how strained and paradoxical is the view of nature which modern science imposes on our thoughts. Wordsworth, to the height of genius, expresses the concrete facts of our apprehension, facts which are distorted in the scientific analysis. Is it not possible that the standardised concepts of science are only valid within narrow limitations, perhaps too narrow for science itself?

Shelley's attitude to science was at the opposite pole to that of Wordsworth. He loved it, and is never tired of expressing in poetry the thoughts which it suggests. It symbolises to him joy, and peace, and illumination. What the hills were to the youth of Wordsworth, a chemical laboratory was to Shelley. It is unfortunate that Shelley's literary critics have, in this respect, so little of Shelley in their own mentality. They tend to treat as a casual oddity of Shelley's nature what was, in fact, part of the main structure of his mind, permeating his poetry through and through. If Shelley had been born a hundred years later, the twentieth century would have seen a Newton among chemists.

For the sake of estimating the value of Shelley's evidence it is important to realise this absorption of his mind in scientific ideas. It can be illustrated by lyric after lyric. I will choose one poem only, the fourth act of his *Prometheus Unbound*. The Earth and the Moon converse together in the language of accurate science. Physical experiments guide his imagery. For example, the Earth's exclamation,

The vaporous exultation not to be confined!

is the poetic transcript of "the expansive force of gases," as it is termed in books on science. Again, take the Earth's stanza,

I spin beneath my pyramid of night,
Which points into the heavens,—dreaming delight,
Murmuring victorious joy in my enchanted sleep;
As a youth lulled in love-dreams faintly sighing,
Under the shadow of his beauty lying,
Which round his rest a watch of light and warmth doth keep.

This stanza could only have been written by someone with a definite geometrical diagram before his inward eye—a diagram which it has often been my business to demonstrate to mathematical classes. As evidence, note especially the last line which gives poetical imagery to the light surrounding night's pyramid. This idea could not occur to anyone without the diagram. But the whole poem and other poems are permeated with touches of this kind.

Now the poet, so sympathetic with science, so absorbed in its ideas, can simply make nothing of the doctrine of secondary qualities which is fundamental to its concepts. For Shelley nature retains its beauty and its colour. Shelley's nature is in its essence a nature of organisms, functioning with the full content of our perceptual experience. We are so used to ignoring the implication of orthodox scientific doctrine, that it is difficult to make evident the criticism upon it which is thereby implied. If anybody could have treated it seriously, Shelley would have done so.

Furthermore Shelley is entirely at one with Wordsworth as to the interfusing Presence in nature. Here is the opening stanza of his poem entitled *Mont Blanc*:

> The everlasting universe of Things
> Flows through the Mind, and rolls its rapid waves,
> Now dark—now glittering—now reflecting gloom—
> Now lending splendour, where from secret springs
> The source of human thought its tribute brings
> Of waters,—with a sound but half its own,
> Such as a feeble brook will oft assume
> In the wild woods, among the Mountains lone,
> Where waterfalls around it leap for ever,
> Where woods and winds contend, and a vast river
> Over its rocks ceaselessly bursts and raves.

Shelley has writen these lines with explicit reference to some form of idealism, Kantian or Berkeleyan or Platonic. But however you construe him, he is here an emphatic witness to a prehensive unification as constituting the very being of nature.

Berkeley, Wordsworth, Shelley are representative of the intuitive refusal seriously to accept the abstract materialism of science.

There is an interesting difference in the treatment of nature by Wordsworth and by Shelley, which brings forward the exact questions we have got to think about. Shelley thinks of nature as changing, dissolving, transforming as it were at a fairy's touch. The leaves fly before the West Wind

> Like ghosts from an enchanter fleeing.

In his poem *The Cloud* it is the transformations of water which excite his imagination. The subject of the poem is the endless, eternal, elusive change of things:

> I change but I cannot die.

This is one aspect of nature, its elusive change: a change not merely to be expressed by locomotion, but a change of inward character. This is where Shelley places his emphasis, on the change of what cannot die.

Wordsworth was born among hills; hills mostly barren of trees, and thus showing the minimum of change with the seasons. He was haunted by the enormous permanences of nature. For him change is an incident which shoots across a background of endurance,

Breaking the silence of the seas
Among the farthest Hebrides.

Every scheme for the analysis of nature has to face these two facts, *change* and *endurance*. There is yet a third fact to be placed by it, *eternality*, I will call it. The mountain endures. But when after ages it has been worn away, it has gone. If a replica arises, it is yet a new mountain. A colour is eternal. It haunts time like a spirit. It comes and it goes. But where it comes, it is the same colour. It neither survives nor does it live. It appears when it is wanted. The mountain has to time and space a different relation from that which colour has. In the previous lecture, I was chiefly considering the relation to space-time of things which, in my sense of the term, are eternal. It was necessary to do so before we can pass to the consideration of the things which endure. . . .

The literature of the nineteenth century, especially its English poetic literature, is a witness to the discord between the aesthetic intuitions of mankind and the mechanism of science. Shelley brings vividly before us the elusiveness of the eternal objects of sense as they haunt the change which infects underlying organisms. Wordsworth is the poet of nature as being the field of enduring permanences carrying within themselves a message of tremendous significance. The eternal objects are also there for him,

The light that never was, on sea or land.

Both Shelley and Wordsworth emphatically bear witness that nature cannot be divorced from its aesthetic values, and that these values arise from the cumulation, in some sense, of the brooding presence of the whole on to its various parts. Thus we gain from the poets the doctrine that a philosophy of nature must concern itself at least with these six notions: change, value, eternal objects, endurance, organism, interfusion. . . . In a certain sense, everything is everywhere at all times. For every location involves an aspect of itself in every other location. Thus every spatio-temporal standpoint mirrors the world.

If you try to imagine this doctrine in terms of our conventional views of space and time, which presuppose simple location, it is a great paradox. But if you think of it in terms of our naïve experience, it is a mere transcript of the obvious facts. You are in a certain place perceiving things. Your perception takes place where you are, and is entirely dependent on how your body is functioning. But this functioning of the body in one place, exhibits for your cognisance an aspect of the distant environment, fading away into the general knowledge that there are things beyond. If this cognisance conveys knowledge of a transcendent world, it must be because the event which is the bodily life unifies in itself aspects of the universe.

# IN BACK OF MAN A WORLD OF NATURE

## Joseph Wood Krutch

*The career of Joseph Wood Krutch (1893–1970) is one that cannot be questioned by today's standards of relevance. Though a distinguished scholar and critic of English literature, he seldom failed to emphasize the connection between his study and the world at large. Toward the end of his career his writing turned to reflections on science and nature. In these areas his remarks are all the more impressive because they are tempered by a wise and humble humanism. This essay, for example, is characteristically deceptive in that it begins modestly, and before it ends on a note of incisive irony it involves the reader in some of the more pressing problems of this age.*

More than twenty-five years ago I first bought a house in the country. My motives were various, but as in the case of many another man who acquires something he never had before, the most compelling was the fact that my wife wanted it. Among the reasons I gave myself to explain my acquiescence, the possibility that I might some day be writing about nature was certainly not one. As a matter of fact I had been leading a double life, in the city and out of it, for almost a quarter of a century before it occurred to me. Then one evening when I was reading with delight about someone else's feeling for his countryside, I said to myself that it would be nice if I could do it too.

Until then I had never realized what had been happening to me. I still thought that I thought the country was a place where one went to get away from the fatigues of the city. Its virtues were still supposed to be largely negative and I had never realized how, in actual fact, I had long been regarding

SOURCE: From *The Best of Two Worlds* by Joseph Wood Krutch. Published by William Sloane Associates. Reprinted by permission of William Morrow and Company, Inc. Copyright 1951, 1953 by Joseph Wood Krutch.

it as the principal scene of those activities for the sake of which I made a living elsewhere, how completely it had become the place where it seemed to me that I must live or have no being. Since then I have seen myself referred to somewhat condescendingly as a "nature writer," and though I did not mind a bit, I was set to wondering just what a nature writer might be and just why or from just what he should be distinguished by his special label.

A great many people, I am afraid, do not even care to know. When they hear the phrase they think of "the birds and the bees"—a useful device in the sex education of children but hardly, they think, an occupation for a grown man. A biologist is all right and so is a sportsman. But a nature writer can hardly expect more than a shrug from the realistic or the robust.

The more literary quote Wordsworth: "Nature never did betray the heart that loved her" or "to me the meanest flower that blows can give thoughts that do often lie too deep for tears." But Darwin, they say, exploded such ideas nearly a century ago. Even Tennyson knew better: "Nature red in tooth and claw." We can't very well learn anything from all that. Ours are desperate times. The only hope for today lies in the study of politics, sociology, and economics. If a nature writer is someone who advocates a return to nature then he must be a sentimentalist and a very old-fashioned one at that.

But why, then, should I have discovered that my weekends and my summers in the country were becoming more and more important, that I spent more and more time looking at and thinking about plants and animals and birds, that I began to feel the necessity of writing about what I had seen and thought? Why, finally, when I saw fifteen free months ahead of me, should I have been sure that I wanted to go back to the southwestern desert to see what an entirely different, only half-known natural environment might have to say to me?

It was more than a mere casual interest in natural history as a hobby. Certainly it was also more than the mere fact that the out-of-doors is healthful and relaxing. I was not merely being soothed and refreshed by an escape from the pressures of urban life. I was seeking for something, and I got at least the conviction that there was something I really was learning. I seemed to be getting a glimpse of some wisdom of which I had less than an inkling before.

Actually, of course, nature writing does flourish quite vigorously today as a separate department of literature even though it does remain to some extent a thing apart, addressed chiefly to a group of readers called "nature lovers" who are frequently referred to by outsiders in the same condescending tone they would use in speaking of prohibitionists, diet cranks, Holy Rollers, or the followers—if there still are any—of the late Mr. Coué. And that makes me sometimes wonder whether its very importance as a "department of literature" does not mean that what was once an almost inevitable motive in most imaginative writing has become something recognizably special, for the reason that most writers of novels and poems and plays no longer find the contemplation of nature relevant to their purposes—at least to the same extent they once did. And I wonder also whether that seeming fact is merely the result of urban living, or whether "merely" is not the wrong term to use in reference to a phenomenon which may mean a great deal more than the mere disappearance from fiction of apostrophes to the mountains.

A generation ago the first page of the epoch-making *Cambridge History of English Literature* listed among the enduring characteristics of the English people, "love of nature." That phrase covers something which has meant a good many different things at different times. But would it have suggested itself to a critic who was concerned only with the most esteemed English or American books of our time? Is there any love of nature—as distinguished from an intellectual approval of the processes of biology—in Bernard Shaw? Does T. S. Eliot find much gladness in contemplating nature? Does Joyce's apostrophe to a river count, and is Hemingway's enthusiasm for the slaughter of animals really to be considered as a modern expression of even that devotion to blood sports which, undoubtedly, really is a rather incongruous aspect of the Anglo-Saxon race's love of nature? In America Robert Frost is almost the only poet whose work is universally recognized as of major importance and in which the loving contemplation of natural phenomena seems a central activity from which the poetry springs.

All this is probably the result of something more than mere fashion. True, ridicule of conventional description in fiction is no new thing, and Mark Twain once tested his theory that readers always skip it by inserting a paragraph which told how a solitary esophagus might have been seen winging its way across the sky. It is true also that observations of natural phenomena do still sometimes get into fiction. *The Grapes of Wrath* begins with the symbolic use of a turtle crossing a highway and nearly everybody knows that at least one tree did grow in Brooklyn. But somewhere along the road they have traveled, most moderns have lost the sense that nature is the most significant background of human life. They see their characters as part of society or, more specifically, members of some profession or slaves at some industry, rather than as part of nature. Neither her appearances nor her ways any longer seem—to use a favorite modern term—as "relevant" as they once did. She is not often, nowadays, invoked to furnish the resolution of an emotional problem.

Whistler was probably the first English-speaking writer ever to say flatly "Nature is wrong." Of course he meant to be shocking and he also meant "artistically wrong"—unpictorial or badly composed. And goodness only knows most contemporary painting is the product of hearty agreement with this dictum. Either natural objects are so distorted in the effort to correct nature's wrongness that they are just barely recognizable, or the artist, refusing to look at nature at all, plays at being God by attempting to create a whole new universe of man-made shapes. In either case the assumption is that modern man is more at home and gets more emotional satisfaction in this world of his own making than in the world which nature gave him.

Only the most extreme forms of the most desperately "experimental" writing go anything like so far. Only the poets of the Dada produce literary analogies of abstract painting. But without being, for the most part, even aware of a theory about what they are doing, many novelists and poets have obviously ceased to feel that the significant physical setting for their characters or sentiments is the fields or woods, or that the intellectual and emotional context of their difficulties and problems is the natural world rather than that of exclusively human concepts. This amounts to a good deal more than a mere loss of faith in the dogma that nature never did betray the heart that loved her. It also amounts to a good deal more than the somewhat romantic Victorian distress in the face of her red tooth and red claw. It means that the writer finds neither God dwelling in the light of setting suns nor a very significant aspect of the problem of evil in nature's often careless cruelty. Faced with either her beauty or her ruthlessness, his reaction is more likely to be only, "so what?" Man's own achievements, follies, and crimes seem to him to lie in a different realm.

As cities grow and daily life becomes of necessity more and more mechanized, we inevitably come to have less to do with, even to see less often and to be less aware of, other things which live; and it comes to seem almost as though all the world outside ourselves were inanimate. In so far as the writer thinks of himself as a secretary of society he has that much justification for treating man as a creature whose most significant environment is that of the machines he has built and, to some small extent, the art which he has created. But in so far as the writer is more than a recorder, in so far as he should see deeper than a secretary is required to see, he might be expected to be aware both of the extent to which the fact that man is an animal sharing the earth with other animals is still significantly true, and of the consequences of the extent to which it has actually become less so. Those consequences, though difficult to assess, are certainly enormous and they did not begin to be fully felt until the twentieth century.

The nineteenth century was deeply concerned with what it called "man's place in nature," and as some of the writers pointed out, that had much more than merely scientific implications. It did not imply only, or even most importantly, that man was descended from the apes and was, therefore, still apelike in many of his characteristics. It meant also that animal life supplied the inescapable context of his life, spiritually as well as physically. It meant that life was an adventure which he shared with all living things, that the only clue to himself was in them. But of that fact many, perhaps most, of the most intelligent and cultivated people of our time are unaware. Having to do almost exclusively with other human beings and with machines, they tend to forget what we are and what we are like. Even the graphic arts are forsaking nature so that even on the walls of our apartments the wheel or the lever are more familiar than the flower or the leaf. And perhaps all this is the real reason why we have tended more and more to think about man and society as though they were machines, why we have mechanistic theories about consciousness and about human behavior in general, why we have begun to think that even the brain is something like an electronic calculating machine. After all, it is only with machines that most people are more than casually familiar. And perhaps it is trying to think in this way that makes us unhappy—nearly everybody seems to agree that we are—because we know in our hearts that we are not machines and grow lonesome in a universe where we are little aware of anything else which is not.

In so far then as nature writing has become a special department of literature, consciously concerned with an expression of the individual writer's awareness of something which most people are not aware of; in so far as it finds readers other than those members of the small group to whom natural history and allied subjects is a hobby like any other unusual interest; then perhaps to that extent what the existence of this department of literature means is simply that it exists principally because works in other departments are no longer concerned with certain truths of fact and feeling which some part of even the general public recognizes as lacking. In any event, at least all of those nature writers who stem ultimately from Thoreau are concerned not only with the aesthetic and hedonistic aspects of the love of nature, but also with what can only be called its moral aspect.

Every now and then the average man looks at a kitten and thinks it is "cute." Or he looks at the stars and they make him feel small. When he thinks either of those things he is being aware of the context of nature, and that is good as far as it goes. But it does not go very far, and there is little in modern life or art to make him go further. What the nature writer is really asking him to do is to explore the meaning of such thoughts and feelings. He is asking him to open his heart and mind to nature as another kind of writer asks him to open them to art or music or literature. Nearly everyone admits that these have something to say that science and sociology cannot say. Nature has something to say that art and literature have not.

But nature writing also implies something more fundamental than that. It also raises the question of the moral consequences of taking that opposite point of view which is now more usual. It raises the question of the effect which forgetting that he is alive may have upon man and his society.

Today the grandest of all disputes is that between those who are determined to manipulate man as though he were a machine and those who hope, on the contrary, to let him grow like an organism. Whether our future is to be totalitarian or free depends entirely upon which side wins the dispute, and the question which side we ourselves are on may in the end depend upon our conception of "man's place in nature." Nearly every taste we cultivate and nearly every choice we make, down to the very decoration of our wall, tends to proclaim where our sympathies lie. Do they suggest a preference for the world which lives and grows or for the world which obeys the laws of the machine? Do they remind us of the natural universe in which every individual thing leads its own individual, unique and rebellious life? Of the universe where nothing, not even one of two leaves, is quite like anything else? Or do they accustom us instead to feel more at home in surroundings where everything suggests only machines and machine parts, which do as they are told and could never have known either joy or desire? In its direct, brutal way even official Marxism recognized that fact when it rested its hopes for a revolution not on the peasants but on the urban proletariat. It is certainly no accident that the totalitarian countries are those which have made mechanistic theories the official philosophy of the state. If man is nothing but a machine, if there are laws of psychology precisely analogous to the laws of physics, then he can be "conditioned" to do and to want whatever his masters decide. Society is then a machine composed of standard parts, and men can be ordered to become whatever cogs and levers the machine happens to require at the moment.

Our forefathers were in no danger of forgetting that they were dependent upon nature and were a part of it. No man who clears forests, grows crops, and tends cattle is. He sees his food coming up out of the ground, and when he would travel further than he can walk, he calls a living, four-footed creature for aid. But unless our civilization should be destroyed as completely as the gloomiest prophets of the atomic age sometimes predict, we shall never again be a rural people and probably we would not want to be. What tools can give us is very much worth having, and the machinery of an industrial society is merely a collection of tools. But our physical as well as our spiritual dependence upon nature is merely obscured, not abolished, and to be unaware of that fact is to be as naïvely obtuse as the child who supposes that cows are no longer necessary because we now get milk from bottles.

In the early stages of mechanized civilization no one thought of mechanics as a threat to the soul of man. The machine was too obviously an alien though useful contrivance. The threat did not arise until we began to be overawed by the lifeless things we had created and to worship, rather than merely to use, our tools. Science and psychology began to talk about the "body machine" and the "brain machine" just about the time when we began to be more and more aware of the artificial rather than the natural environment. When men lived most intimately with things which were alive they thought of themselves as living. When they began, on the contrary, to live most intimately with dead things, they began to suppose that they, too, were dead. And once men were thought of as machines, governments began inevitably to be thought of as merely a method of making the machines operate productively. An engineer does not consider the tastes, the preferences, the desires, or the possible individual talents of the mechanical units he employs.

Why should a government concerned with body machines and mechanically conditioned reflexes do so either?

It is highly improbable that we shall ever again lead what a Thoreau would be willing to call a "natural life." Moreover, since not every man has the temperament of a camper, it is not likely that most people will take even periodically to the woods. But there are other ways in which even the urban dweller not fortunate enough to have his country home can remain aware, at least in the important back of his mind, of the fact that this true place is somewhere in nature.

The more complicated life becomes, the more important is the part that symbols play, and all the arts, including the arts of architecture and decoration, are the products of symbolic acts. To plant a garden, a window box, or even to cultivate a house plant is to perform a sort of ritual and thereby to acknowledge, even in the middle of a city, one's awareness that our real kinship is with life, not with mechanism. To hang a picture or to choose a design or a color may be, only a little more remotely, the same thing.

It may be upon such rituals that our fate will ultimately depend. Organisms are rebellious, individual, and self-determining. It is the machine which is manageable and obedient, which always does the expected and behaves as it is told to behave. You can plan for it as you cannot plan for anything which is endowed with a life of its own, because nothing which is alive ever wholly surrenders its liberties. Man will not surrender his unless he forgets that he too belongs among living things.

This is not, in its emphasis at least, quite the same lesson which others at other times have deducted. From nature as from everything else worth studying one gets what at the moment one needs, sometimes perhaps only what at the moment one would like to have. The Wordsworthians, for instance, got the conviction that some pantheistic God who approved of the same thing that they approved of really did exist. Many of us who look back at them are inclined to suspect that what they were contemplating was only a projection of themselves and that Wordsworth might have profited by a little more objective observation—a little more of what he called the scientist's willingness to "botanize upon his mother's grave"—instead of devoting so much of his time to reading into clouds or mountains or sunsets what he liked to find there.

As for myself I settle for what Wordsworth would have considered a good deal less than what he managed to get. What I think I find myself most positively assured of is not that man is divine but only that he is at least not a machine, that he is like an animal, not like even the subtlest electronic contraption. Today in a good many different ways most of us are willing to settle for less than most men at most times got, and that less seems quite enough for me.

In other respects, also, the demands I make upon nature before I will call her sum total "good" are less exorbitant than those which others have made. I do not think that her every prospect pleases or that only man is vile. No one who actually looks at nature rather than at something his fancy reads into her can ever fail to realize that she represents some ultimate things-as-they-are, not some ideal of things-as-he-thinks-they-ought-to-be. There is in her what we call cruelty and also, even more conspicuously, what we call grotesqueness, even what we call comedy. If she warns the so-called realist how limited his conception of reality is, she is no less likely to bring the sentimentalist back, literally, to earth.

How much of the cruelty, the grotesqueness, or the sublimity any given man will see depends no doubt to some considerable extent upon his own temperament, and I suppose it is some indication of mine when I confess that what I see most often and relish the most is, first, the intricate marvel; second, the comedy. To be reminded that one is very much like other members of the animal kingdom is often funny enough, though it is never, like being compared to a machine, merely humiliating. I do not too much mind being somewhat like a cat, a dog, or even a frog, but I resent having it said that even an electronic calculator is like me.

Not very long ago I was pointing out to a friend the courtship of two spiders in a web just outside my door. Most people know that the male is often much smaller than his mate, and nearly everybody knows by now that the female of many species sometimes eats her husband. Both of these things were true of the common kind beside my door, and the insignificant male was quite obviously torn between ardor and caution. He danced forward and then darted back. He approached now from one side and now from the other. He would and he wouldn't.

My friend, no nature student and not much given to observing such creatures, was gratifyingly interested. Presently he could contain himself no longer. "You know," he said thoughtfully, "there is only one difference between that spider and a human male. The spider knows it's dangerous."

That, I maintain, both is and ought to be as much grist for a nature writer's mill as a sunset or a bird song.

from **ON THE ORIGIN OF SPECIES BY MEANS OF NATURAL SELECTION**

*Charles Darwin*

*Beyond a doubt, the publication of Darwin's* **On the Origin of Species by Means of Natural Selection** *(1859) marked the successful negotiation of a turning point in the history of man. Evolution as an idea had been present for many years before that date, but it took Darwin (1809–1881), a dedicated scientist with a profound confidence in the goodness of life, to establish the thought in the human psyche. Though he saw that the economy of nature dictates that all living creatures be involved in an endless struggle for survival, he was more impressed with the completeness of the picture that emerges than he was with the cruelty that is necessary to the overall design.*

Several writers have misapprehended or objected to the term *Natural Selection.* Some have even imagined that natural selection induces variability, whereas it implies only the preservation of such variations as arise and are beneficial to the being under its conditions of life. No one objects to agriculturists speaking of the potent effects of man's selection; and in this case the individual differences given by nature, which man for some object selects, must of necessity first occur. Others have objected that the term selection implies conscious choice in the animals which become modified; and it has even been urged that, as plants have no volition, natural selection is not applicable to them! In the literal sense of the word, no doubt, natural selection is a false term; but who ever objected to chemists speaking of the elective affinities of the various elements?—and yet an acid cannot strictly be said to

SOURCE: Charles Darwin, *On the Origin of Species by Means of Natural Selection*, 6th ed. (New York and London, 1872).

elect the base with which it in preference combines. It has been said that I speak of natural selection as an active power or Deity; but who objects to an author speaking of the attraction of gravity as ruling the movements of the planets? Every one knows what is meant and is implied by such metaphorical expressions; and they are almost necessary for brevity. So again it is difficult to avoid personifying the word *Nature;* but I mean by Nature, only the aggregate action and product of many natural laws, and by laws the sequence of events as ascertained by us. With a little familiarity such superficial objections will be forgotten.

We shall best understand the probable course of natural selection by taking the case of a country undergoing some slight physical change, for instance, of climate. The proportional numbers of its inhabitants will almost immediately undergo a change, and some species will probably become extinct. We may conclude, from what we have seen of the intimate and complex manner in which the inhabitants of each country are bound together, that any change in the numerical proportions of the inhabitants, independently of the change of climate itself, would seriously affect the others. If the country were open on its borders, new forms would certainly immigrate, and this would likewise seriously disturb the relations of some of the former inhabitants. Let it be remembered how powerful the influence of a single introduced tree or mammal has been shown to be. But in the case of an island, or of a country partly surrounded by barriers, into which new and better adapted forms could not freely enter, we should then have places in the economy of nature which would assuredly be better filled up, if some of the original inhabitants were in some manner modified; for, had the area been open to immigration, these same places would have been seized on by intruders. In such cases, slight modifications, which in any way favoured the individuals of any species, by better adapting them to their altered conditions, would tend to be preserved; and natural selection would have free scope for the work of improvement.

We have good reason to believe . . . that changes in the conditions of life give a tendency to increased variability; and in the foregoing cases the conditions have changed, and this would manifestly be favourable to natural selection, by affording a better chance of the occurrence of profitable variations. Unless such occur, natural selection can do nothing. Under the term of "variations," it must never be forgotten that mere individual differences are included. As man can produce a great result with his domestic animals and plants by adding up in any given direction individual differences, so could natural selection, but far more easily from having incomparably longer time for action. Nor do I believe that any great physical change, as of climate, or any unusual degree of isolation to check immigration, is necessary in order that new and unoccupied places should be left, for natural selection to fill up by improving some of the varying inhabitants. For as all the inhabitants of each country are struggling together with nicely balanced forces, extremely slight modifications in the structure or habits of one species would often give it an advantage over others; and still further modifications of the same kind would often still further increase the advantage, as long as the species continued under the same conditions of life and profited by similar means of subsistence and defence. No country can be named in which all the native inhabitants are now so perfectly adapted to each other and to the physical conditions under which they live, that none of them could be still better adapted or improved; for in all countries, the natives have been so far conquered by naturalised productions, that they have allowed some foreigners to take firm possession of the land. And as foreigners have thus in every country beaten some of the natives, we may safely conclude that the natives might have been modified with advantage, so as to have better resisted the intruders.

As man can produce, and certainly has produced, a great result by his methodical and unconscious means of selection, what may not natural selection effect? Man can act only on external and visible characters: Nature, if I may be allowed to personify the natural preservation or survival of the fittest, cares nothing for appearances, except insofar as they are useful to any being. She can act on every internal organ, on every shade of constitutional difference, on the whole machinery of life. Man selects only for his own good: Nature only for that of the being which she tends. Every selected character is fully exercised by her, as is implied by the fact of their selection. Man keeps the natives of many climates in the same country; he seldom exercises each selected character in some peculiar and fitting manner; he feeds a long- and a short-beaked pigeon on the same food; he does not exercise a long-backed or long-legged quadruped in any peculiar manner; he exposes sheep with long and short wool to the same climate. He does not allow the most vigorous males to struggle for the females. He does not rigidly destroy all inferior animals, but protects during each varying season, as far as lies in his power, all his productions. He often begins his selection by some half-monstrous form; or at least by some modification prominent enough to catch the eye or to be plainly useful to him. Under nature, the slightest differences of structure or constitution may well turn the nicely balanced scale in the struggle for life, and so be preserved. How fleeting are the wishes and efforts of man! how short his time! and consequently how poor will be his results, compared with those accumulated by Nature during the whole geological periods! Can we wonder, then, that Nature's productions should be far "truer" in character than man's productions; that they should be infinitely better adapted to the most complex conditions of life, and should plainly bear the stamp of far higher workmanship?

It may metaphorically be said that natural selection is daily and hourly scrutinising, throughout the world, the slightest variations; rejecting those that are bad, preserving and adding up all that are good; silently and insensibly working, *whenever and wherever opportunity offers,* at the improvement of each organic being in relation to its organic and inorganic conditions of life. We see nothing of these slow changes in progress, until the hand of time has marked the lapse of ages, and then so imperfect is our view into long-past geological ages, that we see only that the forms of life are now different from what they formerly were.

*        *        *

It is interesting to contemplate a tangled bank, clothed with many plants of many kinds, with birds singing on the bushes, with various insects flitting about, and with worms crawling through the damp earth, and to reflect that these elaborately constructed forms, so different from each other, and dependent upon each other in so complex a manner, have all been produced by laws acting around us. These laws, taken in the largest sense, being Growth with Reproduction; Inheritance which is almost implied by reproduction; Variability from the indirect and direct action of the conditions of life, and from use and disuse; a Ratio of Increase so high as to lead to a Struggle for Life, and as a consequence to Natural Selection, entailing Divergence of Character and the Extinction of less-improved forms. Thus, from the war of nature, from famine and death, the most exalted object which we are capable of conceiving, namely, the production of the higher animals, directly follows. There is grandeur in this view of life, with its several powers, having been originally breathed by the Creator into a few forms or into one; and that, whilst this planet has gone cycling on according to the fixed law of gravity, from so simple a beginning endless forms most beautiful and most wonderful have been, and are being evolved.

# THE WILL IN NATURE

*Arthur Schopenhauer*

*For Arthur Schopenhauer (1788–1860) the only reality is will. In man it is a self-conscious force, but in nature it is unconscious and is known only by way of our intuition. Schopenhauer also understood the world to be created by will, and for him nature exists as a malignant illusion, which inveigles man into reproducing and perpetuating life. His solution was to oppose the "will to live" (ego) with the moral law of compassion, which was based on the intuition of the essential identity of all beings. Whether or not one agrees with this thesis (Schopenhauer's father died a lunatic), there is no doubt that many people subscribe to his vision of nature in which everyone and everything is constantly in a state of strife.*

The one will . . . always seeks the highest possible objectification, and has therefore in this case given up the lower grades of its manifestation after a conflict, in order to appear in a higher grade, and one so much the more powerful. No victory without conflict: since the higher Idea or objectification of will can only appear through the conquest of the lower, it endures the opposition of these lower Ideas, which, although brought into subjection, still constantly strive to obtain an independent and complete expression of their being. The magnet that has attracted a piece of iron carries on a perpetual conflict with gravitation, which, as the lower objectification of will, has a prior right to the matter of the iron; and in this constant battle the magnet indeed grows stronger, for the opposition excites it, as it were, to greater effort. In the same way every manifestation of the will, including that which expresses itself in

SOURCE: Arthur Schopenhauer, *The World as Will and Idea*, trans. R. B. Haldane and J. Kemp (London, 1883), vol. I.

the human organism, wages a constant war against the many physical and chemical forces which, as lower Ideas, have a prior right to that matter. Thus the arm falls which for a while, overcoming gravity, we have held stretched out; thus the pleasing sensation of health, which proclaims the victory of the Idea of the self-conscious organism over the physical and chemical laws, which originally governed the humours of the body, is so often interrupted, and is indeed always accompanied by greater or less discomfort, which arises from the resistance of these forces, and on account of which the vegetative part of our life is constantly attended by slight pain. Thus also digestion weakens all the animal functions, because it requires the whole vital force to overcome the chemical forces of nature by assimilation. Hence also in general the burden of physical life, the necessity of sleep, and, finally, of death; for at last these subdued forces of nature, assisted by circumstances, win back from the organism, wearied even by the constant victory, the matter it took from them, and attain to an unimpeded expression of their being. . . .

Thus everywhere in nature we see strife, conflict, and alternation of victory, and in it we shall come to recognise more distinctly that variance with itself which is essential to the will. Every grade of the objectification of will fights for the matter, the space, and the time of the others. The permanent matter must constantly change its form; for under the guidance of causality, mechanical, physical, chemical, and organic phenomena, eagerly striving to appear, wrest the matter from each other, for each desires to reveal its own Idea. This strife may be followed through the whole of nature; indeed nature exists only through it. Yet this strife itself is only the revelation of that variance with itself which is essential to the will. This universal conflict becomes most distinctly visible in the animal kingdom. For animals have the whole of the vegetable kingdom for their food, and even within the animal kingdom every beast is the prey and the food of another; that is, the matter in which its Idea expresses itself must yield itself to the expression of another Idea, for each animal can only maintain its existence by the constant destruction of some other. Thus the will to live everywhere preys upon itself, and in different forms is its own nourishment, till finally the human race, because it subdues all the others, regards nature as a manufactory for its use. Yet even the human race . . . reveals in itself with most terrible distinctness this conflict, this variance with itself of the will. Meanwhile we can recognise this strife, this subjugation, just as well in the lower grades of the objectification of will. Many insects (especially ichneumon-flies) lay their eggs on the skin, and even in the body of the larvae of other insects, whose slow destruction is the first work of the newly hatched brood. The young hydra, which grows like a bud out of the old one, and afterwards separates itself from it, fights while it is still joined to the old one for the prey that offers itself, so that the one snatches it out of the mouth of the other. But the bulldog-ant of Australia affords us the most extraordinary example of this kind; for if it is cut in two, a battle begins between the head and the tail. The head seizes the tail with its teeth, and the tail defends itself bravely by stinging the head: the battle may last for half an hour, until they die or are dragged away by other ants. This contest takes place every time the experiment is tried. On the banks of the Missouri one sometimes sees a mighty oak the stem and branches of which are so encircled, fettered, and interlaced by a gigantic wild vine, that it withers as if choked. The same thing shows itself in the lowest grades; for example, when water and carbon are changed into vegetable sap, or vegetables or bread into blood by organic assimilation; and so also in every case in which animal secretion takes place, along with the restriction of chemical forces to a subordinate mode of activity. This also occurs in unorganised nature, when, for example, crystals in process of formation meet, cross, and mutually disturb each other to such an extent that they are unable to assume the pure crystalline form, so that almost every cluster of crystals is an image of such a conflict of will at this low grade of objectification; or again, when a magnet forces its magnetism upon iron, in order to express its Idea in it; or when galvanism overcomes chemical affinity, decomposes the closest combinations, and so entirely suspends the laws of chemistry that the acid of a decomposed salt at the negative pole must pass to the positive pole without combining with the alkalies through which it goes on its way, or turning red the litmus paper that touches it. On a large scale it shows itself in the relation between the central body and the planet, for although the planet is in absolute dependence, yet it always resists, just like the chemical forces in the organism; hence arises the constant tension between centripetal and centrifugal force, which keeps the globe in motion, and is itself an example of that universal essential conflict of the manifestation of will which we are considering. . . .

Will is the thing-in-itself, the inner content, the essence of the world. Life, the visible world, the phenomenon, is only the mirror of the will. Therefore life accompanies the will as inseparably as the shadow accompanies the body; and if will exists, so will life, the world, exist. Life is, therefore, assured to the will to live; and so long as we are filled with the will to live we need have no fear for our existence, even in the presence of death. It is true we see the individual come into being and pass away; but the individual is only phenomenal, exists only for the knowledge which is bound to . . . the principle of individuation. Certainly, for this kind of knowledge, the individual receives his life as a gift, rises out of nothing, then suffers the loss of this gift through death, and returns again to nothing. But we desire to consider life philosophically . . . and in this sphere we shall find that neither the will, the thing-in-itself in all phenomena, nor the subject of knowing, that which perceives all phenomena, is affected at all by birth or by death. Birth and death belong merely to the phenomenon of will, thus to life; and it is essential to this to exhibit itself in individuals which come into being and pass away, as fleeting phenomena appearing in the form of time—phenomena of that which in itself knows no time, but must exhibit itself precisely in the way we have said, in order to objectify its peculiar nature. Birth and death belong in like manner to life, and hold the balance as reciprocal conditions of each other, or, if one likes the expression, as poles of the whole phenomenon of life. The wisest of all mythologies, the Indian, expresses this by giving to the very god that symbolises destruction, death (as Brahma, the most sinful and lowest god of the Trimurti, symbolises generation, coming into being, and Vishnu maintaining or preserving), by giving, I say, to Siva as an attribute not only the necklace of skulls, but also the lingam, the symbol of generation, which appears here as the counterpart of death, thus signifying that generation and death are essentially correlatives, which reciprocally neutralise and annul each other. It was precisely the same sentiment that led the Greeks and Romans to adorn their costly sarcophagi, just as we see them now, with feasts, dances, marriages, the chase, fights of wild beasts, bacchanalians, &c.; thus with representations of the full ardour of life, which they place before us not only in such revels and sports, but also in sensual groups, and even go so far as to represent the sexual intercourse of satyrs and goats. Clearly the aim was to point in the most impressive manner away from the death of the mourned individual to the immortal life of nature, and thus to indicate, though without abstract knowledge, that the whole of nature is the phenomenon and also the fulfilment of the will to live. The form of this phenomenon is time, space, and causality, and by means of these individuation, which carries with it that the individual must come into being and pass away. But this no more affects the will to live, of whose manifestation the individual is, as it were, only a particular example or specimen, than the death of an individual injures the whole of nature. For it is not the individual, but only the species that Nature cares for, and for the preservation of which she so earnestly strives, providing for it with the utmost prodigality through the vast surplus of the seed and the great strength of the fructifying impulse. The individual, on the contrary, neither has nor can have any value for Nature, for her kingdom is infinite time and infinite space, and in these infinite multiplicity of possible individuals. Therefore she is always ready to let the individual fall, and hence it is not only exposed to destruction in a thousand ways by the most insignificant accident, but originally destined for it, and conducted towards it by Nature herself from the moment it has served its end of maintaining the species.

*    *    *

Dogmas change and our knowledge is deceptive; but Nature never errs, her procedure is sure, and she never conceals it. Everything is entirely in Nature, and Nature is entire in everything. She has her centre in every brute. It has surely found its way into existence, and it will surely find its way out of it. In the meantime it lives, fearless and without care, in the presence of annihilation, supported by the consciousness that it is Nature herself, and imperishable as she is. Man alone carries about with him, in abstract conceptions, the certainty of his death; yet this can only trouble him very rarely, when for a single moment some occasion calls it up to his imagination. Against the mighty voice of Nature reflection can do little. In man, as in the brute which does not think, the certainty that springs from his inmost consciousness that he himself is Nature, the world, predominates as a lasting frame of mind; and on account of this no man is observably disturbed by the thought of certain and never-distant death, but lives as if he would live for ever.

# from NATURAL THEOLOGY

## William Paley

*William Paley (1743–1805) was a major influence on Cambridge University at the turn of the nineteenth century. The following excerpt is from his* Natural Theology *(1802) in which Paley gives proof of the existence of God from the design apparent in natural phenomena. Much of this work was an attempt to controvert the theory of adaptation of the organism to its circumstances by use. Of course, his position was weakened by subsequent evolutionary discoveries. More interesting for our purposes is the way in which Paley's theological unitarianism enters into his understanding of nature. Even today, there are those who appear to believe in a watchmaker God.*

In crossing a heath, suppose I pitched my foot against a *stone,* and were asked how the stone came to be there; I might possibly answer, that, for any thing I knew to the contrary, it had lain there for ever: nor would it perhaps be very easy to show the absurdity of this answer. But suppose I had found a *watch* upon the ground, and it should be inquired how the watch happened to be in that place; I should hardly think of the answer which I had before given, that, for any thing I knew, the watch might have always been there. Yet why should not this answer serve for the watch as well as for the stone? why is it not as admissible in the second case, as in the first? For this reason, and no other, viz. that, when we come to inspect the watch, we perceive (what we could not discover with the stone) that its several parts are framed and put together for a purpose, e.g. that they are so formed and adjusted as to produce motion, and that motion so regulated as to point out the hour of the day; that, if the different parts had been differently shaped from what they

SOURCE: *The Works of William Paley, D.D.* (London, 1828), vol. II.

are, of a different size from what they are, or placed after any other manner, or in any other order, than that in which they are placed, either no motion at all would have been carried on in the machine, or none which would have answered the use that is now served by it. To reckon up a few of the plainest of these parts, and of their offices, all tending to one result:—We see a cylindrical box containing a coiled elastic spring, which, by its endeavour to relax itself, turns round the box. We next observe a flexible chain (artificially wrought for the sake of flexure), communicating the action of the spring from the box to the fusee. We then find a series of wheels, the teeth of which catch in, and apply to, each other, conducting the motion from the fusee to the balance, and from the balance to the pointer; and at the same time, by the size and shape of those wheels, so regulating that motion, as to terminate in causing an index, by an equable and measured progression, to pass over a given space in a given time. We take notice that the wheels are made of brass in order to keep them from rust; the springs of steel, no other metal being so elastic; that over the face of the watch there is placed a glass, a material employed in no other part of the work, but in the room of which, if there had been any other than a transparent substance, the hour could not be seen without opening the case. This mechanism being observed (it requires indeed an examination of the instrument, and perhaps some previous knowledge of the subject, to perceive and understand it; but being once, as we have said, observed and understood), the inference, we think, is inevitable, that the watch must have had a maker: that there must have existed, at some time, and at some place or other, an artificer or artificers who formed it for the purpose which we find it actually to answer; who comprehended its construction, and designed its use.

*     *     *

Every indication of contrivance, every manifestation of design, which existed in the watch, exists in the works of nature; with the difference, on the side of nature, of being greater and more, and that in a degree which exceeds all computation. I mean that the contrivances of nature surpass the contrivances of art, in the complexity, subtility, and curiosity of the mechanism; and still more, if possible, do they go beyond them in number and variety; yet, in a multitude of cases, are not less evidently mechanical, not less evidently accommodated to their end, or suited to their office, than are the most perfect productions of human ingenuity.

# NATURE

## John Stuart Mill

*This selection is from a longer work by John Stuart Mill (1806–1873) entitled* Three Essays on Religion. *In it Mill talks of nature in both senses, as phenomena and as human behavior, and tries to correct certain beliefs that are founded on the relation of the two. Here we are only concerned with his objective evaluation of the phenomenal universe and have avoided those parts of the discussion that are concerned with morality. The general points, however, can be briefly summarized.*

*1. The doctrine that man should follow nature is a truism since man has no other course—all actions concur with one or a number of the so-called natural laws, which are both physical and mental.*

*2. In the sense that to follow nature is to imitate nature, the doctrine is both irrational and immoral. It is irrational because "all human action whatever, consists in altering, and all useful action in improving, the spontaneous course of nature." It is immoral because very often imitating nature involves man in abhorrent behavior. In fact, says Mill, "anyone who endeavored in his actions to imitate the natural course of things would be universally seen and acknowledged to be the wickedest of men."*

*The essay was written in 1854. Even today it provides a strong antidote for those who cloud the issue by being excessively pious about their love for nature.*

The consciousness that whatever a man does to improve his condition is in so much a censure and a thwarting of the spontaneous order of Nature, has in all ages caused new and unprecedented attempts at improvement to

SOURCE: John Stuart Mill, *Three Essays on Religion*, ed. Helen Taylor (London, 1874).

be generally at first under a shade of religious suspicion; as being in any case uncomplimentary, and very probably offensive to the powerful beings (or, when polytheism gave place to monotheism, to the all-powerful Being) supposed to govern the various phenomena of the universe, and of whose will the course of nature was conceived to be the expression. Any attempt to mould natural phenomena to the convenience of mankind might easily appear an interference with the government of those superior beings: and though life could not have been maintained, much less made pleasant, without perpetual interferences of the kind, each new one was doubtless made with fear and trembling, until experience had shown that it could be ventured on without drawing down the vengeance of the Gods. The sagacity of priests showed them a way to reconcile the impunity of particular infringements with the maintenance of the general dread of encroaching on the divine administration. This was effected by representing each of the principal human inventions as the gift and favour of some God. The old religions also afforded many resources for consulting the Gods, and obtaining their express permission for what would otherwise have appeared a breach of their prerogative. When oracles had ceased, any religion which recognized a revelation afforded expedients for the same purpose. The Catholic religion had the resource of an infallible Church, authorized to declare what exertions of human spontaneity were permitted or forbidden; and in default of this, the case was always open to argument from the Bible whether any particular practice had expressly or by implication been sanctioned. The notion remained that this liberty to control Nature was conceded to man only by special indulgence, and as far as required by his necessities; and there was always a tendency, though a diminishing one, to regard any attempt to exercise power over nature, beyond a certain degree, and a certain admitted range, as an impious effort to usurp divine power, and dare more than was permitted to man. The lines of Horace in which the familiar arts of shipbuilding and navigation are reprobated as *vetitum nefas,* indicate even in that sceptical age a still unexhausted vein of the old sentiment. The intensity of the corresponding feeling in the middle ages is not a precise parallel, on account of the superstition about dealing with evil spirits with which it was complicated: but the imputation of prying into the secrets of the Almighty long remained a powerful weapon of attack against unpopular inquirers into nature; and the charge of presumptuously attempting to defeat the designs of Providence, still retains enough of its original force to be thrown in as a makeweight along with other objections when there is a desire to find fault with any new exertion of human forethought and contrivance. No one, indeed, asserts it to be the intention of the Creator that the spontaneous order of the creation should not be altered, or even that it should not be altered in any new way. But there still exists a vague notion that though it is very proper to control this or the other natural phenomenon, the general scheme of nature is a model for us to imitate: that with more or less liberty in details, we should on the whole be guided by the spirit and general conception of nature's own ways: that they are God's work, and as such perfect; that man cannot rival their unapproachable excellence, and can best show his skill and piety by attempting, in however imperfect a way, to reproduce their likeness; and that if not the whole, yet some particular parts of the spontaneous order of nature, selected according to the speaker's predilections, are in a peculiar sense, manifestations of the Creator's will; a sort of finger posts pointing out the direction which things in general, and therefore our voluntary actions, are intended to take. Feelings of this sort, though repressed on ordinary occasions by the contrary current of life, are ready to break out whenever custom is silent, and the native promptings of the mind have nothing opposed to them but reason: and appeals are continually made to them by rhetoricians, with the effect, if not of convincing opponents, at least of making those who already hold the opinion which the rhetorician desires to recommend, better satisfied with it. For in the present day it probably seldom happens that any one is persuaded to approve any course of action because it appears to him to bear an analogy to the divine government of the world, though the argument tells on him with great force, and is felt by him to be a great support, in behalf of anything which he is already inclined to approve.

If this notion of imitating the ways of Providence as manifested in Nature, is seldom expressed plainly and downrightly as a maxim of general application, it also is seldom directly contradicted. Those who find it on their path, prefer to turn the obstacle rather than to attack it, being often themselves not free from the feeling, and in any case afraid of incurring the charge of impiety by saying anything which might be held to disparage the works of the Creator's power. They therefore, for the most part, rather endeavour to show, that they have as much right to the religious argument as their opponents, and that if the course they recommend seems to conflict with some part of the ways of Providence, there is some other part with which it agrees better than what is contended for on the other side. In this mode of dealing with the great a priori fallacies, the progress of improvement clears away particular errors while the causes of errors are still left standing, and very little weakened by each conflict: yet by a long series of such partial victories precedents are accumulated, to which an appeal may be made against these powerful prepossessions, and which afford a growing hope that the misplaced feeling, after having so often learnt to recede, may some day be compelled to an unconditional surrender. For however offensive the proposition may appear to many religious persons, they should be willing to look in the face the undeniable fact, that the order of nature, insofar as unmodified by man, is such as no being, whose attributes are justice and benevolence, would have made, with the intention that his rational creatures should follow it as an example. If made wholly by such a Being, and not partly by beings of very different qualities, it could only be as a designedly imperfect work, which man, in his limited sphere, is to exercise justice and benevolence in amending. The best persons have always held it to be the essence of religion, that the paramount duty of man upon earth is to amend himself: but all except monkish quietists have annexed to this in their inmost minds (though seldom willing to enunciate the obligation with the same clearness) the additional religious duty of amending the world, and not solely the human part of it but the material; the order of physical nature.

In considering this subject it is necessary to divest ourselves of certain preconceptions which may justly be called natural prejudices, being grounded on feelings which, in themselves natural and inevitable, intrude into matters with which they ought to have no concern. One of these feelings is the astonishment, rising into awe, which is inspired (even independently of all religious sentiment) by any of the greater natural phenomena. A hurricane; a mountain precipice; the desert; the ocean, either agitated or at rest; the solar system, and the great cosmic forces which hold it together; the boundless firmament, and to an educated mind any single star; excite feelings which make all human enterprises and powers appear so insignificant, that to a mind thus occupied it seems insufferable presumption in so puny a creature as man to look critically on things so far above him, or dare to measure himself against the grandeur of the universe. But a litttle interrogation of our own consciousness will suffice to convince us, that what makes these phenomena so impressive is simply their vastness. The enormous extension in space and time, or the enormous power they exemplify, constitutes their sublimity; a feeling in all cases, more allied to terror than to any moral emotion. And though the vast scale of these phenomena may well excite wonder, and sets at defiance all idea of rivalry, the feeling it inspires is of a totally different character from admiration of excellence. Those in whom awe produces admiration may be aesthetically developed, but they are morally uncultivated. It is one of the endowments of the imaginative part of our mental nature that conceptions of greatness and power, vividly realized, produce a feeling which though in its higher degrees closely bordering on pain, we prefer to most of what are accounted pleasures. But we are quite equally capable of experiencing this feeling toward maleficent power; and we never experience it so strongly towards most of the powers of the universe, as when we have most present to our consciousness a vivid sense of their capacity of inflicting evil. Because these natural powers have what we cannot imitate, enormous might, and overawe us by that one attribute, it would be a great error to infer that their other attributes are such as we ought to emulate, or that we should be justified in using our small powers after the example which Nature sets us with her vast forces.

For, how stands the fact? That next to the greatness of these cosmic forces, the quality which most forcibly strikes every one who does not avert his eyes from it, is their perfect and absolute recklessness. They go straight to their end, without regarding what or whom they crush on the road. Optimists, in their attempts to prove that "whatever is, is right," are obliged to maintain, not that Nature ever turns one step from her path to avoid trampling us into destruction, but that it would be very unreasonable in us to expect that she should. Pope's "Shall gravitation cease when you go by?" may be a just rebuke to any one who should be so silly as to expect common human morality from nature. But if the question were between two men, instead of between a man and a natural phenomenon, that triumphant apostrophe would be thought a rare piece of impudence. A man who should persist in hurling stones or firing cannon when another man "goes by," and having killed him should urge a similar plea in exculpation, would very deservedly be found guilty of murder.

In sober truth, nearly all the things which men are hanged or imprisoned for doing to one another, are nature's every day performances. Killing, the most criminal act recognized by human laws, Nature does once to every being that lives; and in a large proportion of cases, after protracted tortures such as only the greatest monsters whom we read of ever purposely inflicted on their living fellow creatures. If, by an arbitrary reservation, we refuse to account anything murder but what abridges a certain term supposed to be allotted to human life, nature also does this to all but a small percentage of lives, and does it in all the modes, violent or insidious, in which the worst human beings take the lives of one another. Nature impales men, breaks them as if on the wheel, casts them to be devoured by wild beasts, burns them to death, crushes them with stones like the first Christian martyr, starves them with hunger, freezes them with cold, poisons them by the quick or slow venom of her exhalations, and has hundreds of other hideous deaths in reserve, such as the ingenious cruelty of a Nabis or a Domitian never surpassed. All this, Nature does with the most supercilious disregard both of mercy and of justice, emptying her shafts upon the best and noblest indifferently with the meanest and worst; upon those who are engaged in the highest and worthiest enterprises, and often as the direct consequence of the noblest acts; and it might almost be imagined as a punishment for them. She mows down those on whose existence hangs the well-being of a whole people, perhaps the prospects of the human race for generations to come, with as little compunction as those whose death is a relief to themselves, or a blessing to those under their noxious influence. Such are Nature's dealings with life. Even when she does not intend to kill, she inflicts the same tortures in apparent wantonness. In the clumsy provision which she has made for that perpetual renewal of animal life, rendered necessary by the prompt termination she puts to it in every individual instance, no human being ever comes into the world but another human being is literally stretched on the rack for hours or days, not unfrequently issuing in death. Next to taking life (equal to it according to a high authority) is taking the means by which we live; and Nature does this too on the largest scale and with the most callous indifference. A single hurricane destroys the hopes of a season; a flight of locusts, or an inundation, desolates a district; a trifling chemical change in an edible root starves a million of people. The waves of the sea, like banditti, seize and appropriate the wealth of the rich and the little all of the poor with the same accompaniments of stripping, wounding, and killing as their human antitypes. Everything in short, which the worst men commit either against life or property is perpetrated on a larger scale by natural agents. Nature has Noyades more fatal than those of Carrier; her explosions of fire damp are as destructive as human artillery; her plague and cholera far surpass the poison cups of the Borgias. Even the love of "order" which is thought to be a following of the ways of Nature, is in fact a contradiction of them. All which people are accustomed to deprecate as "disorder" and its consequences, is precisely a counterpart of Nature's ways. Anarchy and the Reign of Terror are overmatched in injustice, ruin, and death, by a hurricane and a pestilence.

But, it is said, all these things are for wise and good ends. On this I must first remark that whether they are so or not, is altogether beside the point. Supposing it true that contrary to appearances these horrors when perpetrated by Nature, promote good ends, still as no one believes that good ends would be promoted by our following the example, the course of Nature cannot be a proper model for us to imitate. Either it is right that we should kill because nature kills; torture because nature tortures; ruin and devastate because nature does the like; or we ought not to consider at all what nature does, but what it is good to do. If there is such a thing as a *reductio ad absurdum,* this surely amounts to one. If it is a sufficient reason for doing one thing, that nature does it, why not another thing? If not all things, why anything? The physical government of the world being full of the things which when done by men are deemed the greatest enormities, it cannot be religious or moral in us to guide our actions by the analogy of the course of nature. This proposition remains true, whatever occult quality of producing good may reside in those facts of nature which to our perceptions are most noxious, and which no one considers it other than a crime to produce artificially.

But, in reality, no one consistently believes in any such occult quality. The phrases which ascribe perfection to the course of nature can only be considered as the exaggerations of poetic or devotional feeling, not intended to stand the test of a sober examination. No one, either religious or irreligious, believes that the hurtful agencies of nature, considered as a whole, promote good purposes, in any other way than by inciting human rational creatures to rise up and struggle against them. If we believe that those agencies were appointed by a benevolent Providence as the means of accomplishing wise purposes which could not be compassed if they did not exist, then everything done by mankind which tends to chain up these natural agencies or to restrict their mischievous operation, from draining a pestilential marsh down to curing the toothache, or putting up an umbrella, ought to be accounted impious; which assuredly nobody does account them, notwithstanding an undercurrent of sentiment setting in that direction which is occasionally perceptible. On the contrary, the improvements on which the civilized part of mankind most pride themselves, consist in more successfully warding off those natural calamities which if we really believe what most people profess to believe, we should cherish as medicines provided for our earthly state by infinite wisdom. Inasmuch too as each generation greatly surpasses its predecessors in the amount of natural evil which it succeeds in averting, our condition, if the theory were true, ought by this time to have become a terrible manifestation of some tremendous calamity, against which the physical evils we have learnt to overmaster, had previously operated as a preservative. Any one, however, who acted as if he supposed this to be the case, would be more likely, I think, to be confined as a lunatic, than reverenced as a saint.

# from THE EDUCATION OF HENRY ADAMS

## Henry Adams

*Because of the sudden and sadly ironic death of his sister, Henry Adams (1838–1918) was convinced of the whimsically cruel behavior of nature. A lively, bright, and innocently reckless girl, she had received a slight wound in a cab accident, contracted tetanus, and died in the grip of a convulsion. Adams was especially impressed by the way nature enjoyed death, the way she "played with it, the horror added to her charm, she liked the torture, and smothered her victim with caresses." Adams was a complex man and possessed an attractively sane, skeptical attitude (his attempt to see himself objectively led to writing his autobiography in the third person); yet this incident in his life may have influenced his reaction to the dynamo at the Chicago Exposition of 1900. The following excerpt from that autobiography describes his reaction to the dynamo.*

Until the Great Exposition of 1900 closed its doors in November, Adams haunted it, aching to absorb knowledge, and helpless to find it. He would have liked to know how much of it could have been grasped by the best-informed man in the world. While he was thus meditating chaos, Langley came by, and showed it to him. At Langley's behest, the Exhibition dropped its superfluous rags and stripped itself to the skin, for Langley knew what to study, and why, and how; while Adams might as well have stood outside in the night, staring at the Milky Way. Yet Langley said nothing new, and taught nothing that one might not have learned from Lord Bacon, three hundred years

SOURCE: Henry Adams, *The Education of Henry Adams* (Boston: Houghton Mifflin Co., 1918). Copyright 1946 by Charles F. Adams. Reprinted by permission of the publisher, Houghton Mifflin Co.

before; but though one should have known the "Advancement of Science" as well as one knew the "Comedy of Errors," the literary knowledge counted for nothing until some teacher should show how to apply it. Bacon took a vast deal of trouble in teaching King James I and his subjects. American or other, towards the year 1620, that true science was the development or economy of forces; yet an elderly American in 1900 knew neither the formula nor the forces; or even so much as to say to himself that his historical business in the Exposition concerned only the economies or developments of force since 1893, when he began the study at Chicago.

Nothing in education is so astonishing as the amount of ignorance it accumulates in the form of inert facts. Adams had looked at most of the accumulations of art in the storehouses called Art Museums; yet he did not know how to look at the art exhibits of 1900. He had studied Karl Marx and his doctrines of history with profound attention, yet he could not apply them at Paris. Langley, with the ease of a great master of experiment, threw out of the field every exhibit that did not reveal a new application of force, and naturally threw out, to begin with, almost the whole art exhibit. Equally, he ignored almost the whole industrial exhibit. He led his pupil directly to the forces. His chief interest was in new motors to make his airship feasible, and he taught Adams the astonishing complexities of the new Daimler motor, and of the automobile, which, since 1893, had become a nightmare at a hundred kilometres an hour, almost as destructive as the electric tram which was only ten years older; and threatening to become as terrible as the locomotive steam-engine itself, which was almost exactly Adams's own age.

Then he showed his scholar the great hall of dynamos, and explained how little he knew about electricity or force of any kind, even of his own special sun, which spouted heat in inconceivable volume, but which, as far as he knew, might spout less or more, at any time, for all the certainty he felt in it. To him, the dynamo itself was but an ingenious channel for conveying somewhere the heat latent in a few tons of poor coal hidden in a dirty engine-house carefully kept out of sight; but to Adams the dynamo became a symbol of infinity. As he grew accustomed to the great gallery of machines, he began to feel the forty-foot dynamos as a moral force, much as the early Christians felt the Cross. The planet itself seemed less impressive, in its old-fashioned, deliberate, annual or daily revolution, than this huge wheel, revolving within arm's-length at some vertiginous speed, and barely murmuring—scarcely humming an audible warning to stand a hair's-breadth further for respect of power —while it would not wake the baby lying close against its frame. Before the end, one began to pray to it; inherited instinct taught the natural expression of man before silent and infinite force. Among the thousand symbols of ultimate energy, the dynamo was not so human as some, but it was the most expressive.

Yet the dynamo, next to the steam-engine, was the most familiar of exhibits. For Adams's objects its value lay chiefly in its occult mechanism. Between the dynamo in the gallery of machines and the engine-house outside, the break of continuity amounted to abysmal fracture for a historian's objects. No more relation could he discover between the steam and the electric current than between the Cross and the cathedral. The forces were interchangeable if not reversible, but he could see only an absolute *fiat* in electricity as in faith. Langley could not help him. Indeed, Langley seemed to be worried by the same trouble, for he constantly repeated that the new forces were anarchical, and specially that he was not responsible for the new rays, that were little short of parricidal in their wicked spirit towards science. His own rays, with which he had doubled the solar spectrum, were altogether harmless and beneficent; but Radium denied its God—or, what was to Langley the same thing, denied the truths of his Science. The force was wholly new.

A historian who asked only to learn enough to be as futile as Langley or Kelvin, made rapid progress under this teaching, and mixed himself up in the tangle of ideas until he achieved a sort of Paradise of ignorance vastly consoling to his fatigued senses. He wrapped himself in vibrations and rays which were new, and he would have hugged Marconi and Branly had he met them, as he hugged the dynamo; while he lost his arithmetic in trying to figure out the equation between the discoveries and the economies of force. The economies, like the discoveries, were absolute, supersensual, occult; incapable of expression in horsepower. What mathematical equivalent could he suggest as the value of a Branly coherer? Frozen air, or the electric furnace, had some scale of measurement, no doubt, if somebody could invent a thermometer adequate to the purpose; but X-rays had played no part whatever in man's consciousness, and the atom itself had figured only as a fiction of thought. In these seven years man had translated himself into a new universe which had no common scale of measurement with the old. He had entered a supersensual world, in which he could measure nothing except by chance collisions of movements imperceptible to his senses, perhaps even imperceptible to his instruments, but perceptible to each other, and so to some known ray at the end of the scale. Langley seemed prepared for anything, even for an indeterminable number of universes interfused—physics stark mad in metaphysics.

# from EVOLUTION AND ETHICS

*Thomas H. Huxley*

*Thomas H. Huxley (1825–1895) was both a biologist and a philos-*
*opher. He worked by induction; and though he recognized certain uni-*
*versal laws, that of evolution being one, he never allowed himself to*
*assign authority for those laws to anything like the orthodox God of*
*theology. Furthermore, his agnostic position led him to believe that man*
*himself was responsible for finding morality in nature. Thus, he came*
*to see man's task in the world as that of subduing nature in order to*
*pursue the higher ends of ethics. Others have arrived at the same con-*
*clusion, but, unlike Huxley, they have not argued with so much knowl-*
*edge of the workings of nature.*

Modern thought is making a fresh start from the base whence Indian
and Greek philosophy set out; and, the human mind being very much what it
was six-and-twenty centuries ago, there is no ground for wonder if it presents
indications of a tendency to move along the old lines to the same results.

We are more than sufficiently familiar with modern pessimism, at least
as a speculation; for I cannot call to mind that any of its present votaries have
sealed their faith by assuming the rags and the bowl of the mendicant Bhikku,
or the wallet of the Cynic. The obstacles placed in the way of sturdy vagrancy
by an unphilosophical police have, perhaps, proved too formidable for philosophical consistency. We also know modern speculative optimism, with its
perfectability of the species, reign of peace, and lion and lamb transformation
scenes; but one does not hear so much of it as one did forty years ago; indeed,
I imagine it is to be met with more commonly at the tables of the healthy and
wealthy, than in the congregations of the wise. The majority of us, I appre-

SOURCE: Thomas H. Huxley, *Evolution and Ethics* (New York: D. Appleton and Company, 1896).

hend, profess neither pessimism nor optimism. We hold that the world is neither so good, nor so bad, as it conceivably might be; and, as most of us have reason, now and again, to discover that it can be. Those who have failed to experience the joys that make life worth living are, probably, in as small a minority as those who have never known the griefs that rob existence of its savor and turn its richest fruits into mere dust and ashes.

Further, I think I do not err in assuming that, however diverse their views on philosophical and religious matters, most men are agreed that the proportion of good and evil in life may be very sensibly affected by human action. I never heard anybody doubt that the evil may be thus increased, or diminished; and it would seem to follow that good must be similarly susceptible of addition or subtraction. Finally, to my knowledge, nobody professes to doubt that, so far forth as we possess a power of bettering things, it is our paramount duty to use it and to train all our intellect and energy to this supreme service of our kind.

Hence the pressing interest of the question, to what extent modern progress in natural knowledge, and, more especially, the general outcome of that progress in the doctrine of evolution, is competent to help us in the great work of helping one another?

The propounders of what are called the "ethics of evolution," when the "evolution of ethics" would usually better express the object of their speculations, adduce a number of more or less interesting facts and more or less sound arguments in favor of the origin of the moral sentiments, in the same way as other natural phenomena, by a process of evolution. I have little doubt, for my own part, that they are on the right track; but as the immoral sentiments have no less been evolved, there is, so far, as much natural sanction for the one as the other. The thief and the murderer follow nature just as much as the philanthropist. Cosmic evolution may teach us how the good and the evil tendencies of man may have come about; but, in itself, it is incompetent to furnish any better reason why what we call good is preferable to what we call evil than we had before. Some day, I doubt not, we shall arrive at an understanding of the evolution of the aesthetic faculty; but all the understanding in the world will neither increase nor diminish the force of the intuition that this is beautiful and that is ugly.

There is another fallacy which appears to me to pervade the so-called "ethics of evolution." It is the notion that because, on the whole, animals and plants have advanced in perfection of organization by means of the struggle for existence and the consequent "survival of the fittest"; therefore men in society, men as ethical beings, must look to the same process to help them towards perfection. I suspect that this fallacy has arisen out of the unfortunate ambiguity of the phrase "survival of the fittest." "Fittest" has a connotation of "best"; and about "best" there hangs a moral flavor. In cosmic nature, however, what is "fittest" depends upon the conditions. Long since, I ventured to point out that if our hemisphere were to cool again, the survival of the fittest might bring about, in the vegetable kingdom, a population of more and more stunted and humbler organisms, until the "fittest" that survived might be nothing but lichens, diatoms, and such microscopic organisms as those which give red snow its color; while, if it became hotter, the pleasant valleys of the Thames and Isis might be uninhabitable by any animated beings save those that flourish in a tropical jungle. They, as the fittest, the best adapted to the changed conditions, would survive.

Men in society are undoubtedly subject to the cosmic process. As among other animals, multiplication goes on without cessation, and involves severe competition for the means of support. The struggle for existence tends to eliminate those less fitted to adapt themselves to the circumstances of their existence. The strongest, the most self-assertive, tend to tread down the weaker. But the influence of the cosmic process on the evolution of society is the greater the more rudimentary its civilization. Social progress means a checking of the cosmic process at every step and the substitution for it of another, which may be called the ethical process; the end of which is not the survival of those who may happen to be the fittest, in respect of the whole of the conditions which obtain, but of those who are ethically the best.

As I have already urged, the practice of that which is ethically best— what we call goodness or virtue—involves a course of conduct which, in all respects, is opposed to that which leads to success in the cosmic struggle for existence. In place of ruthless self-assertion it demands self-restraint; in place of thrusting aside, or treading down, all competitors, it requires that the individual shall not merely respect, but shall help his fellows; its influence is directed, not so much to the survival of the fittest, as to the fitting of as many as possible to survive. It repudiates the gladiatorial theory of existence. It demands that each man who enters into the enjoyment of the advantages of a polity shall be mindful of his debt to those who have laboriously constructed it; and shall take heed that no act of his weakens the fabric in which he has been permitted to live. Laws and moral precepts are directed to the end of curbing the cosmic process and reminding the individual of his duty to the community, to the protection and influence of which he owes, if not existence itself, at least the life of something better than a brutal savage.

It is from neglect of these plain considerations that the fanatical individualism of our time attempts to apply the analogy of cosmic nature to society. Once more we have a misapplication of the stoical injunction to follow nature; the duties of the individual to the state are forgotten, and his tendencies to self-assertion are dignified by the name of rights. It is seriously debated whether the members of a community are justified in using their combined strength to constrain one of their number to contribute his share to the maintenance of it; or even to prevent him from doing his best to destroy it. The struggle for existence which has done such admirable work in cosmic nature, must, it appears, be equally beneficent in the ethical sphere. Yet if that which I have insisted upon is true; if the cosmic process has no sort of relation to moral ends; if the imitation of it by man is inconsistent with the first principles of ethics; what becomes of this surprising theory?

Let us understand, once for all, that the ethical progress of society depends, not on imitating the cosmic process, still less in running away from it, but in combating it. It may seem an audacious proposal thus to put the microcosm against the macrocosm and to set man to subdue nature to his higher ends; but I venture to think that the great intellectual difference between the ancient times with which we have been occupied and our day, lies in the solid foundation we have acquired for the hope that such an enterprise may meet with a certain measure of success.

The history of civilization details the steps by which men have succeeded in building up an artificial world within the cosmos. Fragile reed as he may be, man, as Pascal says, is a thinking reed: there lies within him a fund of energy operating intelligently and so far akin to that which pervades the universe, that it is competent to influence and modify the cosmic process. In virtue of his intelligence, the dwarf bends the Titan to his will. In every family, in every polity that has been established, the cosmic process in man has been restrained and otherwise modified by law and custom; in surrounding nature, it has been similarly influenced by the art of the shepherd, the agriculturist, the artisan. As civilization has advanced, so has the extent of this interference increased; until the organized and highly developed sciences and arts of the present day have endowed man with a command over the course of non-human nature greater than that once attributed to the magicians. The most impressive, I might say startling, of these changes have been brought about in the course of the last two centuries; while a right comprehension of the process of life and of the means of influencing its manifestations is only just dawning upon us. We do not yet see our way beyond generalities; and we are befogged by the obtrusion of false analogies and crude anticipations. But Astronomy, Physics, Chemistry, have all had to pass through similar phases, before they reached the stage at which their influence became an important factor in human affairs. Physiology, Psychology, Ethics, Political Science, must submit to the same ordeal. Yet it seems to me irrational to doubt that, at no distant period, they will work as great a revolution in the sphere of practice.

The theory of evolution encourages no millennial anticipations. If, for millions of years, our globe has taken the upward road, yet, some time, the summit will be reached and the downward route will be commenced. The most daring imagination will hardly venture upon the suggestion that the power and the intelligence of man can ever arrest the procession of the great year.

Moreover, the cosmic nature born with us and, to a large extent, necessary for our maintenance, is the outcome of millions of years of severe training, and it would be folly to imagine that a few centuries will suffice to subdue its masterfulness to purely ethical ends. Ethical nature may count upon having to reckon with a tenacious and powerful enemy as long as the world lasts. But, on the other hand, I see no limit to the extent to which intelligence and will, guided by sound principles of investigation, and organized in common effort, may modify the conditions of existence, for a period longer than that now covered by history. And much may be done to change the nature of man himself. The intelligence which has converted the brother of the wolf into the faithful guardian of the flock ought to be able to do something towards curbing the instincts of savagery in civilized men.

But if we may permit ourselves a larger hope of abatement of the essential evil of the world than was possible to those who, in the infancy of exact knowledge, faced the problem of existence more than a score of centuries ago, I deem it an essential condition of the realization of that hope that we should cast aside the notion that the escape from pain and sorrow is the proper object of life.

We have long since emerged from the heroic childhood of our race, when good and evil could be met with the same "frolic welcome"; the attempts to escape from evil, whether Indian or Greek, have ended in flight from the battlefield; it remains to us to throw aside the youthful overconfidence and the no less youthful discouragement of nonage. We are grown men, and must play the man

> strong in will
> To strive, to seek, to find, and not to yield,

cherishing the good that falls in our way, and bearing the evil, in and around us, with stout hearts set on diminishing it. So far, we all may strive in one faith towards one hope:

> It may be that the gulfs will wash us down,
> It may be we shall touch the Happy Isles,
>
> . . . but something ere the end,
> Some work of noble note may yet be done.

# WAS THE WORLD MADE FOR MAN?

## Samuel Langhorne Clemens

Alfred Russell Wallace's revival of the theory that this earth is at the center of the stellar universe, and is the only habitable globe, has aroused great interest in the world.

—*Literary Digest*

For ourselves we do thoroughly believe that man, as he lives just here on this tiny earth, is in essence and possibilities the most sublime existence in all the range of non-divine being—the chief love and delight of God.

—Chicago *Interior* (Presb.)

*The wit of Samuel L. Clemens or Mark Twain (1835–1910) is well known for its shrewd penetration into the inanities of the American scene. Few have been his equal in the precise art of comic burlesque; yet beneath the humor, one is aware of his intense desire for social justice and a pervasive equalitarian attitude. For the most part, his subject is vanity. In the following selection, he strips the mask of anthropocentricity from the human animal.*

I seem to be the only scientist and theologian still remaining to be heard from on this important matter of whether the world was made for man or not. I feel that it is time for me to speak.

I stand almost with the others. They believe the world was made for man, I believe it likely that it was made for man; they think there is proof, astronomical mainly, that it was made for man, I think there is evidence only,

SOURCE: From *Letters from the Earth* by Mark Twain, pp. 211–216. Edited by Bernard DeVoto. Copyright © 1962 by The Mark Twain Co. Reprinted by permission of Harper & Row, Publishers, Inc.

not proof, that it was made for him. It is too early, yet, to arrange the verdict, the returns are not all in. When they are all in, I think they will show that the world was made for man; but we must not hurry, we must patiently wait till they are all in.

Now as far as we have got, astronomy is on our side. Mr. Wallace has clearly shown this. He has clearly shown two things: that the world was made for man, and that the universe was made for the world—to stiddy it, you know. The astronomy part is settled, and cannot be challenged.

We come now to the geological part. This is the one where the evidence is not all in, yet. It is coming in, hourly, daily, coming in all the time, but naturally it comes with geological carefulness and deliberation, and we must not be impatient, we must not get excited, we must be calm, and wait. To lose our tranquillity will not hurry geology; nothing hurries geology.

It takes a long time to prepare a world for man, such a thing is not done in a day. Some of the great scientists, carefully ciphering the evidences furnished by geology, have arrived at the conviction that our world is prodigiously old, and they may be right, but Lord Kelvin is not of their opinion. He takes a cautious, conservative view, in order to be on the safe side, and feels sure it is not so old as they think. As Lord Kelvin is the highest authority in science now living, I think we must yield to him and accept his view. He does not concede that the world is more than a hundred million years old. He believes it is that old, but not older. Lyell believed that our race was introduced into the world 31,000 years ago, Herbert Spencer makes it 32,000. Lord Kelvin agrees with Spencer.

Very well. According to these figures it took 99,968,000 years to prepare the world for man, impatient as the Creator doubtless was to see him and admire him. But a large enterprise like this has to be conducted warily, painstakingly, logically. It was foreseen that man would have to have the oyster. Therefore the first preparation was made for the oyster. Very well, you cannot make an oyster out of whole cloth, you must make the oyster's ancestor first. This is not done in a day. You must make a vast variety of invertebrates, to start with—belemnites, trilobites, Jebusites, Amalekites, and that sort of fry, and put them to soak in a primary sea, and wait and see what will happen. Some will be a disappointment—the belemnites, the Ammonites and such; they will be failures, they will die out and become extinct, in the course of the nineteen million years covered by the experiment, but all is not lost, for the Amalekites will fetch the homestake; they will develop gradually into encrinites, and stalactites, and blatherskites, and one thing and another as the mighty ages creep on and the Archaean and the Cambrian Periods pile their lofty crags in the primordial seas, and at last the first grand stage in the preparation of the world for man stands completed, the oyster is done. An oyster has hardly any more reasoning power than a scientist has; and so it is reasonably certain that this one jumped to the conclusion that the nineteen million years was a preparation for *him*; but that would be just like an oyster, which is the most conceited animal there is, except man. And anyway, this one could not know, at that early date, that he was only an incident in a scheme, and that there was some more to the scheme, yet.

The oyster being achieved, the next thing to be arranged for in the preparation of the world for man was fish. Fish, and coal—to fry it with. So the Old Silurian seas were opened up to breed the fish in, and at the same time the great work of building Old Red Sandstone mountains eighty thousand feet high to cold-storage their fossils in was begun. This latter was quite indispensable, for there would be no end of failures again, no end of extinctions—millions of them—and it would be cheaper and less trouble to can them in the rocks than keep tally of them in a book. One does not build the coal beds and eighty thousand feet of perpendicular Old Red Sandstone in a brief time—no, it took twenty million years. In the first place, a coal bed is a slow and troublesome and tiresome thing to construct. You have to grow prodigious forests of tree-ferns and reeds and calamites and such things in a marshy region; then you have to sink them under out of sight and let them rot; then you have to turn the streams on them, so as to bury them under several feet of sediment, and the sediment must have time to harden and turn to rock; next you must grow another forest on top, then sink it and put on another layer of sediment and harden it; then more forest and more rock, layer upon layer, three miles deep—ah, indeed it is a sickening slow job to build a coal-measure and do it right!

So the millions of years drag on; and meantime the fish culture is lazying along and frazzling out in a way to make a person tired. You have developed ten thousand kinds of fishes from the oyster; and come to look, you have raised nothing but fossils, nothing but extinctions. There is nothing left alive and progressive but a ganoid or two and perhaps a half a dozen asteroids. Even the cat wouldn't eat such.

Still, it is no great matter; there is plenty of time, yet, and they will develop into something tasty before man is ready for them. Even a ganoid can be depended on for that, when he is not going to be called on for sixty million years.

The Paleozoic time limit having now been reached, it was necessary to begin the next stage in the preparation of the world for man, by opening up the Mesozoic Age and instituting some reptiles. For man would need reptiles. Not to eat, but to develop himself from. This being the most important detail of the scheme, a spacious liberality of time was set apart for it—thirty million years. What wonders followed! From the remaining ganoids and asteroids and alkaloids were developed by slow and steady and painstaking culture those stupendous saurians that used to prowl about the steamy world in those remote ages, with their snaky heads reared forty feet in the air and sixty feet of body and tail racing and thrashing after. All gone, now, alas—all extinct, except the little handful of Arkansawrians left stranded and lonely with us here upon this far-flung verge and fringe of time.

Yes, it took thirty million years and twenty million reptiles to get one that would stick long enough to develop into something else and let the scheme proceed to the next step.

Then the pterodactyl burst upon the world in all his impressive solemnity and grandeur, and all Nature recognized that the Cenozoic threshold was crossed and a new Period open for business, a new stage begun in the preparation of the globe for man. It may be that the pterodactyl thought the thirty million years had been intended as a preparation for himself, for there was nothing too foolish for a pterodactyl to imagine, but he was in error, the preparation was for man. Without doubt the pterodactyl attracted great attention, for even the least observant could see that there was the making of a bird in him. And so it turned out. Also the makings of a mammal, in time. One thing we have to say to his credit, that in the matter of picturesqueness he was the triumph of his Period; he wore wings and had teeth, and was a starchy and wonderful mixture altogether, a kind of long-distance premonitory symptom of Kipling's marine:

> 'E isn't one o' the reg'lar Line, nor 'e isn't one of the crew,
> 'E's a kind of a giddy harumfrodite—soldier an' sailor too!

From this time onward for nearly another thirty million years the preparation moved briskly. From the pterodactyl was developed the bird; from the bird the kangaroo, from the kangaroo the other marsupials; from these the mastodon, the megatherium, the giant sloth, the Irish elk, and all that crowd that you make useful and instructive fossils out of—then came the first great Ice Sheet, and they all retreated before it and crossed over the bridge at Bering Strait and wandered around over Europe and Asia and died. All except a few, to carry on the preparation with. Six Glacial Periods with two million years between Periods chased these poor orphans up and down and about the earth, from weather to weather—from tropic swelter at the poles to Arctic frost at the equator and back again and to and fro, they never knowing what kind of weather was going to turn up next; and if they ever settled down anywhere the whole continent suddenly sank under them without the least notice and they had to trade places with the fishes and scramble off to where the seas had been, and scarcely a dry rag on them; and when there was nothing else doing a volcano would let go and fire them out from wherever they had located. They led this unsettled and irritating life for twenty-five million years, half the time afloat, half the time aground, and always wondering what it was all for, they never suspecting, of course, that it was a preparation for man and had to be done just so or it wouldn't be any proper and harmonious place for him when he arrived.

And at last came the monkey, and anybody could see that man wasn't far off, now. And in truth that was so. The monkey went on developing for close upon five million years, and then turned into a man—to all appearances.

Such is the history of it. Man has been here 32,000 years. That it took a hundred million years to prepare the world for him is proof that that is what it was done for. I suppose it is. I dunno. If the Eiffel Tower were now representing the world's age, the skin of paint on the pinnacle-knob at its summit would represent man's share of that age; and anybody would perceive that that skin was what the tower was built for. I reckon they would, I dunno.

# III THE CRISIS

# INTRODUCTION

By now only the most obdurate intelligence could remain unaware of the critical condition of the environment. Once again it has been demonstrated that eloquence is the familiar by-product of fear. Despite the apparent lack of positive behavior, there is little doubt that America has experienced a formidable assault waged by the seers of doom and the pessimistic prophets of environmental disaster. There exists, in fact, the distinct possibility of the wolf being overmarketed or, at least, the possibility of the wolf returning to his habitat in the world of fairy tale. America has been so capable in its construction of defense mechanisms against the super-sell that it now is in danger of creating yet another example of the well-known phenomenon referred to as the credibility gap. The impending ecological disaster can, it seems, be hard to believe.

Reluctance to overreact, of course, can be an indication of health and not weakness. Panic and fear do have the awkward habit of fouling the lines of action. However, in this situation, the problems for the writer can become extremely complex. On the one hand, he must communicate urgency as well as information; and on the other, he must somehow avoid the difficulties already mentioned. It goes without saying that he cannot merely stand by and watch.

It was with these difficulties in mind that the editors selected the articles that appear in this section. Ever since that space drama of the spring of 1970 when three astronauts raced the accumulation of deadly carbon oxides in their capsule back to the earth, there has been no doubt that the minds and hearts of America can be moved. The failure in the life-support systems on that occasion provided the world with a most graphic example of what is happening even more subtly in the biosphere with every passing moment. What the writer must do to ensure the same kind of happy ending is to keep the struggle alive in the mind of his fellow man. It is ironical that the whole issue might be resolved by the pen and not by some new marvelous invention, but irony or no, the possibility of such a fate is entirely real.

# from SILENT SPRING

*Rachel Carson*

*Earlier we were impressed with the evocative beauty of Rachel
Carson's prose. In this selection, her love for life is somewhat muted
by the dismay and disappointment she feels as she witnesses man's
apparently insane engagement in biocide. The subject is death; the
weapon, insecticide; and the malignant deity behind it all, chemical
ingenuity.*

## I. A Fable for Tomorrow

There was once a town in the heart of America where all life seemed to
live in harmony with its surroundings. The town lay in the midst of a checkerboard of prosperous farms, with fields of grain and hillsides of orchards where,
in spring, white clouds of bloom drifted above the green fields. In autumn,
oak and maple and birch set up a blaze of color that flamed and flickered
across a backdrop of pines. Then foxes barked in the hills and deer silently
crossed the fields, half hidden in the mists of the fall mornings.

Along the roads, laurel, viburnum and alder, great ferns and wildflowers
delighted the traveler's eye through much of the year. Even in winter the roadsides were places of beauty, where countless birds came to feed on the berries
and on the seed heads of the dried weeds rising above the snow. The
countryside was, in fact, famous for the abundance and variety of its bird life,
and when the flood of migrants was pouring through in spring and fall
people traveled from great distances to observe them. Others came to fish
the streams, which flowed clear and cold out of the hills and contained shady

pools where trout lay. So it had been from the days many years ago when the first settlers raised their houses, sank their wells, and built their barns.

Then a strange blight crept over the area and everything began to change. Some evil spell had settled on the community: mysterious maladies swept the flocks of chickens; the cattle and sheep sickened and died. Everywhere was a shadow of death. The farmers spoke of much illness among their families. In the town the doctors had become more and more puzzled by new kinds of sickness appearing among their patients. There had been several sudden and unexplained deaths, not only among adults but even among children, who would be stricken suddenly while at play and die within a few hours.

There was a strange stillness. The birds, for example—where had they gone? Many people spoke of them, puzzled and disturbed. The feeding stations in the backyards were deserted. The few birds seen anywhere were moribund; they trembled violently and could not fly. It was a spring without voices. On the mornings that had once throbbed with the dawn chorus of robins, catbirds, doves, jays, wrens, and scores of other bird voices there was now no sound; only silence lay over the fields and woods and marsh.

On the farms the hens brooded, but no chicks hatched. The farmers complained that they were unable to raise any pigs—the litters were small and the young survived only a few days. The apple trees were coming into bloom but no bees droned among the blossoms, so there was no pollination and there would be no fruit.

The roadsides, once so attractive, were now lined with browned and withered vegetation as though swept by fire. These, too, were silent, deserted by all living things. Even the streams were now lifeless. Anglers no longer visited them, for all the fish had died.

In the gutters under the eaves and between the shingles of the roofs, a white granular powder still showed a few patches; some weeks before it had fallen like snow upon the roofs and the lawns, the fields and streams.

No witchcraft, no enemy action had silenced the rebirth of new life in this stricken world. The people had done it themselves.

This town does not actually exist, but it might easily have a thousand counterparts in America or elsewhere in the world. I know of no community that has experienced all the misfortunes I describe. Yet every one of these disasters has actually happened somewhere, and many real communities have already suffered a substantial number of them. A grim specter has crept upon us almost unnoticed, and this imagined tragedy may easily become a stark reality we all shall know. . . .

## II. The Obligation to Endure

The history of life on earth has been a history of interaction between living things and their surroundings. To a large extent, the physical form and the habits of the earth's vegetation and its animal life have been molded by the environment. Considering the whole span of earthly time, the opposite effect, in which life actually modifies its surroundings, has been relatively slight. Only within the moment of time represented by the present century has one species—man—acquired significant power to alter the nature of his world.

During the past quarter century this power has not only increased to one of disturbing magnitude but it has changed in character. The most alarming of all man's assaults upon the environment is the contamination of air, earth, rivers, and sea with dangerous and even lethal materials. This pollution is for the most part irrecoverable; the chain of evil it initiates not only in the world that must support life but in living tissues is for the most part irreversible. In this now universal contamination of the environment, chemicals are the sinister and little-recognized partners of radiation in changing the very nature of the world—the very nature of its life. Strontium 90, released through nuclear explosions into the air, comes to earth in rain or drifts down as fallout, lodges in soil, enters into the grass or corn or wheat grown there, and in time takes up its abode in the bones of a human being, there to remain until his death. Similarly, chemicals sprayed on croplands or forests or gardens lie long in soil, entering into living organisms, passing from one to another in a chain of poisoning and death. Or they pass mysteriously by underground streams until they emerge and, through the alchemy of air and sunlight, combine into new forms that kill vegetation, sicken cattle, and work unknown harm on those who drink from once pure wells. As Albert Schweitzer has said, "Man can hardly even recognize the devils of his own creation."

It took hundreds of millions of years to produce the life that now inhabits the earth—eons of time in which that developing and evolving and diversifying life reached a state of adjustment and balance with its surroundings. The environment, rigorously shaping and directing the life it supported, contained elements that were hostile as well as supporting. Certain rocks gave out dangerous radiation; even within the light of the sun, from which all life draws its energy, there were shortwave radiations with power to injure. Given time—time not in years but in millennia—life adjusts, and a balance has been reached. For time is the essential ingredient; but in the modern world there is no time.

The rapidity of change and the speed with which new situations are created follow the impetuous and heedless pace of man rather than the deliberate pace of nature. Radiation is no longer merely the background radiation of rocks, the bombardment of cosmic rays, the ultraviolet of the sun that have existed before there was any life on earth; radiation is now the unnatural creation of man's tampering with the atom. The chemicals to which life is asked to make its adjustment are no longer merely the calcium and silica and copper and all the rest of the minerals washed out of the rocks and carried in rivers to the sea; they are the synthetic creations of man's inventive mind, brewed in his laboratories, and having no counterparts in nature.

To adjust to these chemicals would require time on the scale that is nature's; it would require not merely the years of a man's life but the life of generations. And even this, were it by some miracle possible, would be futile, for the new chemicals come from our laboratories in an endless stream; almost five hundred annually find their way into actual use in the United States alone. The figure is staggering and its implications are not easily grasped—five hundred new chemicals to which the bodies of men and animals are required somehow to adapt each year, chemicals totally outside the limits of biologic experience.

Among them are many that are used in man's war against nature. Since the mid-1940's over two hundred basic chemicals have been created for use in killing insects, weeds, rodents, and other organisms described in the modern vernacular as "pests"; and they are sold under several thousand different brand names.

These sprays, dusts, and aerosols are now applied almost universally to farms, gardens, forests, and homes—nonselective chemicals that have the power to kill every insect, the "good" and the "bad," to still the song of birds and the leaping of fish in the streams, to coat the leaves with a deadly film, and to linger on in soil—all this though the intended target may be only a few weeds or insects. Can anyone believe it is possible to lay down such a barrage of poisons on the surface of the earth without making it unfit for all life? They should not be called "insecticides," but "biocides."

The whole process of spraying seems caught up in an endless spiral. Since DDT was released for civilian use, a process of escalation has been going on in which ever more toxic materials must be found. This has happened because insects, in a triumphant vindication of Darwin's principle of the survival of the fittest, have evolved super races immune to the particular insecticide used, hence a deadlier one has always to be developed—and then a deadlier one than that. It has happened also because, for reasons to be described later, destructive insects often undergo a "flareback," or resurgence, after spraying, in numbers greater than before. Thus the chemical war is never won, and all life is caught in its violent crossfire.

Along with the possibility of the extinction of mankind by nuclear war, the central problem of our age has therefore become the contamination of man's total environment with such substances of incredible potential for harm —substances that accumulate in the tissues of plants and animals and even penetrate the germ cells to shatter or alter the very material of heredity upon which the shape of the future depends.

Some would-be architects of our future look toward a time when it will be possible to alter the human germ plasm by design. But we may easily be doing so now by inadvertence, for many chemicals, like radiation, bring about gene mutations. It is ironic to think that man might determine his own future by something so seemingly trivial as the choice of an insect spray.

All this has been risked—for what? Future historians may well be amazed by our distorted sense of proportion. How could intelligent beings seek to control a few unwanted species by a method that contaminated the entire environment and brought the threat of disease and death even to their own kind? Yet this is precisely what we have done. We have done it, moreover, for reasons that collapse the moment we examine them. We are told that the enormous and expanding use of pesticides is necessary to maintain farm production. Yet is our real problem not one of *overproduction*? Our farms, despite measures to remove acreages from production and to pay farmers *not* to produce, have yielded such a staggering excess of crops that the American taxpayer in 1962 is paying out more than one billion dollars a year as the total carrying cost of the surplus-food storage program. And is the situation helped when one branch of the Agriculture Department tries to reduce production while another states, as it did in 1958, "It is believed generally that reduction of crop acreages under provisions of the Soil Bank will stimulate interest in use of chemicals to obtain maximum production on the land retained in crops."

All this is not to say there is no insect problem and no need of control. I am saying, rather, that control must be geared to realities, not to mythical situations, and that the methods employed must be such that they do not destroy us along with the insects.

The problem whose attempted solution has brought such a train of disaster in its wake is an accompaniment of our modern way of life. Long before the age of man, insects inhabited the earth—a group of extraordinarily varied and adaptable beings. Over the course of time since man's advent, a small percentage of the more than half a million species of insects have come into conflict with human welfare in two principal ways: as competitors for the food supply and as carriers of human disease.

Disease-carrying insects become important where human beings are crowded together, especially under conditions where sanitation is poor, as in time of natural disaster or war or in situations of extreme poverty and deprivation. Then control of some sort becomes necessary. It is a sobering fact, however, as we shall presently see, that the method of massive chemical control has had only limited success, and also threatens to worsen the very conditions it is intended to curb.

Under primitive agricultural conditions the farmer had few insect problems. These arose with the intensification of agriculture—the devotion of immense acreages to a single crop. Such a system set the stage for explosive increases in specific insect populations. Single-crop farming does not take advantage of the principles by which nature works; it is agriculture as an engineer might conceive it to be. Nature has introduced great variety into the landscape, but man has displayed a passion for simplifying it. Thus he undoes the built-in checks and balances by which nature holds the species within bounds. One important natural check is a limit on the amount of suitable habitat for each species. Obviously then, an insect that lives on wheat can build up its population to much higher levels on a farm devoted to wheat than on one in which wheat is intermingled with other crops to which the insect is not adapted.

The same thing happens in other situations. A generation or more ago, the towns of large areas of the United States lined their streets with the noble elm tree. Now the beauty they hopefully created is threatened with complete destruction as disease sweeps through the elms, carried by a beetle that would have only limited chance to build up large populations and to spread from tree to tree if the elms were only occasional trees in a richly diversified planting.

Another factor in the modern insect problem is one that must be viewed against a background of geologic and human history: the spreading of thousands of different kinds of organisms from their native homes to invade new territories. This worldwide migration has been studied and graphically described by the British ecologist Charles Elton in his recent book *The Ecology of Invasions*. During the Cretaceous Period, some hundred million years ago, flooding seas cut many land bridges between continents and living things found themselves confined in what Elton calls "colossal separate nature reserves." There, isolated from others of their kind, they developed many new species. When some of the land masses were joined again, about fifteen million years ago, these species began to move out into new territories—a movement that is not only still in progress but is now receiving considerable assistance from man.

The importation of plants is the primary agent in the modern spread of species, for animals have almost invariably gone along with the plants, quarantine being a comparatively recent and not completely effective innovation. The United States Office of Plant Introduction alone has introduced almost 200,000 species and varieties of plants from all over the world. Nearly half of the 180 or so major insect enemies of plants in the United States are accidental imports from abroad, and most of them have come as hitchhikers on plants.

In new territory, out of reach of the restraining hand of the natural enemies that kept down its numbers in its native land, an invading plant or animal is able to become enormously abundant. Thus it is no accident that our most troublesome insects are introduced species.

These invasions, both the naturally occurring and those dependent on human assistance, are likely to continue indefinitely. Quarantine and massive chemical campaigns are only extremely expensive ways of buying time. We are faced, according to Dr. Elton, "with a life-and-death need not just to find new technological means of suppressing this plant or that animal"; instead we need the basic knowledge of animal populations and their relations to their surroundings that will "promote an even balance and damp down the explosive power of outbreaks and new invasions."

Much of the necessary knowledge is now available but we do not use it. We train ecologists in our universities and even employ them in our governmental agencies but we seldom take their advice. We allow the chemical death rain to fall as though there were no alternative, whereas in fact there are many, and our ingenuity could soon discover many more if given opportunity.

Have we fallen into a mesmerized state that makes us accept as inevitable that which is inferior or detrimental, as though having lost the will or the vision to demand that which is good? Such thinking, in the words of the ecologist Paul Shepard, "idealizes life with only its head out of water, inches above the limits of toleration of the corruption of its own environment . . . Why should we tolerate a diet of weak poisons, a home in insipid surroundings, a circle of acquaintances who are not quite our enemies, the noise of motors with just enough relief to prevent insanity? Who would want to live in a world which is just not quite fatal?"

Yet such a world is pressed upon us. The crusade to create a chemically sterile, insect-free world seems to have engendered a fanatic zeal on the part of many specialists and most of the so-called control agencies. On every hand there is evidence that those engaged in spraying operations exercise a ruthless power. "The regulatory entomologists . . . function as prosecutor, judge and jury, tax assessor and collector and sheriff to enforce their own orders," said Connecticut entomologist Neely Turner. The most flagrant abuses go unchecked in both state and federal agencies.

It is not my contention that chemical insecticides must never be used. I do contend that we have put poisonous and biologically potent chemicals indiscriminately into the hands of persons largely or wholly ignorant of their potentials for harm. We have subjected enormous numbers of people to contact with these poisons, without their consent and often without their knowledge. If the Bill of Rights contains no guarantee that a citizen shall be secure against lethal poisons distributed either by private individuals or by public officials, it is surely only because our forefathers, despite their considerable wisdom and foresight, could conceive of no such problem.

I contend, furthermore, that we have allowed these chemicals to be used with little or no advance investigation of their effect on soil, water, wildlife, and man himself. Future generations are unlikely to condone our lack of prudent concern for the integrity of the natural world that supports all life.

There is still very limited awareness of the nature of the threat. This is an era of specialists, each of whom sees his own problem and is unaware of or intolerant of the larger frame into which it fits. It is also an era dominated by industry, in which the right to make a dollar at whatever cost is seldom challenged. When the public protests, confronted with some obvious evidence of damaging results of pesticide applications, it is fed little tranquilizing pills of half truth. We urgently need an end to these false assurances, to the sugar coating of unpalatable facts. It is the public that is being asked to assume the risks that the insect controllers calculate. The public must decide whether it wishes to continue on the present road, and it can do so only when in full possession of the facts. In the words of Jean Rostand, "The obligation to endure gives us the right to know."

# ECO-CATASTROPHE

*Paul Ehrlich*

*The following selection is a "pretty grim scenario" of a future to which a great number of creatures have already subscribed. Brown pelicans face extinction off the shores of California, the osprey and the American egret . . . and so it goes. Why? Because by the time we got after DDT its relations had disappeared into industrial processes where they lived and polluted under the initials PCB's and by the time we get to the producers of paints and plastics. . . . Paul Ehrlich is the Malthus of the 1970's. His book* The Population Bomb *(1968) has had a profound impact on America.*

## I.

The end of the ocean came late in the summer of 1979, and it came even more rapidly than the biologists had expected. There had been signs for more than a decade, commencing with the discovery in 1968 that DDT slows down photosynthesis in marine plant life. It was announced in a short paper in the technical journal, *Science,* but to ecologists it smacked of doomsday. They knew that all life in the sea depends on photosynthesis, the chemical process by which green plants bind the sun's energy and make it available to living things. And they knew that DDT and similar chlorinated hydrocarbons had polluted the entire surface of the earth, including the sea.

But that was only the first of many signs. There had been the final gasp of the whaling industry in 1973, and the end of the Peruvian anchovy fishery in 1975. Indeed, a score of other fisheries had disappeared quietly from

SOURCE: Paul Ehrlich, "Eco-Catastrophe," *Ramparts,* September 1969. Reprinted by permission of the author.

overexploitation and various eco-catastrophes by 1977. The term *eco-catastrophe* was coined by a California ecologist in 1969 to describe the most spectacular of man's attacks on the systems which sustain his life. He drew his inspiration from the Santa Barbara offshore oil disaster of that year, and from the news which spread among naturalists that virtually all of the Golden State's seashore bird life was doomed because of chlorinated hydrocarbon interference with its reproduction. Eco-catastrophes in the sea became increasingly common in the early 1970's. Mysterious "blooms" of previously rare microorganisms began to appear in offshore waters. Red tides—killer outbreaks of a minute single-celled plant—returned to the Florida Gulf coast and were sometimes accompanied by tides of other exotic hues.

It was clear by 1975 that the entire ecology of the ocean was changing. A few types of phytoplankton were becoming resistant to chlorinated hydrocarbons and were gaining the upper hand. Changes in the phytoplankton community led inevitably to changes in the community of zooplankton, the tiny animals which eat the phytoplankton. These changes were passed on up the chains of life in the ocean to the herring, plaice, cod and tuna. As the diversity of life in the ocean diminished, its stability also decreased.

Other changes had taken place by 1975. Most ocean fishes that returned to fresh water to breed, like the salmon, had become extinct, their breeding streams so dammed up and polluted that their powerful homing instinct only resulted in suicide. Many fishes and shellfishes that bred in restricted areas along the coasts followed them as onshore pollution escalated.

By 1977 the annual yield of fish from the sea was down to thirty million metric tons, less than one-half the per capita catch of a decade earlier. This helped malnutrition to escalate sharply in a world where an estimated fifty million people per year were already dying of starvation. The United Nations attempted to get all chlorinated hydrocarbon insecticides banned on a worldwide basis, but the move was defeated by the United States. This opposition was generated primarily by the American petrochemical industry, operating hand in glove with its subsidiary, the United States Department of Agriculture. Together they persuaded the government to oppose the U.N. move—which was not difficult since most Americans believed that Russia and China were more in need of fish products than was the United States. The United Nations also attempted to get fishing nations to adopt strict and enforced catch limits to preserve dwindling stocks. This move was blocked by Russia, who, with the most modern electronic equipment, was in the best position to glean what was left in the sea. It was, curiously, on the very day in 1977 when the Soviet Union announced its refusal that another ominous article appeared in *Science*. It announced that incident solar radiation had been so reduced by worldwide air pollution that serious effects on the world's vegetation could be expected.

## II.

Apparently it was a combination of ecosystem destabilization, sunlight reduction, and a rapid escalation in chlorinated hydrocarbon pollution from massive Thanodrin applications which triggered the ultimate catastrophe. Seventeen huge Soviet-financed Thanodrin plants were operating in underdeveloped countries by 1978. They had been part of a massive Russian "aid offensive" designed to fill the gap caused by the collapse of America's ballyhooed "Green Revolution."

It became apparent in the early 70's that the Green Revolution was more talk than substance. Distribution of high yield "miracle" grain seeds had caused temporary local spurts in agricultural production. Simultaneously, excellent weather had produced record harvests. The combination permitted bureaucrats, especially in the United States Department of Agriculture and the Agency for International Development (AID), to reverse their previous pessimism and indulge in an outburst of optimistic propaganda about staving off famine. They raved about the approaching transformation of agriculture in the underdeveloped countries (UDCs). The reason for the propaganda reversal was never made clear. Most historians agree that a combination of utter ignorance of ecology, a desire to justify past errors, and pressure from agro-industry (which was eager to sell pesticides, fertilizers, and farm machinery to the UDCs and agencies helping the UDCs) was behind the campaign. Whatever the motivation, the results were clear. Many concerned people, lacking the expertise to see through the Green Revolution drivel, relaxed. The population-food crisis was "solved."

But reality was not long in showing itself. Local famine persisted in northern India even after good weather brought an end to the ghastly Bihar famine of the mid-60's. East Pakistan was next, followed by a resurgence of general famine in northern India. Other foci of famine rapidly developed in Indonesia, the Philippines, Malawi, the Congo, Egypt, Colombia, Ecuador, Honduras, the Dominican Republic, and Mexico.

Everywhere hard realities destroyed the illusion of the Green Revolution. Yields dropped as the progressive farmers who had first accepted the new seeds found that their higher yields brought lower prices—effective demand (hunger plus cash) was not sufficient in poor countries to keep prices up. Less progressive farmers, observing this, refused to make the extra effort required to cultivate the "miracle" grains. Transport systems proved inadequate to bring the necessary fertilizer to the fields where the new and extremely fertilizer-sensitive grains were being grown. The same systems were also inadequate to move produce to markets. Fertilizer plants were not built fast enough, and most of the underdeveloped countries could not scrape together funds to purchase supplies, even on concessional terms. Finally, the inevitable happened, and pests began to reduce yields in even the most carefully cultivated fields. Among the first were the famous "miracle rats" which invaded Philippine "miracle rice" fields in early 1969. They were quickly followed by many insects and viruses, thriving on the relatively pest-susceptible new grains, encouraged by the vast and dense plantings, and rapidly acquiring resistance to the chemicals used against them. As chaos spread until even the most obtuse agriculturists and economists realized that the Green Revolution had turned brown, the Russians stepped in.

In retrospect it seems incredible that the Russians, with the American mistakes known to them, could launch an even more incompetent program of aid to the underdeveloped world. Indeed, in the early 1970's there were cynics in the United States who claimed that outdoing the stupidity of American foreign aid would be physically impossible. Those critics were, however, unaware that the Russians had been busily destroying their own environment for many years. The virtual disappearance of sturgeon from Russian rivers caused a great shortage of caviar by 1970. A standard joke among Russian scientists at that time was that they had created an artificial caviar which was indistinguishable from the real thing—except by taste. At any rate the Soviet Union, observing with interest the progressive deterioration of relations between the UDCs and the United States, came up with a solution. It had recently developed what it claimed was the ideal insecticide, a highly lethal chlorinated hydrocarbon complexed with a special agent for penetrating the external skeletal armor of insects. Announcing that the new pesticide, called Thanodrin, would truly produce a Green Revolution, the Soviets entered into negotiations with various UDCs for the construction of massive Thanodrin factories. The USSR would bear all the costs; all it wanted in return were certain trade and military concessions.

It is interesting now, with the perspective of years, to examine in some detail the reasons why the UDCs welcomed the Thanodrin plan with such open arms. Government officials in these countries ignored the protests of their own scientists that Thanodrin would not solve the problems which plagued them. The governments now knew that the basic cause of their problems was overpopulation, and that these problems had been exacerbated by the dullness, daydreaming, and cupidity endemic to all governments. They knew that only population control and limited development aimed primarily at agriculture could have spared them the horrors they now faced. They knew it, but they were not about to admit it. How much easier it was simply to accuse the Americans of failing to give them proper aid; how much simpler to accept the Russian panacea.

And then there was the general worsening of relations between the United States and the UDCs. Many things had contributed to this. The situation in America in the first half of the 1970's deserves our close scrutiny. Being more dependent on imports for raw materials than the Soviet Union, the United States had, in the early 1970's, adopted more and more heavy-handed policies in order to insure continuing supplies. Military adventures in Asia and Latin America had further lessened the international credibility of the United States as a great defender of freedom—an image which had begun to deteriorate rapidly during the pointless and fruitless Viet-Nam conflict. At home, acceptance of the carefully manufactured image lessened dramatically, as even the more romantic and chauvinistic citizens began to understand the role of the military and the industrial system in what John Kenneth Galbraith had aptly named "The New Industrial State."

At home in the USA the early 70's were traumatic times. Racial violence grew and the habitability of the cities diminished, as nothing substantial was done to ameliorate either racial inequities or urban blight. Welfare rolls grew as automation and general technological progress forced more and more people into the category of "unemployable." Simultaneously a taxpayers' revolt occurred. Although there was not enough money to build the schools, roads, water systems, sewage systems, jails, hospitals, urban transit lines, and all the other amenities needed to support a burgeoning population, Americans refused to tax themselves more heavily. Starting in Youngstown, Ohio in 1969 and followed closely by Richmond, California, community after community was forced to close its schools or curtail educational operations for lack of funds. Water supplies, already marginal in quality and quantity in many places by 1970, deteriorated quickly. Water rationing occurred in 1723 municipalities in the summer of 1974, and hepatitis and epidemic dysentery rates climbed about 500 per cent between 1970–1974.

III.

Air pollution continued to be the most obvious manifestation of environmental deterioration. It was, by 1972, quite literally in the eyes of all Americans. The year 1973 saw not only the New York and Los Angeles smog disasters, but also the publication of the Surgeon General's massive report on air pollution and health. The public had been partially prepared for the worst by the publicity given to the U.N. pollution conference held in 1972. Deaths in the late 60's caused by smog were well known to scientists, but the public had ignored them because they mostly involved early demise of the old and sick rather than people dropping dead on the freeways. But suddenly our citizens were faced with nearly 200,000 corpses and massive documentation that they could be the next to die from respiratory disease. They were not ready for that scale of disaster. After all, the U.N. conference had not predicted that accumulated air pollution would make the planet uninhabitable until almost 1990. The population was terrorized as TV screens became filled with scenes of horror from the disaster areas. Especially vivid was NBC's coverage of hundreds of unattended people choking out their lives outside of

New York's hospitals. Terms like nitrogen oxide, acute bronchitis, and cardiac arrest began to have real meaning for most Americans.

The ultimate horror was the announcement that chlorinated hydrocarbons were now a major constituent of air pollution in all American cities. Autopsies of smog disaster victims revealed an average chlorinated hydrocarbon load in fatty tissue equivalent to 26 parts per million of DDT. In October, 1973, the Department of Health, Education and Welfare announced studies which showed unequivocally that increasing death rates from hypertension, cirrhosis of the liver, liver cancer, and a series of other diseases had resulted from the chlorinated hydrocarbon load. They estimated that Americans born since 1946 (when DDT usage began) now had a life expectancy of only 49 years, and predicted that if current patterns continued, this expectancy would reach 42 years by 1980, when it might level out. Plunging insurance stocks triggered a stock market panic. The president of . . . a major pesticide producer went on television to "publicly eat a teaspoonful of DDT" (it was really powdered milk) and announce that HEW had been infiltrated by Communists. Other giants of the petro-chemical industry, attempting to dispute the indisputable evidence, launched a massive pressure campaign on Congress to force HEW to "get out of agriculture's business." They were aided by the agro-chemical journals, which had decades of experience in misleading the public about the benefits and dangers of pesticides. But by now the public realized that it had been duped. The Nobel Prize for medicine and physiology was given to Drs. J. L. Radomski and W. B. Deichmann, who in the late 1960's had pioneered in the documentation of the long-term lethal effects of chlorinated hydrocarbons. A Presidential Commission with unimpeachable credentials directly accused the agro-chemical complex of "condemning many millions of Americans to an early death." The year 1973 was the year in which Americans finally came to understand the direct threat to their existence posed by environmental deterioration.

And 1973 was also the year in which most people finally comprehended the indirect threat. Even the president of Union Oil Company and several other industrialists publicly stated their concern over the reduction of bird populations which had resulted from pollution by DDT and other chlorinated hydrocarbons. Insect populations boomed because they were resistant to most pesticides and had been freed, by the incompetent use of those pesticides, from most of their natural enemies. Rodents swarmed over crops, multiplying rapidly in the absence of predatory birds. The effect of pests on the wheat crop was especially disastrous in the summer of 1973, since that was also the year of the great drought. Most of us can remember the shock which greeted the announcement by atmospheric physicists that the shift of the jet stream which had caused the drought was probably permanent. It signalled the birth of the Midwestern desert. Man's air-polluting activities had by then caused gross changes in climatic patterns. The news, of course, played hell with commodity and stock markets. Food prices skyrocketed, as savings were poured into hoarded canned goods. Official assurances that food supplies would remain ample fell on deaf ears, and even the government showed signs of nervousness when California migrant field workers went on strike again in protest against the continued use of pesticides by growers. The strike burgeoned into farm burning and riots. The workers, calling themselves "The Walking Dead," demanded immediate compensation for their shortened lives, and crash research programs to attempt to lengthen them.

It was in the same speech in which President Edward Kennedy, after much delay, finally declared a national emergency and called out the National Guard to harvest California's crops, that the first mention of population control was made. Kennedy pointed out that the United States would no longer be able to offer any food aid to other nations and was likely to suffer food shortages herself. He suggested that, in view of the manifest failure of the Green Revolution, the only hope of the UDCs lay in population control. His statement, you will recall, created an uproar in the underdeveloped countries. Newspaper editorials accused the United States of wishing to prevent small countries from becoming large nations and thus threatening American hegemony. Politicians asserted that President Kennedy was a "creature of the giant drug combine" that wished to shove its pills down every woman's throat.

Among Americans, religious opposition to population control was very slight. Industry in general also backed the idea. Increasing poverty in the UDCs was both destroying markets and threatening supplies of raw materials. The seriousness of the raw material situation had been brought home during the Congressional Hard Resources hearings in 1971. The exposure of the ignorance of the cornucopian economists had been quite a spectacle—a spectacle brought into virtually every American's home in living color. Few would forget the distinguished geologist from the University of California who suggested that economists be legally required to learn at least the most elementary facts of geology. Fewer still would forget that an equally distinguished Harvard economist added that they might be required to learn some economics, too. The overall message was clear: America's resource situation was bad and bound to get worse. The hearings had led to a bill requiring the Departments of State, Interior, and Commerce to set up a joint resource procurement council with the express purpose of "insuring that proper consideration of American resource needs be an integral part of American foreign policy."

Suddenly the United States discovered that it had a national consensus: population control was the only possible salvation of the underdeveloped world. But that same consensus led to heated debate. How could the UDCs be persuaded to limit their populations, and should not the United States lead the way by limiting its own? Members of the intellectual community wanted America to set an example. They pointed out that the United States was in the midst of a new baby boom: her birth rate, well over twenty per thousand per year, and her growth rate of over one per cent per annum were among the very highest of the developed countries. They detailed the deterioration of the American physical and psychic environments, the growing health threats, the impending food shortages, and the insufficiency of funds for desperately needed public works. They contended that the nation was clearly unable or unwilling to properly care for the people it already had. What possible reason could there be, they queried, for adding any more? Besides, who would listen to requests by the United States for population control when that nation did not control her own profligate reproduction?

Those who opposed population controls for the U.S. were equally vociferous. The military-industrial complex, with its all-too-human mixture of ignorance and avarice, still saw strength and prosperity in numbers. Baby food magnates, already worried by the growing nitrate pollution of their products, saw their market disappearing. Steel manufacturers saw a decrease in aggregate demand and slippage for that holy of holies, the Gross National Product. And military men saw, in the growing population-food-environment crisis, a serious threat to their carefully nurtured Cold War. In the end, of course, economic arguments held sway, and the "inalienable right of every American couple to determine the size of its family," a freedom invented for the occasion in the early 70's, was not compromised.

The population control bill, which was passed by Congress early in 1974, was quite a document, nevertheless. On the domestic front, it authorized an increase from 100 to 150 million dollars in funds for "family planning" activities. This was made possible by a general feeling in the country that the growing army on welfare needed family planning. But the gist of the bill was a series of measures designed to impress the need for population control on the UDCs. All American aid to countries with overpopulation problems was required by law to consist in part of population control assistance. In order to receive any assistance each nation was required not only to accept the population control aid, but also to match it according to a complex formula. "Overpopulation" itself was defined by a formula based on U.N. statistics, and the UDCs were required not only to accept aid, but also to show progress in reducing birth rates. Every five years the status of the aid program for each nation was to be reevaluated.

The reaction to the announcement of this program dwarfed the response to President Kennedy's speech. A coalition of UDCs attempted to get the U.N. General Assembly to condemn the United States as a "genetic aggressor." Most damaging of all to the American cause was the famous "25 Indians and a dog" speech by Mr. Shankarnarayan, Indian Ambassador to the U.N. Shankarnarayan pointed out that for several decades the United States, with less than six per cent of the people of the world, had consumed roughly fifty per cent of the raw materials used every year. He described vividly America's contribution to worldwide environmental deterioration, and he scathingly denounced the miserly record of United States foreign aid as "unworthy of a fourth-rate power, let alone the most powerful nation on earth."

It was the climax of his speech, however, which most historians claim once and for all destroyed the image of the United States. Shankarnarayan informed the assembly that the average American family dog was fed more animal protein per week than the average Indian got in a month. "How do you justify taking fish from protein-starved Peruvians and feeding them to your animals?" he asked. "I contend," he concluded, "that the birth of an American baby is a greater disaster for the world than that of 25 Indian babies." When the applause had died away, Mr. Sorensen, the American representative, made a speech which said essentially that "other countries look after their own self-interest, too." When the vote came, the United States was condemned.

## IV.

This condemnation set the tone of U.S.–UDC relations at the time the Russian Thanodrin proposal was made. The proposal seemed to offer the masses in the UDCs an opportunity to save themselves and humiliate the United States at the same time; and in human affairs, as we all know, biological realities could never interfere with such an opportunity. The scientists were silenced, the politicians said yes, the Thanodrin plants were built, and the results were what any beginning ecology student could have predicted. At first Thanodrin seemed to offer excellent control of many pests. True, there was a rash of human fatalities from improper use of the lethal chemical, but as Russian technical advisors were prone to note, these were more than compensated for by increased yields. Thanodrin use skyrocketed throughout the underdeveloped world. The Mikoyan design group developed a dependable, cheap, agricultural aircraft which the Soviets donated to the effort in large numbers. MIG sprayers became even more common in UDCs than MIG interceptors.

Then the troubles began. Insect strains with cuticles resistant to Thanodrin penetration began to appear. And as streams, rivers, fish culture ponds and onshore waters became rich in Thanodrin, more fisheries began to disappear. Bird populations were decimated. The sequence of events was standard for broadcast use of a synthetic pesticide: great success at first, followed by removal of natural enemies and development of resistance by the pest. Populations of crop-eating insects in areas treated with Thanodrin made steady comebacks and soon became more abundant than ever. Yields plunged while farmers in their desperation increased the Thanodrin dose and shortened the time between treatments. Death from Thanodrin poisoning became common. The first violent incident occurred in the Canete Valley of

Peru, where farmers had suffered a similar chlorinated hydrocarbon disaster in the mid-50's. A Russian advisor serving as an agricultural pilot was assaulted and killed by a mob of enraged farmers in January, 1978. Trouble spread rapidly during 1978, especially after the word got out that two years earlier Russia herself had banned the use of Thanodrin at home because of its serious effects on ecological systems. Suddenly Russia, and not the United States, was the *bête noir* in the UDCs. "Thanodrin parties" became epidemic, with farmers, in their ignorance, dumping carloads of Thanodrin concentrate into the sea. Russian advisors fled, and four of the Thanodrin plants were leveled to the ground. Destruction of the plants in Rio and Calcutta led to hundreds of thousands of gallons of Thanodrin concentrate being dumped directly into the sea.

Mr. Shankarnarayan again rose to address the U.N., but this time it was Mr. Potemkin, representative of the Soviet Union, who was on the hot seat. Mr. Potemkin heard his nation described as the greatest mass killer of all time as Shankarnarayan predicted at least thirty million deaths from crop failure due to overdependence on Thanodrin. Russia was accused of "chemical aggression," and the General Assembly, after a weak reply by Potemkin, passed a vote of censure.

It was in January, 1979, that huge blooms of a previously unknown variety of diatom were reported off the coast of Peru. The blooms were accompanied by a massive die-off of sea life and of the pathetic reminder of the birds which had once feasted on the anchovies of the area. Almost immediately another huge bloom was reported in the Indian ocean, centering around the Seychelles, and then a third in the South Atlantic off the African coast. Both of these were accompanied by spectacular die-offs of marine animals. Even more ominous were growing reports of fish and bird kills at oceanic points where there were no spectacular blooms. Biologists were soon able to explain the phenomena: the diatom had evolved an enzyme which broke down Thanodrin; that enzyme also produced a breakdown product which interfered with the transmission of nerve impulses, and was therefore lethal to animals. Unfortunately, the biologists could suggest no way of repressing the poisonous diatom bloom in time. By September, 1979, all important animal life in the sea was extinct. Large areas of coastline had to be evacuated, as windrows of dead fish created a monumental stench.

But stench was the least of man's problems. Japan and China were faced with almost instant starvation from a total loss of the seafood on which they were so dependent. Both blamed Russia for their situation and demanded immediate mass shipments of food. Russia had none to send. On October 13, Chinese armies attacked Russia on a broad front. . . .

## V.

A pretty grim scenario. Unfortunately, we're a long way into it already. Everything mentioned as happening before 1970 has actually occurred; much of the rest is based on projections of trends already appearing. Evidence that pesticides have long-term lethal effects on human beings has started to accumulate, and recently Robert Finch, Secretary of the Department of Health, Education and Welfare, expressed his extreme apprehension about the pesticide situation. Simultaneously the petrochemical industry continues its unconscionable poison-peddling. For instance, Shell Chemical has been carrying on a high-pressure campaign to sell the insecticide Azodrin to farmers as a killer of cotton pests. They continue their program even though they know that Azodrin is not only ineffective, but often *increases* the pest density. They have covered themselves nicely in an advertisement which states, "Even if an overpowering migration [sic] develops, the flexibility of Azodrin lets you regain control fast. Just increase the dosage according to label recommendations." It's a great game—get people to apply the poison and kill the natural enemies of the pests. Then blame the increased pests on "migration" and sell even more pesticide!

Right now fisheries are being wiped out by overexploitation, made easy by modern electronic equipment. The companies producing the equipment know this. They even boast in advertising that only their equipment will keep fishermen in business until the final kill. Profits must obviously be maximized in the short run. Indeed, Western society is in the process of completing the rape and murder of the planet for economic gain. And, sadly, most of the rest of the world is eager for the opportunity to emulate our behavior. But the underdeveloped peoples will be denied that opportunity—the days of plunder are drawing inexorably to a close.

Most of the people who are going to die in the greatest cataclysm in the history of man have already been born. More than three and a half billion people already populate our moribund globe, and about half of them are hungry. Some ten to twenty million will starve to death *this year*. In spite of this, the population of the earth will increase by seventy million souls in 1969. For mankind has artificially lowered the death rate of the human population, while in general birth rates have remained high. With the input side of the population system in high gear and the output side slowed down, our fragile planet has filled with people at an incredible rate. It took several million years for the population to reach a total of two billion people in 1930, *while a second two billion will have been added by 1975!* By that time some experts feel that food shortages will have escalated the present level of world hunger and starvation into famines of unbelievable proportions. Other experts, more optimistic, think the ultimate food-population collision will not occur until the decade of the 1980's. Of course, more massive famine may be avoided if other events cause a prior rise in the human death rate.

Both worldwide plague and thermonuclear war are made more probable as population growth continues. These, along with famine, make up the trio of potential "death rate solutions" to the population problem—solutions in which the birth rate–death rate imbalance is redressed by a rise in the death rate rather than by a lowering of the birth rate. Make no mistake about it, *the imbalance will be redressed.* The shape of the population growth curve is one familiar to the biologist. It is the outbreak part of an outbreak-crash sequence. A population grows more rapidly in the presence of abundant resources, finally runs out of food or some other necessity, and crashes to a low level or extinction. Man is not only running out of food, he is also destroying the life support systems of the Spaceship Earth. The situation was recently summarized very succinctly: "It is the top of the ninth inning. Man, always a threat at the plate, has been hitting Nature hard. It is important to remember, however, that NATURE BATS LAST."

# SOIL AND FRESH WATER: DAMAGED GLOBAL FABRIC

## Barry Commoner

*Barry Commoner is the director of the famed Center for the Biology of Natural Systems at Washington University. Perhaps more than any other contemporary scientist, Dr. Commoner has a profound understanding of the environmental crisis in its many and varied forms. This brief article speaks of some of the stresses that our technological society is placing on the living systems, which support all life, including our own. He pleads for a technology that has at its foundations ecological principles rather than purely economic motives.*

With startling suddenness environmental pollution has jumped to the top of the agenda of public concern. A short time ago the condition of the environment was largely a subject for discussion among scientists; although some of us did venture from our laboratories to alert the public and legislators to the problem, until recently the response was one of polite attention, but little demand for remedial action. Now, suddenly, things are different. Anti-pollution picket lines have appeared before industrial plants, and legislators are being pressed for new laws to limit pollution. Citizens have taken legal action against polluters. Student groups have organized to defend the environment from attack; environmental teach-ins will be held on most of the nation's campuses this month. Important state-wide environmental conferences have been organized under the sponsorship of the governors of California, Vermont, and New Hampshire. The President, in his State of the Union message last January, designated environmental quality as the major domestic issue for the 1970s.

SOURCE: Barry Commoner, "Soil and Fresh Water: Damaged Global Fabric," *Environment*, April 1970, vol. 12, no. 3. Reprinted by permission of Barry Commoner. Copyright © 1970 by Barry Commoner.

The immediate reasons for this intense concern are not difficult to detect, for they constantly assail our senses: Our eyes smart from smog, our ears throb with the noise of automobiles, aircraft, and construction tools; we are assailed by the odors of polluted waters and the sight of mounting heaps of rubbish.

Less apparent than the *fact* of pollution is what can be done about it. The problems are enormous in size: Cities are running out of places to dump garbage and rubbish, and a lake as large as Erie has been nearly totally polluted. The problems are bewildering in their complexity: If we expand sewage treatment facilities, we only increase the pollution due to rotting masses of algae; if we incinerate garbage, we intensify air pollution; if we attempt to control smog by means of exhaust devices which reduce waste fuel emission, we worsen the pollution due to nitrogen oxides.

The degradation of the environment in which we live has become a pervasive, intractable, discouraging problem. It clashes noisomely with the magnificent progress of the age, with the marvelous competence of our new machines, with the rising productivity of our factories and our farms, with the new inventions that have revolutionized communications and management. It raises perplexing questions about the human value of our technological competence.

Why has a society which is so enriched by the progress of technology now become so impoverished in the quality of the life which that technology supports? What are the causes of this dismaying paradox? What can be done to resolve it?

The following thesis may provide some useful insights into these problems: We are in an environmental crisis which threatens the survival of this nation, and of the world as a suitable place of human habitation. Environmental pollution is not to be regarded as an unfortunate but incidental by-product of the growth of population, the intensification of production, or of technological progress. It is, rather, an intrinsic feature of the very technology which we have developed to enhance productivity. Our technology is enormously successful in producing material goods, but too often is disastrously incompatible with natural environmental systems. Yet, the survival of all living things—including man—the quality of life, and the continued success of all human activities—including technology, industry, and agriculture—depends on the integrity of the complex web of biological processes which comprise the environment—the earth's ecosystem. And what man is now doing on the earth violates this fundamental requisite of human existence. With tragic perversity we have linked much of our productive economy to precisely those features of technology which violate the environment that supports us.

Moreover, the technologies which threaten the stability of the environment are now so massively embedded in our system of industrial and agricultural production that any effort to make them conform to the demands of the environment will involve serious economic dislocations. If environmental pollution is a sign of major incompatibilities between our system of productivity and the environmental system that supports it, then, if we are to survive, we must successfully confront these economic obligations, however severe and challenging to our social concepts they may be.

Finally, unless we start now with a *fundamental* attack on the environmental crisis, we will find ourselves, in a decade, locked into an irreversible, self-destructive course.

All living things including man, and all human activities on the surface of the earth, including all of our technology, industry, and agriculture, are dependent on the great interwoven cyclical processes followed by the four elements that make up the major portion of living things and the environment: carbon, oxygen, hydrogen, and nitrogen. All of these cycles are driven by the action of living things: green plants convert carbon dioxide into food, fiber, and fuel; at the same time they produce oxygen, so that the total oxygen supply in our atmosphere is the product of plant activity. Plants also convert inorganic nutrients into foodstuffs. Animals, basically, live on plant-produced food; in turn they regenerate the inorganic materials—carbon dioxide, nitrates, and phosphates—which must support plant life. Also involved are myriads of microorganisms in the soil and water. Altogether this vast web of biological interactions generates the very physical system in which we live: the soil and the air. It maintains the purity of surface waters and by governing the movement of water in the soil and its evaporation into the air regulates the weather. This is the environment. It is a place created by living things, maintained by living things, and through the marvelous reciprocities of biological evolution is essential to the support of living things.

The city clearly illustrates our dependence on the environment, for it is sustained by essential links to a number of ecological systems. Thus the purity of the water which is delivered to the city dweller is achieved by biological processes in some distant waterway. In turn, the city releases its sewage wastes to nearby waterways, imposing on their self-cleansing biological systems a strain which must be overcome if local beaches are to be usable and if downstream communities are to have pure water for their own needs. The city must have air to support its life; but it discharges massive wastes into it, so that the purity of the air becomes, again, dependent on natural processes in the weather system. Similarly, the city is linked to the land by its requirements for food, as well as water. Organic wastes, such as garbage, which in nature sustain the fertility of the soil, are carelessly thrust back into the environment by the city as landfill or through incineration. And the city has its own internal ecological problems; rats and other vermin live on waste, and in their proliferation contribute to human misery. In effect, the city, like all other human activities, is part of a huge ecological system, and the quality of the urban environment depends on the successful functioning of that system.

In sum, the environment makes up a huge, enormously complex living machine—the ecosphere—and *every* human activity depends on the integrity and proper functioning of that machine. Without the photosynthetic activity of green plants there would be no oxygen for our smelters and furnaces, let alone to support human and animal life. Without the action of plants and animals in aquatic systems, we can have no pure water to supply agriculture, industry, and the cities. Without the biological processes that have gone on in the soil for thousands of years, we would have neither food crops, oil, nor coal. This machine is our biological capital, the basic apparatus on which our total productivity depends. If we destroy it, our most advanced technology will come to naught and any economic and political system which depends on it will founder. Yet, the major threat to the integrity of this biological capital is technology itself.

Consider, for example, the technological process by which modern cities dispose of human wastes—sewage treatment. This technology makes use of the ecological processes which occur in surface waters, such as lakes and rivers. This is the cycle which links aquatic animals to their organic wastes; these wastes to the oxygen-requiring microorganisms that convert them into inorganic nitrate, phosphate, and carbon dioxide; the inorganic nutrients to the algae which photosynthetically reconvert them into organic substances (thereby also adding to the oxygen content of the water and so providing support for the animals and the organisms of decay); and algal organic matter to the chain of animals which feed on it, thus completing the cycle.

This cycle works continuously, so long as it is not too heavily stressed; when that happens it breaks down and stops. Thus, if the load of organic waste imposed on the system becomes too great, the demand of the bacteria of decay for oxygen may exceed the limited oxygen content of the water. When the oxygen content falls to zero, the bacteria die, the biological cycle breaks down, and organic debris accumulate. Similarly, if the inorganic nutrient level (especially nitrate and phosphate) of the water becomes so great as to stimulate the rapid growth of algae, the dense algal population cannot be long sustained because of the intrinsic limitations of photosynthetic efficiency. As the thickness of the algal layer in the water increases, the light required for photosynthesis that can reach the lower parts of the algal layer becomes sharply diminished, so that any strong overgrowth of algae very quickly dies back, again releasing organic debris. These are examples of a fundamental attribute of all ecological systems—that under stress they may undergo dramatic overgrowths and equally dramatic collapse.

Modern sewage treatment technology is intended to convert the noxious organic materials of human wastes into innocuous inorganic materials. This technology reflects an excellent understanding of *part* of the aquatic cycle: that given sufficient oxygen, aquatic microorganisms can convert organic matter to innocuous inorganic products which are readily carried off in surface waters. By domesticating such microorganisms in artificially aerated sewage plants we can indeed convert nearly all of the organic matter of sewage into inorganic products and discharge them to rivers and lakes.

So far, so good; the fatal stress of an overburden of organic matter on the stability of the aquatic cycle is avoided. But given the circularity of the process, it is evident that now a new stress must appear, this time the impact of excessive inorganic nutrients on the growth of algae. And given the tendency of algal populations to excessive growth under heavy nutrient stimulation we ought to expect trouble at this point. And indeed the trouble has come— but it has been largely unexpected. Only in the last decade, when the effects of algal overgrowths had already largely destroyed the self-purifying capability of an ecosystem as massive as Lake Erie was the phenomenon recognized as a serious limitation on the technology of sewage treatment. In effect, the modern system of sewage technology has failed in its stated aim of reducing the organic oxygen demand on surface waters because it did not take into account the circularity of the ecological system on which it intruded. Because of this circularity the inorganic products of sewage treatment were themselves reconverted to organic nutrients by the algae, which on their death simply reimposed the oxygen demand that the treatment was supposed to remove on the lakes and rivers. This failure can be attributed, therefore, to a simple violation of a fundamental principle of ecology. The price that we pay for this defect is the nearly catastrophic pollution of our surface waters.

Some features of the impending catastrophe are predictable. For example, the Spilhaus Report estimates that by 1980 the oxygen demand due to municipal wastes will equal the oxygen content of the total flow of all the U.S. river systems in the summer months.[1] To this we must add the fact that oxygen-demanding wastes from livestock feedlots are now larger than the total municipal waste output. And, finally, we must also add pollution due to algal overgrowths stimulated by phosphate from synthetic detergents and by nitrate leaching from heavily fertilized soil.

The effects of these changes on human health are less predictable but of great potential danger. Excess nitrate in drinking water can cause a fatal disease in infants, methemoglobinemia, and at lower levels may, according to recent studies, cause chronic deficiencies in infant development. Moreover, with the death of algal overgrowths and the resultant accumulation of organic matter, microorganisms normally restricted to the soil—many of which are pathogenic in man and animals—may begin to grow in waterways, raising the possibility of the outbreak of new types of water-borne disease.

The overall lesson is clear: These modern technological practices—sewage treatment, synthetic detergents, feedlot techniques, intensive use of

[1] National Academy of Sciences—National Research Council, *Waste Management and Control*, Washington, D.C.: National Research Council Publication 1400, 1966.

inorganic nitrogen fertilizer—tend to break down the natural biological cycles of the soil and water. The resultant ecological backlash deprives us of the pure water needed by homes and industries and creates hazards to health.

Another example of technology-generated environmental deterioration is air pollution due to automotive exhaust fumes. This problem is the direct outcome of the technological *improvement* of gasoline engines: the development of the modern high-compression engine. Such engines operate at higher temperatures than older ones; at these elevated temperatures the oxygen and nitrogen of the air taken into the engine tend to combine rapidly, with the resultant production of nitrogen oxides. Once released into the air nitrogen oxides are activated by sunlight. They then react with waste hydrocarbon fuel forming eventually the notorious PAN—the toxic agent of the smog made famous by Los Angeles.

The present smog control technique—reduction of waste fuel emission— by diminishing the interaction of nitrogen oxides with hydrocarbon wastes, enhances the level of airborne nitrogen oxides—which are themselves toxic substances. In the air nitrogen oxides are readily converted to nitrates, which are then brought down by rain and snow to the land and surface waters. There they add to the growing burden of nitrogen fertilizer, which is itself an important aspect of water pollution. What is surprising is the amount of nitrogen oxides that are generated by automotive traffic: it amounts to more than one-third of the nitrogen contained in the fertilizer currently employed on U.S. farms. One calculation shows that farms in New Jersey receive about 25 pounds of nitrogen fertilizer per year (a significant amount in agricultural practice) from the trucks and cars that travel the New Jersey highways. Another recent study shows that in the heavily populated Eastern section of the country, the nitrate content of local rainfall is proportional to the local rate of gasoline consumption. Thus, the emergence of a new technology—the modern gasoline engine—is itself responsible for most of the smog problem and for an appreciable part of the pollution of surface waters with nitrate.

Again we see the endless web of environmental processes at work. Get the engines too hot—for the sake of generating the power needed to drive a huge car at destructive speeds—and you set off a chain of events that keeps kids in Los Angeles off the playground, sends older people to a premature death, and in passing, adds to the already excessive burden of water pollutants.

As a final example of the intrinsic failure of a technology which bears a considerable responsibility for the present pollution of the environment, we may look at the current status of the insecticide problem. Recent reports from Asia, Africa, Latin America, and California show that, with awesome regularity, major outbreaks of insect pests have been *induced* by the use of modern contact-killing insecticides, because such insecticides kill the natural predator and parasitic insects which ordinarily keep the spread of insect pests under control. At the same time there is now increasing evidence that synthetic insecticides are responsible for declining populations of birds and fish. Because of such hazards and the still poorly understood danger to man, widespread use of DDT is being banned.

I have cited these examples in order to illustrate the point that major problems of environmental pollution arise, not out of some minor inadequacies in our new technologies—but because of the very success of these technologies in accomplishing their designed aims. A modern sewage treatment plant causes algal overgrowths and resultant pollution *because* it produces, as it is designed to do, so much plant nutrient in its effluent. Modern, highly concentrated nitrogen fertilizers result in the drainage of nitrate pollutants into streams and lakes just *because* they succeed in the aim of raising the nutrient level of the soil. The modern high-compression gasoline engine contributes to smog and nitrate *because* it successfully meets its design criterion—the development of a high level of power. Modern synthetic insecticides kill birds, fish, and useful insects just *because* they are successful in being absorbed by insects, and killing them, as they are intended to do.

This is some of the tragic destruction that lies hidden in the changing environment—costs that do not appear as entries in the balance sheets of industry and agriculture. These are some of the great debts which must be paid if the environment is to be saved from ultimate destruction. The debts are so embedded in every feature of the economy that it is almost impossible to calculate them.

Their scale, at least, can be sensed from a recent survey which I have made relevant to the state of California.[2] There, according to one study, it will cost $5 billion over a 50-year period, and continuing at $100 million per year, for a water quality control system which will transfer pollutants now entering the San Francisco Bay area to the ocean—where they are likely to cause new pollution problems. To get rid of the smog that now envelops Los Angeles and an increasing number of California metropolitan areas—which surely must be done if these places are to survive—there is the cost of replacing present cars by essentially emissionless vehicles. Smog damage has already been detected in California pine forests; what will it cost if the state's magnificent forests begin to die, unleashing enormous flood problems? How shall we determine the cost of the urban spread which has covered the richest soil in the state—soil that may yet need to be restored to agriculture when the state is forced to limit intensive, pollution-generating fertilization, and new lands have to be used to sustain food production? What is the price of those massive walls of concrete, the freeways, which slice across the land, disrupting drainage patterns and upsetting the delicate balance of forces that keeps the land from sliding into ravines? Against the value of the new real estate

[2] Kaiser Engineers Report, *San Francisco Bay-Delta Water Quality Control Program*, March 1969.

developments on landfills in San Francisco Bay calculate the cost of the resulting changes in tidal movement which have decreased the dilution of the polluting nutrients by fresh water from the sea and have worsened the algal overgrowths. Or balance against the value of the offshore oil the cost of a constant risk of beach and ocean pollution until the offending wells are pumped dry. Finally, figure what it will cost to restore the natural fertility of the soil in central California to keep nitrogen in the soil, where it belongs, and to develop a new, more mixed form of agriculture that will make it possible to get rid of most insecticides and make better use of the natural biological controls.

This is the staggering size of one state's environmental problems. Similar problems burden every state in the union. Most of Lake Erie has been lost to pollution. In Illinois every major river has been overloaded with fertilizer drainage and has lost its powers of self-purification. Automobile smog hangs like a pall over even Denver and Phoenix. Every major city is experiencing worsening air pollution. The entire nation is in the grip of the environmental crisis.

Nor can we forget that in an important sense, the environmental crisis is worldwide. Apart from the environmental deterioration that we in the technologically developed nations are experiencing, there is in the rest of the world a related crisis—the coming collision between the world's rising population and its limited supplies of food. The world will not survive *this* crisis without massive technical support from those nations which have the capability. This will in turn place new, massive demands on our own industrial and agricultural capacities—which will only intensify our own environmental crisis if we do not, at the same time, rectify the basic ecological faults in our present system of productivity.

What is to be done? What *can* be done? Although we are on a path which can only lead to self-destruction, I am also convinced that we have not yet passed the point of no return. We have time—perhaps a generation—in which to save the environment from the final effects of the violence we have already done to it, and to save ourselves from our own suicidal folly. But this is a very short time to achieve the massive environmental repair that is needed. We will need to start, now, on a new path. And the first action is to recognize how badly we have gone wrong in the use of the environment and to mobilize every available resource for the huge task of saving it.

An immediate essential is that we recognize the urgency of *timely* action. No estimate of the time course of such massive and complex processes as those which are degrading the environment can be accurate. But the urgency of the problem requires us to try. My own estimate is that if we are to avoid environmental catastrophe by the 1980s we will need to begin the vast process of correcting the *fundamental* incompatibilities of major technologies with the demands of the ecosystem. This means that we will need to put into operation essentially emissionless versions of automotive vehicles, power plants, refineries, steel mills, and chemical plants. Agricultural technology will need to find ways of sustaining productivity without breaking down the natural soil cycle or disrupting the natural control of destructive insects. Sewage and garbage treatment plants will need to be designed to return organic waste to the soil, where, in nature, it belongs. Vegetation will need to be massively reintroduced into urban areas. Housing and urban sanitary facilities will need to be drastically improved. All of these will demand serious economic adjustments, and our economic and social system will need to be prepared to accommodate them.

Obviously, such massive undertakings in the 1980s and thereafter will be impossible unless we prepare for them in advance. The tragic fact is that we are, today, almost totally unprepared. In some cases we lack the needed basic scientific information; in nearly all cases we lack the technological means. This will take time to achieve. And time is needed, too, for large-scale pilot programs to test the new technologies and to work out the multiple interrelations among them. For at no cost can we afford to forget that the environmental system is an integrated whole, and that our system of technology must be compatible, as a whole, with the environment.

Thus, I believe that we have, as of now, a single decade in which to design the fundamental changes in technology that we must put into effect in the 1980s—if we are to survive. We will need to seize on the decade of the 70s as a period of grace—a decade which must be used for a vast pilot program to guide the coming reconstruction of the nation's system of productivity. This is the urgency of the environmental crisis—we must determine now to develop in the next decade, the new means of our salvation.

How can we possibly accomplish this enormous task? None of us is wise enough to offer a blueprint. Nevertheless, the roles and responsibilities of the various segments of our society can, in a general way, be delineated.

Clearly, an initial responsibility must be undertaken by the community of scientists, engineers, and technologists. New priorities need to be set for basic research and technological development; new types of skills will need to be taught and learned. These needs have already been sensed by our students and by many of their teachers. In the last year there has been a sharp awakening in many university classrooms and laboratories to the urgency of the environmental crisis. At the December 1969 annual meeting of the nation's largest scientific organization, the American Association for the Advancement of Science, nearly every major symposium—about 40 in all—was devoted to some aspect of the environmental crisis. At the 1968 meeting there were perhaps a dozen such sessions; a few years ago they were rare. In 1970 we shall witness the self-mobilization of the vast human resources of the scientific and academic community for the task of learning how to survive the environmental crisis.

We are enormously fortunate that our young people have become particularly sensitive to the threat of environmental catastrophe. Nor is this surprising, for *they* are the first generation in human history to carry strontium-90 in their bones and DDT in their fat; their bodies will record, in time, the effects of the new environmental insults on human health. It is they who face the frightful task of seeking humane knowledge in a world which has, with cunning perversity, transformed the power that knowledge generates into an instrument of catastrophe. And in the coming months I think that our young people will demonstrate that they are, in fact, equal to this task—as their environmental teach-ins and ecological actions begin to mobilize the knowledge of our schools and universities and the civic zeal of our communities for a real attack on the environmental crisis.

# THE TRAGEDY OF THE COMMONS

## Garrett Hardin

*In this essay Garrett Hardin focuses our attention on one of those notions which is so much a part of the general fabric of being human that it is uncritically accepted—the notion of the commons. His main target for attack is our unwillingness to control population growth. However, the commons idea can be recognized in other areas of human thought and behavior as well. In this tightly argued paper the author also applies his thesis in passing to such subjects as property rights, morality, the psychology of guilt and anxiety, and background music in department stores.*

At the end of a thoughtful article on the future of nuclear war, Jerome Wiesner and Herbert York[1] concluded that: "Both sides in the arms race are . . . confronted by the dilemma of steadily increasing military power and steadily decreasing national security. *It is our considered professional judgment that this dilemma has no technical solution.* If the great powers continue to look for solutions in the area of science and technology only, the result will be to worsen the situation."

I would like to focus your attention not on the subject of the article (national security in a nuclear world) but on the *kind* of conclusion they reached, namely that there is no "technical solution" to the problem. An implicit and almost universal assumption of discussions published in professional and semi-popular scientific journals is that the problem under discussion has a "technical

SOURCE: Garrett Hardin, "The Tragedy of the Commons," *Science*, vol. 162, 13 December 1968, pp. 1243–1248. Copyright 1968 by the American Association for the Advancement of Science.

[1] J. B. Wiesner and H. F. York, *Scientific American*, 211 (April 1964): 27.

solution." A technical solution may be defined as one that requires a change only in the techniques of the natural sciences, demanding little or nothing in the way of change in human values or ideas of morality.

In our day (though not in earlier times) technical solutions are always welcome. Because of previous failures in prophecy it takes courage to assert that a desired technical solution is not possible. Wiesner and York exhibited this courage; publishing in a science journal they insisted that the solution to the problem was not to be found in the natural sciences. They cautiously qualified their statement with the phrase, "It is our considered professional judgment. . . ." Whether they were right or not is the concern of the present article. Rather, the concern here is with the important concept of a class of human problems which can be called "no technical solution problems"; and more specifically, with the identification and discussion of one of these.

It is easy to show that the class is not a null class. Recall the game of tick-tack-toe. Consider the problem, "How can I win the game of tick-tack-toe?" It is well known that I cannot, if I assume (in keeping with the conventions of game theory) that my opponent understands the game perfectly. Put another way, there is no "technical solution" to the problem. I can win only by giving a radical meaning to the word "win." I can hit my opponent over the head; or I can drug him; or I can falsify the records. Every way in which I "win" involves, in some sense, an abandonment of the *game,* as we intuitively understand it. (I can also, of course, openly abandon the game—refuse to play it. This is what most adults do.)

The class of "No technical solution problems" has members. It is the thesis of the present article that the "population problem," as conventionally conceived, is a member of this class. How it is conventionally conceived needs some comment. I think it is fair to say that most people who anguish over the population problem are trying to find a way to avoid the evils of overpopulation without relinquishing any of the privileges they now enjoy. They think that farming the seas or developing new strains of wheat will solve the problem —technologically. I shall try to show here that the solution they seek cannot be found. The population problem cannot be solved in a technical way, any more than can the problem of winning the game of tick-tack-toe.

## What Shall We Maximize?

Population, as Malthus said, naturally tends to grow "geometrically"; or as we would say, exponentially. In a finite world this means that the per capita share of the world's goods must steadily decrease. Is ours a finite world?

A fair defense can be put forward for the view that the world is infinite; or that we don't know that it isn't. But, in terms of the practical problems we must face in the next few generations with the foreseeable technology, it is clear that we will greatly increase human misery if we do not, during the immediate future, assume that the world *available to the terrestrial human population* is finite. "Space" is no escape.[2]

A finite world can support only a finite population; therefore, population growth must eventually equal zero. (The case of perpetual wide fluctuations above and below zero is a trivial variant that need not be discussed.) When this condition is met, what will be the situation of mankind? Specifically, can Bentham's goal of "the greatest good for the greatest number" be realized?

No—for two reasons, each sufficient by itself. The first is a theoretical one. It is not mathematically possible to maximize for two (or more) variables at the same time. This was clearly stated by von Neumann and Morgenstern,[3] but the principle is implicit in the theory of partial differential equations, dating back at least to D'Alembert (1717–1783).

The second reason springs directly from biological facts. To live, any organism must have a source of energy (e.g., food). This energy is utilized for two purposes: mere maintenance, and work. For man, maintenance of life requires about 1600 kilocalories a day ("maintenance calories"). Anything that he does over and above merely staying alive will be defined as work, and is supported by "work calories," which he takes in. Work calories are used not only for what we call work in common speech; they are also required for all forms of enjoyment, from gormandizing and automobile racing to playing music and writing poetry. If our goal is to maximize population it is obvious what we must do: we must make the work calories per person approach as close to zero as possible. No gourmet meals, no vacations, no sports, no music, no literature, no art. . . . I think everyone will grant, without argument or proof, that maximizing population does not maximize goods. Bentham's goal is impossible.

In reaching this conclusion I have made the usual assumption that it is the acquisition of energy that is the problem. The appearance of atomic energy has led some to question this assumption. Given an infinite source of energy, however, population growth still produces an inescapable problem. The problem of the acquisition of energy is replaced by the problem of its dissipation, as J. H. Fremlin has so wittily shown.[4] The arithmetic signs in the analysis are, as it were, reversed; but Bentham's goal is still unattainable.

The optimum population is, then, less than a maximum. The difficulty of defining the optimum is enormous; so far as I know no one has seriously tackled this problem. Reaching an acceptable and stable solution will surely require more than one generation of hard analytical work; and much persuasion.

[2] G. Hardin, *J. Heredity*, 50 (1959): 68; S. von Hoernor, *Science*, 137 (1962): 18.

[3] J. von Neumann and O. Morgenstern, *Theory of Games and Economic Behavior*, Princeton: Princeton Univ. Press, 1947, p. 11.

[4] J. H. Fremlin, *New Scientist*, no. 415 (1964): 285.

We want the maximum good per person; but what is "good"? To one person it is wilderness, to another it is ski lodges for thousands. To one it is estuaries to nourish ducks for hunters to shoot at; to another it is factory land. Comparing one good with another is, we usually say, impossible because goods are incommensurable. Incommensurables cannot be compared.

Theoretically this may be true; but in real life *incommensurables are commensurable*. All that is needed is a criterion of judgment and a system of weighting. In nature the criterion is survival. Is it better for a species to be small and hideable, or large and powerful? Natural selection commensurates the incommensurables. The compromise achieved depends on a natural weighting of the values of the variables.

Man must imitate this process. There is no doubt that in fact he already does, but unconsciously. It is when the hidden decisions are made explicit that the arguments begin. The problem for the years ahead is to work out an acceptable theory of weighting. Synergistic effects, nonlinear variation, and difficulties in discounting the future make the intellectual problem difficult, but not (in principle) insoluble.

Has any cultural group solved this practical problem at the present time, even on an intuitive level? One simple fact proves that none has: there is no prosperous population in the world today that has, and has had for some time, a growth rate of zero. Any people that has intuitively identified its optimum point will soon reach it, after which its growth rate becomes and remains zero.

Of course, a positive growth rate might be taken as evidence that a population is below its optimum. It is widely recognized, however that, by any reasonable standards, the most rapidly growing populations on earth today are (in general) the most miserable. This association (which need not be invariable) casts doubt on the optimistic assumption that the positive growth rate of a population is evidence that it has yet to reach its optimum.

We can make little progress in working toward optimum population size until we explicitly exorcize the spirit of Adam Smith in the field of practical demography. In economic affairs, *The Wealth of Nations* (1776) popularized the "invisible hand," the idea that an individual who "intends only his own gain," is, as it were, "led by an invisible hand to promote . . . the public interest."[5] Adam Smith did not assert that this was invariably true, and perhaps neither did any of his followers. But he contributed to a dominant tendency of thought that has ever since interfered with positive action based on rational analysis, namely the tendency to assume that decisions reached individually will, in fact, be the best decisions for an entire society. If this assumption is correct it justifies the continuance of our present policy of *laissez-faire* in reproduction. If it is correct we can assume that men will control their

[5] Adam Smith, *The Wealth of Nations*, New York: Modern Library, 1937, p. 423.

individual fecundity so as to produce the optimal population. If the assumption is not correct, we need to reexamine our individual freedoms to see which ones are defensible.

## Tragedy of Freedom in a Commons

The rebuttal to the "invisible hand" in population control is to be found in a "scenario" first sketched in a little known pamphlet[6] in 1833 by a mathematical amateur named William Forster Lloyd (1794–1852). We may well call it "The Tragedy of the Commons," using the word "tragedy" as the philosopher Whitehead used it:[7] "The essence of dramatic tragedy is not unhappiness. It resides in the solemnity of the remorseless working of things." He then goes on to say: "This inevitableness of destiny can only be illustrated in terms of human life by incidents which in fact involve unhappiness. For it is only by them that the futility of escape can be made evident in the drama."

The tragedy of the commons develops in this way. Picture a pasture open to all. It is to be expected that each herdsman will try to keep as many cattle as possible on the commons. Such an arrangement may work reasonably satisfactorily for centuries because tribal wars, poaching, and disease keep the numbers of both man and beast well below the "carrying capacity" of the land. Finally, however, comes the day of reckoning, i.e., the day when the long-desired social stability becomes a reality. At this point, the inherent logic of the commons remorselessly generates tragedy.

As a rational being each herdsman seeks to maximize his gain. Explicitly or implicitly, more or less consciously, he asks: "What is the utility *to me* of adding one more animal to my herd?" This utility has two components:

1. A positive component, which is a function of the increment of one animal. Since the herdsman receives all the proceeds from the sale of the additional animal, the positive utility is nearly $+1$.

2. A negative component, which is a function of the additional overgrazing created by one more animal. But since the effects of overgrazing are shared by all the herdsmen, the negative utility for any particular decision-making herdsman is only a fraction of $-1$.

Adding together the component partial utilities, the rational herdsman concludes that the only sensible course for him to pursue is to add another animal to his herd. And another; and another . . . But this is the conclusion reached by each and every rational herdsman sharing a commons. Therein is the tragedy. Each man is *locked in* to a system that compels him to increase his herd without limit—in a world that is limited. Ruin is the destination toward which all men rush, each pursuing his own best interest in a society that

[6] W. F. Lloyd, *Two Lectures on the Checks to Population*, Oxford, 1833.
[7] A. N. Whitehead, *Science and the Modern World*, New York: Mentor, 1948, p. 17.

believes in the freedom of the commons. *Freedom in a commons brings ruin to all.*

Some would say that this is platitudinous, that is, a truth known to all. Would that it were! In a sense it was learned thousands of years ago, but natural selection favors the forces of psychological denial. The individual benefits *as an individual* from his ability to deny the truth even though society as a whole, of which he is a part, suffers. Education can counteract the natural tendency to do the wrong thing, but the inexorable succession of generations requires that the basis for this knowledge be constantly refreshed.

A simple incident that occurred a few years ago in Leominster, Massachusetts, shows how perishable the knowledge is. During the Christmas shopping season the parking meters downtown were covered with plastic bags that bore tags reading: "Do not open until after Christmas. Free parking courtesy of the mayor and city council." In other words, facing the prospect of an increased demand for already scarce space, the city fathers reinstituted the system of the commons. (Cynically, we suspect that they gained more votes than they lost by this retrogressive act.)

In an approximate way, the logic of the commons has been understood for a long time, perhaps since the discovery of agriculture or the invention of private property in real estate. But it is understood mostly only in special cases, which are not sufficiently generalized. Even at this late date, cattlemen leasing national land on the western ranges demonstrate no more than an ambivalent understanding, constantly pressuring federal authorities to increase the head-count to the point where overgrazing produces erosion and weed-dominance. Similarly, the oceans of the world continue to suffer from the survival of the philosophy of the commons. Maritime nations still respond automatically to the shibboleth of the "freedom of the seas." Professing to believe in the "inexhaustible resources of the oceans," they bring species after species of fish and whales closer to extinction.[8]

The National Parks present another instance of the working out of the tragedy of the commons. At present, they are open to all, without limit. The Parks themselves are limited in extent—there is only one Yosemite Valley —while population seems to grow without limit. The values that visitors seek in the Parks are steadily eroded. Plainly, we must soon cease to treat the Parks as commons or they will be of no value to anyone.

What shall we do? We have several options. We might sell them off as private property. We might keep them as public property, but allocate the right to enter them. The allocation might be on the basis of wealth, using an auction system. It might be on the basis of merit, as defined by some agreed-upon standards. It might be by lottery. Or it might be on a first-come, first-served basis, administered to long queues. These, I think, are all the

---

[8] S. McVay, *Scientific American,* 216 (August 1966): 13.

reasonable possibilities. They are all objectionable. *But we must choose*—or acquiesce in the destruction of the commons that we call our National Parks.

## Pollution

In a reverse way, the tragedy of the commons reappears in problems of pollution. Here it is not a question of taking something out of the commons, but of putting something in—sewage, or chemical, radioactive and heat wastes into water; noxious and dangerous fumes into the air; and distracting and unpleasant advertizing signs into the line of sight. The utility calculations are much the same as before. The rational man finds that his share of the cost of the wastes he discharges into the commons is less than the cost of purifying his wastes before releasing them. Since this is true for everyone, we are locked into a system of "fouling our own nest," so long as we behave only as independent, rational, free-enterprizers.

The tragedy of the commons as a foodbasket is averted by private property, or something formally like it. But the air and waters surrounding us cannot readily be fenced, and so the tragedy of the commons as a cesspool must be prevented by different means, by coercive laws or taxing devices that make it cheaper for the polluter to treat his pollutants than to discharge them untreated. We have not progressed as far with the solution of this problem as we have with the first. Indeed, our particular concept of private property, which deters us from exhausting the positive resources of the earth, favors pollution. The owner of a factory on the bank of a stream—whose "property" extends to the middle of the stream—often has difficulty seeing why it is not his natural "right" to muddy the waters flowing past his door. The law, always behind the times, requires elaborate stitching and fitting to mold it to this newly perceived aspect of the commons.

The pollution problem is a consequence of population. It did not much matter how a lonely American frontiersman disposed of his waste. "Flowing water purifies itself every ten miles," my grandfather used to say, and the myth was near enough to the truth when he was a boy, for there weren't too many people. But as population became denser the natural chemical and biological recycling processes became overloaded, calling for a redefinition of property rights.

## How Legislate Temperance?

Analysis of the pollution problem as a function of population density uncovers a not generally recognized principle of morality, namely: *the morality of an act is a function of the state of the system at the time it is performed.*[9]

---

[9] J. Fletcher, *Situation Ethics*, Philadelphia: Westminster Press, 1966.

Using the commons as a cesspool does not harm the general public under frontier conditions, because there is no public; the same behavior in a metropolis is unbearable. A hundred and fifty years ago a plainsman could kill an American bison, cut out only the tongue for his dinner, and discard the rest of the animal. He was not in any important sense being wasteful. Today, with only a few thousand bison left, we would be appalled at such behavior.

In passing, it is worth noting that the morality of an act cannot be determined from a photograph. One does not know whether a man killing an elephant or setting fire to the grassland is harming others until one knows the total system in which his act appears. "One picture is worth a thousand words," said an ancient Chinese; but it may take 10,000 words to validate it. It is as tempting to ecologists as it is to reformers in general to try to persuade others via the photographic shortcut. But the guts of an argument can't be photographed: they must be presented rationally—in words.

That morality is system-sensitive escaped the attention of most codifiers of ethics in the past. "Thou shalt not . . ." is the form of traditional ethical directives, which make no allowance for particular circumstances. (Christ did; and his continued existence was unbearable to the Establishment.) The laws of our society follow the pattern of ancient ethics, and therefore are poorly suited to governing a complex, crowded, changeable world. Our epicyclic solution is to augment statutory law with administrative law. Since it is practically impossible to spell out all the conditions under which it is safe to burn trash in the back yard or run an automobile without smog-control, by law we delegate the details to bureaus. The result is administrative law, which is rightly feared for an ancient reason: *Quis custodiet ipsos custodes?*—"Who shall watch the watchers themselves?" John Adams said we must have "a government of laws and not men." Bureau administrators, trying to evaluate the morality of acts in the total system, are singularly liable to corruption, producing a government by men, not laws.

Prohibition is easy to legislate (though not necessarily to enforce!); but how do we legislate temperance? Experience indicates that it can be accomplished best through the mediation of administrative law. We limit possibilities unnecessarily if we suppose that the sentiment of *Quis custodiet* denies us the use of administrative law. We should rather retain the phrase as a perpetual reminder of fearful dangers we cannot avoid. The great challenge facing us now is to invent the corrective feedbacks that are needed to keep custodians honest. We must find ways to legitimate the needed authority of both the custodians and the corrective feedbacks.

## Freedom to Breed Is Intolerable

The tragedy of the commons is involved in population problems in another way. In a world governed solely by the principle of "dog eat dog"

—if indeed there ever was such a world—it would not be a matter of public concern how many children a family had. Parents who bred too exuberantly would leave fewer descendants, not more, because they would be unable to care adequately for their children. David Lack and others have found that such a negative feedback demonstrably controls the fecundity of birds.[10] But men are not birds, and have not acted like them for millennia, at least.

If each human family were dependent only on its own resources; if the children of improvident parents starved to death; if, thus, overbreeding brought its own "punishment" to the germ line—then there would be no public interest in controlling the breeding of families. But our society is deeply committed to the welfare state,[11] and hence confronted with another aspect of the tragedy of the commons.

In a welfare state, how shall we deal with the family, the religion, the race, or the class (or indeed any distinguishable and cohesive group) that adopts overbreeding as a policy to secure its own aggrandizement?[12] To couple the concept of freedom to breed with the belief that everyone born has an equal right to the commons is to lock the world into a tragic course of action.

Unfortunately this is just the course of action that is being pursued by the United Nations. In late 1967 some thirty nations agreed to the following:

> The Universal Declaration of Human Rights describes the family as the natural and fundamental unit of society. It follows that any choice and decision with regard to the size of the family must irrevocably rest with the family itself, and cannot be made by anyone else.[13]

It is painful to have to deny categorically the validity of this "right"; denying it, one feels as uncomfortable as a resident of Salem who denied the reality of witches in the seventeenth century. At the present time, in liberal quarters, something like a taboo acts to inhibit criticism of the United Nations. There is a feeling that the U.N. is "our last and best hope," that we shouldn't find fault with it; we shouldn't play into the hands of the arch-conservatives. Let us not forget, however, what Robert Louis Stevenson said: "The truth that is suppressed by friends is the readiest weapon of the enemy." If we love the truth we must openly deny the validity of the Universal Declaration of Human Rights, even though it is promoted by the United Nations. We should also join with Kingsley Davis[14] in attempting to get Planned Parenthood to see the error of its ways in embracing the same tragic ideal.

[10] D. Lack, *The Natural Regulation of Animal Numbers*, Oxford: Clarendon Press, 1954.

[11] H. Girvetz, *From Wealth to Welfare*, Stanford: Stanford University Press, 1950.

[12] G. Hardin, *Persp. Biol. Med.*, 6 (1963): 366.

[13] U Thant, *Intern. Planned Parenthood News*, no. 168, February 1968, p. 3.

[14] K. Davis, *Science*, 158 (1967) 730.

## Conscience Is Self-eliminating

It is a mistake to think that we can control the breeding of mankind *in the long run* by an appeal to conscience. Charles Galton Darwin made this point when he spoke on the centennial of the publication of his grandfather's great book. The argument is straightforward and Darwinian.

People vary. Confronted with appeals to limit breeding, some people will undoubtedly respond to the plea more than others. Those who have more children will produce a larger fraction of the next generation than those with more susceptible consciences. The difference will be accentuated, generation by generation. In C. G. Darwin's words:

> It may well be that it would take hundreds of generations for the progenitive instinct to develop in this way, but if it should do so, nature would have taken her revenge, and the variety *Homo contracipiens* would become extinct and would be replaced by *Homo progenitivus*.[15]

The argument assumes that conscience or the desire for children (no matter which) is hereditary—but hereditary *only in the most general formal sense*. The result will be the same whether the attitude is transmitted via germ cells, or exosomatically, to use A. J. Lotka's term. (If one denies the latter possibility as well as the former, then what's the point of education?) The argument has here been stated in the context of the population problem, but it applies equally well to any instance in which society appeals to an individual exploiting a commons to restrain himself for the general good—by means of his conscience. To make such an appeal is to set up a selective system that works toward the elimination of conscience from the race.

## Pathogenic Effects of Conscience

The long-term disadvantage of an appeal to conscience should be enough to condemn it; but it has serious short-term disadvantages as well.

If we ask a man who is exploiting a commons to desist "in the name of conscience," what are we saying to him? What does he hear—not only at the moment but also in the wee small hours of the night when, half asleep, he remembers not merely the words we used but also the nonverbal communication cues we gave him unawares? Sooner or later, consciously or subconsciously he senses that he has received two communications, and that they are contradictory:

1. (Intended communication) "If you don't do as we ask, we will openly condemn you for not acting like a responsible citizen."

---

[15] S. Tax (ed.), *Evolution After Darwin*, Chicago: Univ. of Chicago Press, 1960, vol. 2, p. 469.

2. (The unintended communication) "If you *do* behave as we ask, we will secretly condemn you for a schlemiel, a sucker, a sap, who can be shamed into standing aside while the rest of us exploit the commons."

In a word, he is damned if he does and damned if he doesn't. He is caught in what Gregory Bateson has called a "double bind." Bateson and his co-workers have made a plausible case for viewing the double bind as an important causative factor in the genesis of schizophrenia.[16] The double bind may not always be so damaging, but it always endangers the mental health of anyone to whom it is applied. "A bad conscience," said Nietzsche, "is a kind of illness."

To conjure up a conscience in others is tempting to anyone who wishes to extend his control beyond the legal limits. Leaders at the highest level succumb to this temptation. Has any President during the past generation failed to call upon labor unions to moderate "voluntarily" their demands for higher wages, or to steel companies to honor "voluntary" guide-lines on prices? I can recall none. The rhetoric used on such occasions is designed to produce feelings of guilt in noncooperators.

For centuries it was assumed without proof that guilt was a valuable, perhaps even an indispensable, ingredient of the civilized life. Now, in this post-Freudian world, we doubt it. Paul Goodman speaks from the modern point of view when he says:

> No good has ever come from feeling guilty, neither intelligence, policy, nor compassion. The guilty do not pay attention to the object but only to themselves, and not even to their own interests, which might make sense, but to their anxieties.[17]

One does not have to be a professional psychiatrist to see the consequences of anxiety. We in the western world are just emerging from a dreadful two-centuries-long Dark Ages of Eros, which was sustained partly by prohibition laws, but perhaps more effectively by the anxiety-generating mechanisms of education. Alex Comfort has told the story well in *The Anxiety Makers*;[18] it is not a pretty one.

Since proof is difficult, we may even concede that the results of anxiety may sometimes, from certain points of view, be desirable. The larger question we should ask is whether, as a matter of policy, we should ever encourage the use of a technique the tendency of which (if not the intention) is psychologically pathogenic. We hear much talk these days of "responsible parenthood"; the coupled words are incorporated into the titles of some

---

[16] G. Bateson, D. D. Jackson, J. Haley, and J. Weakland, *Behav. Sci.*, 1 (1956): 251.

[17] P. Goodman, *New York Review of Books*, 10 (23 May 1968): 22.

[18] A. Comfort, *The Anxiety Makers*, London: 1967.

organizations devoted to birth control. Some people have proposed massive propaganda campaigns to instill responsibility into the nation's (or the world's) breeders. But what is the meaning of the word "responsibility" in this context? Is it not merely a synonym for the word "conscience"? When we use the word "responsibility" in the absence of substantial sanctions are we not trying to browbeat a free man in a commons into acting against his own interest? "Responsibility" is a verbal counterfeit for a substantial *quid pro quo*. It is an attempt to get something for nothing.

If the word *responsibility* is to be used at all, I suggest that it be in the sense Charles Frankel uses it.[19] "Responsibility," says this philosopher, "*is the product of definite social arrangements.*" Notice that Frankel calls for social arrangements—not propaganda.

## Mutual Coercion, Mutually Agreed Upon

The social arrangements that produce responsibility are arrangements that create coercion, of some sort. Consider bank-robbing. The man who takes money from a bank acts as if the bank were a commons. How do we prevent such action? Certainly not by trying to control his behavior solely by a verbal appeal to his sense of responsibility. Rather than rely on propaganda we follow Frankel's lead and insist that a bank is not a commons; we seek the definite social arrangements that will keep it from becoming a commons. That we thereby infringe on the freedom of would-be robbers we neither deny nor regret.

The morality of bank-robbing is particularly easy to understand because we accept complete prohibition of this activity. We are willing to say "Thou shalt not rob banks," without providing for exceptions. But temperance also can be created by coercion. Taxing is a good coercive device. To keep downtown shoppers temperate in their use of parking space we introduce parking meters for short periods, and traffic fines for longer ones. We need not actually forbid a citizen to park as long as he wants to; we need merely make it increasingly expensive for him to do so. Not prohibition, but carefully biased options are what we offer him. A Madison Avenue man might call this *persuasion;* I prefer the greater candor of the word *coercion.*

Coercion is a dirty word to most liberals now, but it need not forever be so. As with the four-letter words, its dirtiness can be cleansed away by exposure to the light, by saying it over and over without apology or embarrassment. To many, the word *coercion* implies arbitrary decisions of distant and irresponsible bureaucrats; but this is not a necessary part of its meaning. The only kind of coercion I recommend is mutual coercion, mutually agreed upon by the majority of the people affected.

[19] C. Frankel, *The Case for Modern Man*, New York: Harper, 1955, p. 203.

To say that we mutually agree to coercion is not to say that we are required to enjoy it, or even to pretend we enjoy it. Who enjoys taxes? We all grumble about them. But we accept compulsory taxes because we recognize that voluntary taxes would favor the conscienceless. We institute and (grumblingly) support taxes and other coercive devices to escape the horror of the commons.

An alternative to the commons need not be perfectly just to be preferable. With real estate and other material goods, the alternative we have chosen is the institution of private property coupled with legal inheritance. Is this system perfectly *just*? As a genetically trained biologist I deny that it is. It seems to me that, if there are to be differences in individual inheritance, legal possession should be perfectly correlated with biological inheritance—that those who are biologically more fit to be the custodians of property and power should legally inherit more. But genetic recombination continually makes a mockery of the doctrine of "like father, like son" implicit in our laws of legal inheritance. An idiot can inherit millions, and a trust fund can keep his estate intact. We must admit that our legal system of private property plus inheritance *is* unjust—but we put up with it because we are not convinced, at the moment, that anyone has invented a better system. The alternative of the commons is too horrifying to contemplate. Injustice is preferable to total ruin.

One of the peculiarities of the warfare between reform and the status quo is that it is thoughtlessly governed by a double standard. Whenever a reform measure is proposed it is often defeated when its opponents triumphantly discover a flaw in it. As Kingsley Davis has pointed out,[20] worshippers of the status quo sometimes imply that no reform is possible without unanimous agreement, an implication contrary to historical fact. As nearly as I can make out, automatic rejection of proposed reforms is based on one of two unconscious assumptions:

1. That the status quo is perfect; or
2. That the choice we face is between reform and no action; if the proposed reform is imperfect, we presumably should take no action at all, while we wait for a perfect proposal.

*But we can never do nothing.* That which we have done for thousands of years is also action. It also produces evils. Once we are aware that the status quo *is* action, we can then compare its discoverable advantages and disadvantages with the predicted advantages and disadvantages of the proposed reform, discounting as best we can for our lack of experience. On the basis of such a comparison, we can make a rational decision, which will not involve the unworkable assumption that only perfect systems are tolerable.

[20] J. D. Roslansky, *Genetics and the Future of Man*, New York: Appleton-Century-Crofts, 1966, p. 177.

## The Recognition of Necessity

Perhaps the simplest summary of this analysis of man's population problems is this: the commons, if justifiable at all, is justifiable only under conditions of low population density. As the human population has increased, the commons has had to be abandoned in one aspect after another.

First we abandoned the commons in food gathering, enclosing farm land and restricting pastures and hunting and fishing areas. These restrictions are still not complete throughout the world.

Somewhat later we saw that the commons as a place for waste disposal would also have to be abandoned. Restrictions on the disposal of domestic sewage are widely accepted in the western world; we are still struggling to close the commons to pollution by automobiles, factories, insecticide sprayers, fertilizing operations, and atomic energy installations.

In a still more embryonic state is our recognition of the evils of the commons in matters of pleasure. There is almost no restriction on the propagation of sound waves in the public medium. The shopping public is assaulted with "mindless music," without its consent. Our government is paying out billions of dollars to create the SST plane, which will disturb fifty thousand people for every one person who is whisked from coast to coast three hours faster. Advertizers muddy the airwaves of radio and TV and pollute the view of travellers. We are a long way from outlawing the commons in matters of pleasure. Is this because our puritan inheritance makes us view pleasure as something of a sin, and pain (i.e., the pollution of advertizing) as the sign of virtue?

Every new enclosure of the commons involves the infringement of somebody's personal liberty. Infringements made in the distant past are accepted because no contemporary complains of a loss. It is the newly proposed infringements that we vigorously oppose: cries of "rights" and "freedom" fill the air. But what does "freedom" mean? When men mutually agreed to pass laws against robbing, mankind became more free, not less so. Individuals locked into the logic of the commons are free only to bring on universal ruin; once they see the necessity of mutual coercion, they become free to pursue other goals. I believe it was Hegel who said, *"Freedom is the recognition of necessity."*

The most important aspect of necessity that we must now recognize, is the necessity of abandoning the commons in breeding. No technical solution can rescue us from the misery of overpopulation. Freedom to breed will bring ruin to all. At the moment, to avoid hard decisions many of us are tempted to propagandize for conscience and "responsible parenthood." The temptation must be resisted, because an appeal to independently acting consciences selects for the disappearance of all conscience in the long run, and an increase in anxiety in the short.

The only way we can preserve and nurture other and more precious freedoms is by relinquishing the freedom to breed, and that very soon. "Freedom is the recognition of necessity"—and it is the role of education to reveal to all the necessity of abandoning the freedom to breed. Only so, can we put an end to this aspect of the tragedy of the commons.

# . . . O AND ALL THE LITTLE BABIES IN THE ALAMEDA GARDENS YES . . .

## Stephanie Mills

*Whereas Garrett Hardin's mind works best with logic, precision, and scholarship, Stephanie Mills prefers to communicate at the more dramatic level of the emotions. A crusader for the Zero Population Growth movement, her voice is filled with passion and tenderness for all living creatures. In the following selection a wide reading in literature complements her intuitive understanding of the life-death cycle, and the result borders on religious joy.*

**I.**

I sink I'd die down over his feet, humbly dumbly, only to washup. Yes, tid. There's where. First. We pass through grass behush the bush to. Whish! A gull. Gulls. Far calls. Coming, far! End here. Us. Then. Finn, again! Take. Bussoftlee, mememormee! Till thousends thee. Lps. The keys to. Given! A way a lone a last a loved a long the

—James Joyce/*Finnegan's Wake*

Rivers flow to the sea, clouds rise from the sea, rain falls to the earth and trickles into rivers. All of life moves, if uninterrupted, in cycles. This was Joyce's vision, and must become ours. Life ever-same and ever-changing once was ours. Or ours to lose? Man interrupts cycles, man changes life with no return to same. Man resists death, and by so doing destroys life.

SOURCE: Stephanie Mills, ". . . O and All the Little Babies in the Alameda Gardens Yes . . . ," *Ecotactics*. Copyright © 1970, by the Sierra Club. Reprinted by permission of Pocket Books/division of Simon & Schuster, Inc.

This is where we are now, in this decade. Faced with a final chance to acknowledge the cycle of life and death and flow with it. Faced with the chance to rejoin nature. But faced with the coming of the worst of all possible worlds as well.

The spectacle of the starving child, the sewer/river, the faceless state confronts us now, glaring at us in our corner. Why are we here now? Why must we, of all generations, live in a make-it-or-break-it era? Our world is crowded and poor. Our neighbors are hungry while we waste our food. Our brother animals are dying, and we poisoned them. Ironically, our poisons seem to affect most brutally those creatures who fly, who soar and laugh at us because we cannot fly without clumsy gadgets. We earth-bound humans are only digits now, crowded into anonymity, our lives and individuality diluted by the presence of rapidly doubling billions like us.

Think of how many a billion is for a moment. Can you? Have you ever *seen* a billion things? Can you imagine a billion human beings? Three, or seven billion human beings? The neighborhood will be getting crowded in 2000.

We find ourselves overpopulated because we attempted to thwart death. Western man, through technology, has lengthened his life span by eliminating many diseases. Since the beginning of this century, missionaries of public health have brought to most of the underdeveloped nations of the world the techniques of achieving longer life. The application of these techniques, by thwarting death temporarily, upset the balance between birth rates and death rates. The growth rates of populations skyrocketed: more human beings lived long enough to produce more children, who had a better chance to survive. Etcetera. Ad infinitum.

## II.

What does it matter? What I hate is death and disease, as you well know. And whether you wish it or not, we're allies, facing them and fighting them together.

—Albert Camus/*The Plague*

These words are uttered by a doctor at the height of an outbreak of bubonic plague. Dr. Rieux has long since recognized that the plague will run its course, in spite of his efforts to combat the disease. Rieux, nonetheless, is dignified by his resistance to the inevitable.

Nobody can deny that the elimination of disease is desirable. There is a case to be made, however, against upsetting a natural balance. For balance, isn't it conceivable that birth control *could* have been employed simultaneously with death control? This is the advantage of hindsight. Yet vigorous population control is still not considered a complementary health measure—and we sink deeper into the morass of too many every day.

Birth control is regarded as tampering with nature. Death control is not. Those people who denounce contraception as interference with a Supreme Will do not, by the same token, denounce typhoid shots. For birth control is regarded as political, not medical. Swallowing Enovoid is certainly more emotionally charged than popping tetracycline. But both actions involve tinkering with the natural order of things.

Acceptance of birth control is absolutely necessary for a humane solution to the population crisis. And it *is* a technical solution. The alternative "natural" solution is to eliminate death control.

There is this to be said of population at this time: the birth rates must go down or the death rates will soon go up. By the Eighties, widespread and cataclysmic death may be caused by hunger, plague, war, or environmental disaster.

Death is finiteness, and Western man rejects finiteness. We long for a limitless supply of everything: air, water, food, wilderness, time, and frontier. But our infinity is linear. We head in a straight, unswerving line for the cosmos, damning any obstacle—even scientific fact—that stands in the way. Sadly enough, as we strive for infinity, we create the irreversible limits. To acquire more electricity and more water, we dam and destroy Glen Canyon for all time. To acquire more food, we deprive the pelican of his—and destroy it, too. For all time. The roster of deaths we have caused in our rush for life is almost endless. And as more of us come into being, *more* death.

More people, we believe, means more power, more consumers, more GNP. Score three. But more people are less individual. More people are less free. A more populous nation becomes necessarily more authoritarian. And beneath the veneer roils a cesspool of chaos.

III.

> Listen! If all must suffer to pay for the eternal harmony, what have children to do with it, tell me please? It's beyond all comprehension why they should suffer, and why they should pay for the harmony.
>
> —Feodor Dostoevski/*The Brothers Karamazov*

Children are the first to die of hunger, children are rebirth, and children are what the concern with population is all about. Children of all ages are the solution.

Population growth intensifies the spate of problems which confront the earth today. Ignoring the population crisis precludes solving the problems of war, hunger, disease, and alienation. Numbers *per se* cause none of these, but mega-behavior and mega-societies will be fraught with all of them. Curiously enough, a recognition of the population problem and its solution—population control—is avoided by all those who have the power to effect the change.

It took a child to perceive the Emperor's nudity, since all the *loyal* subjects had their perceptions filtered politically.

Our emperors would parade their new clothes of environment, simply by announcing that the problems of environment and population will be solved. Thus far, the clothes are invisible, for meaningful solutions to the population-environment crisis must be drastic indeed. They require a revolution in consciousness.

Solving the population problem will require a reorientation of child-bearing attitudes. To encourage such a reorientation, alternative satisfactions must be provided. New modalities of family living such as communes and kibbutzes might be tried. Certainly women's roles must be expanded to encompass much, much more than the production of children. A cultural inversion must take place. The "old maid," not the mother of twelve, must be made the heroine. The childless couple should be applauded, not pitied. And the adopted baby should become the "real" baby (as he always has been).

A danger inherent in the population problem is that the state may finally assume control of reproduction if the individual doesn't. Consider a state so powerful that it controls the reproduction of its citizens. Consider, also, a nation so overpopulated that it can't survive unless drastic steps are taken to alleviate the whole complex of environmental problems.

The opportunity to assume individual responsibility is still ours, but not, perhaps, for long. If individuals abdicate this responsibility, if individuals refuse to act in their enlightened self-interest, then the state will surely take a hand in individual affairs, sooner than later.

Perhaps there is an element of self-fulfilling prophecy in such a warning. Perhaps such fears should not be voiced. Perhaps. Yet we must realize our role in the continuum of evolution. And such realization may lead us to some interesting questions. Is it our turn to become extinct as a species? Is there any point in undertaking a serious attempt to survive? Is there any wisdom in challenging what would seem the inevitable?

These questions are unanswerable; yet the image of Dr. Rieux, fighting impersonal annihilation, is inspiring. Resisting death is humane. But the death we must resist is more than the death of mankind. It is the death of the ecosystem.

## IV.

In gloomy times of bloody confusion
Ordered disorder
Planful wilfulness
Dehumanized humanity
When there is no end to the unrest in our cities:
In such a world, a world like a slaughterhouse—
Summoned by rumors of threatening deeds of violence

To prevent the brute strength of the shortsighted people
From shattering its own tools and
Trampling its own bread-basket to pieces—
We wish to reintroduce
God.

—Bertold Brecht/Saint Joan of the Stockyards

Not a bad idea. It all depends, however, on which god is introduced. Will it be the anthropocentric god of Genesis, or a dryad? Or can this god/godliness be so personal, so innate, that it has no name?

It's not enough to survive, hard as that alone may be. It may not be worth it to survive in a world devoid of humane beings, a world in which man's only aspiration is for biological existence. Quality of life is the concern, and life has no quality without some experience of god. The experience may not even be describable. Can you describe the wisdom of the ecosystem, the flash of awareness that comes when you perceive how the planet functions? Every organism relates to every other. God is an inadequate word.

Man's aspirations so far have been guided by the god of Genesis. For the most part, we have been proud of our subjugation of the planet. Now we are finding that our aspirations have been misguided and destructive. Where can humanity direct its aspirations, now that we see the futility of damming, grading, eroding, overbreeding?

To aspire to survival and to aspire to humanity are the paths. They are one and the same. For openers, we can turn to the humanity within us, and must to survive. All the logic, precision, and practicality in the world can't save us if we lose our own souls.

The prescription is nothing less than a revolution in consciousness. We are beginning to see it now, and must participate. It takes more than lock-jawed resolution to save a world for all creatures. It takes love and joy. There can be no survival without passion. Passion for humanity, love of the earth, joy of existence, and hope for the future. A very wise man has said that "Pessimism has no survival value." Nor hate, nor elitism, nor puritanism.

> . . . O and the sea the sea crimson sometimes like fire and the glorious sunsets and the fig-trees in the Alameda gardens Yes and all the queer little streets and pink and blue and yellow houses and the rosegardens and the jessamine and geraniums and cactuses and Gibraltar as a girl where I was a Flower of the mountain yes when I put the rose in my hair like the Andalusian girls used or shall I wear red yes and how he kissed me under the Moorish wall and I thought well as well him as another and then I asked him with my eyes to ask again yes and then he asked me would I yes to say yes my mountain flower and first I put my arms around him yes and drew him down to me so he could feel my breasts all perfume yes and his heart was going like mad and yes I said yes I will Yes.

—Joyce, again/Ulysses

# ON TOTALITARIANISM AND MODERN ARCHITECTURE

## Norman Mailer

*Norman Mailer is certainly one of the most energetic and colorful writers in America today. He began as a novelist and short story writer and recently has revived the art of the personal essay. His political writings during the Kennedy and Johnson administrations have had an immense influence on the voting public. The selection that follows is taken from a collection entitled* The Idol and the Octopus *(1968). It represents Mailer's imaginative talents at their forceful best.*

Totalitarianism has haunted the twentieth century, haunted the efforts of intellectuals to define it, of politicians to withstand it, and of rebels to find a field of war where it could be given battle. Amoeboid, insidious, totalitarianism came close to conquering the world twenty years ago. In that first large upheaval the Nazis sang of blood and the deep roots of blood and then proceeded to destroy the essential intuition of the primitive, the umbilical idea that death and the appropriate totems of burial are as essential to life as life itself.

That first wave of totalitarianism was a tide which moved in two directions at once. It broke upon the incompatible military force of Russia and of America. But it was an ocean of plague. It contaminated whatever it touched. If Russia had been racing into totalitarianism before the war, it was pervasively totalitarian after the war, in the last half-mad years of Stalin's court. And America was altered from a nation of venture, exploitation, bigotry, initiative, strife, social justice and social injustice, into a vast central swamp of tasteless toneless authority whose dependable heroes were drawn from FBI men, doctors, television entertainers, corporation executives, and athletes who could cooperate with public-relations men.

SOURCE: Norman Mailer, "On Totalitarianism and Modern Architecture," *The Idol and the Octopus* (New York: Dell Publishing Co., 1968). Reprinted by permission of the author and the author's agent, Scott Meredith Literary Agency, Inc., 580 Fifth Avenue, New York, N. Y. 10036.

One must recognize the features of the plague. If it appeared first in Nazi Germany as a political juggernaut, and in the Soviet Union as a psychosis in ideology, totalitarianism has slipped into America with no specific political face. There are liberals who are totalitarian, and conservatives, radicals, rightists, fanatics, hordes of the well-adjusted. Totalitarianism has come to America with no concentration camps and no need for them, no political parties and no desire for new parties, no, totalitarianism has slipped into the body cells and psyche of each of us. It has been transported, modified, codified, and inserted into each of us by way of the popular arts, the social crafts, the political crafts, and the corporate techniques. It sits in the image of the commercials on television which use phallic and vaginal symbols to sell products which are otherwise useless for sex, it is heard in the jargon of educators, in the synthetic continuums of prose with which public-relations men learn to enclose the sense and smell of an event, it resides in the taste of frozen food, the pharmaceutical odor of tranquilizers, the planned obsolescence of automobiles, the lack of workmanship in the mass, it lives in the boredom of a good mind, in the sexual excess of lovers who love each other into apathy, it is the livid passion which takes us to sleeping pills, the mechanical action in every household appliance which breaks too often, it vibrates in the sound of an air conditioner or the flicker of fluorescent lighting. And it proliferates in that new architecture which rests like an incubus upon the American landscape, that new architecture which cannot be called modern because it is not architecture but opposed to architecture. Modern architecture began with the desire to use the building materials of the twentieth century—steel, glass, reinforced concrete—and such techniques as cantilevered structure to increase the sculptural beauty of buildings while enlarging their function. It was the first art to be engulfed by the totalitarians, who distorted the search of modern architecture for simplicity and converted it to monotony. The essence of totalitarianism is that it beheads. It beheads individuality, variety, dissent, extreme possibility, romantic faith, it blinds vision, deadens instinct, it obliterates the past. Since it is also irrational, it puts up buildings with flat roofs and huge expanses of glass in northern climates and then suffocates the inhabitants with super-heating systems. Since totalitarianism is a cancer within the body of history, it obliterates distinctions. It makes factories look like college campuses. It makes the new buildings on college campuses look like factories. It depresses the average American with the unconscious recognition that he is installed in a gelatin of totalitarian environment which is bound to deaden his most individual efforts. This new architecture, this totalitarian architecture, destroys the past. There is no trace of the forms which lived in the centuries before us, none of their arrogance, their privilege, their aspirations, their canniness, their creations, their vulgarities. We are left with less and less sense of the lives of men and women who came before us. So we are less able to judge the sheer psychotic values of the present.

# THE MYTH OF THE PEACEFUL ATOM

*Richard Curtis and Elizabeth Hogan*

*Centuries ago man sadly realized that he would perform almost any insane act for the sake of his own vanity. Yet today the prime mover of all sins can still be seen constructing nuclear power plants in populated areas around the country. Despite the massive evidence to the contrary, our prideful myopia has convinced us that the key to a peaceful future is the atom, especially as it is utilized in nuclear energy plants. The following essay is a strong attack on that particular "myth." In a larger sense, it is a criticism of all behavior that blindly challenges the uncertainties of our world with partially perfected technology.*

"What is past is past, and the damage we may already have done to future generations cannot be rescinded, but we cannot shirk the compelling responsibility to determine if the course we are following is one we should be following."

So said Senator Thruston B. Morton of Kentucky on February 29, 1968, upon introducing into Congress a resolution calling for comprehensive review of federal participation in the atomic energy power program. Admitting he had been remiss in informing himself on this "grave danger," Morton said he had now looked more deeply into nuclear power safety and was "dismayed at some of the things I have found—warnings and facts from highly qualified people who firmly believe that we have moved too fast and without proper safeguards into an atomic power age."

Senator Morton's resolution on nuclear power was by no means the only one before Congress in 1968. Indeed, more than two dozen legislators urged

SOURCE: Richard Curtis and Elizabeth Hogan, "The Myth of the Peaceful Atom," *Natural History*, March 1969. Reprinted by permission of the authors.

investigation and reevaluation of this program. This fact may come as a surprise to much of the public, for the belief is widespread that the nuclear reactors being built to generate electricity for our cities are safe, reliable, and pollution free. But a rapidly growing number of physicists, biologists, engineers, public health officials, and even staff members of the Atomic Energy Commission itself—the government bureau responsible for regulation of this force—have been expressing serious misgivings about the planned proliferation of nuclear power plants. In fact, some have indicated that nuclear power, which Supreme Court Justices William O. Douglas and Hugo L. Black described as "the most deadly, the most dangerous process that man has ever conceived," represents the gravest pollution threat yet to our environment.

As of June, 1968, 15 commercial nuclear power plants were operating or operable within the United States, producing about one per cent of our current electrical output. The government, however, has been promoting a plan by which 25 per cent of our electric power will be generated by the atom by 1980, and half by the year 2000. To meet this goal, 87 more plants are under construction or on the drawing boards. Although atomic power and reactor technology are still imperfect sciences, saturated with hazards and unknowns, these reactors are going up in close proximity to heavy population concentrations. Most of them will be of a size never previously attempted by scientists and engineers. They are, in effect, gigantic nuclear experiments.

As most readers will recall, atomic reactors are designed to use the tremendous heat generated by splitting atoms. They are fueled with a concentrated form of uranium stored in thin-walled tubes bound together to form subassemblies. These are placed in the reactor's core, separated by control rods that absorb neutrons and thus help regulate chain reactions of splitting atoms. When the rods are withdrawn, the chain reactions intensify, producing enormous quantities of heat. Coolant circulated through the fuel elements in the reactor core carries the heat away to heat-exchange systems, where water is brought to a boil. The resultant steam is employed to turn electricity-generating turbines.

Stated in this condensed fashion, the process sounds innocuous enough. Unfortunately, however, heat is not the only form of energy produced by atomic fission. Another is radioactivity. During the course of operation, the fuel assemblies and other components in the reactor's core become intensely radioactive. Some of the fission by-products have been described as a million to a billion times more toxic than any known industrial chemical. Some 200 radioactive isotopes are produced as by-products of reactor operation, and the amount of just one of them, strontium-90, accumulated in a reactor of even modest (100–200 megawatt) size, after it has been operative for six months, is equal to what would be produced by the explosion of a bomb 190 times more powerful than the one dropped on Hiroshima.

Huge concentrations of radioactive material are also to be found in nuclear fuel-reprocessing plants. Because the intense radioactivity in a reactor core eventually interferes with the fuel's efficiency, the spent fuel assemblies must be removed from time to time and replaced by new, uncontaminated ones. The old ones are transported to reprocessing plants where the contaminants are separated from the salvageable fuel as well as from plutonium, a valuable by-product. Since no satisfactory means have been found for neutralizing or for safely releasing into the environment the radioactive liquid containing the contaminants, it must be stored until it is no longer dangerous. Thus, reprocessing plants and storage areas are immense repositories of "hot" and "dirty" material. Furthermore, routes between nuclear power plants and the reprocessing facility carry traffic bearing high quantities of such material.

Even from this glimpse it will be apparent that public and environmental safety depend on the flawless containment of radioactivity every step of the way. For, owing to the incredible potency of fission products, even the slightest leakage is harmful and a massive release would be catastrophic. The fundamental question, then, is how heavily can we rely on human wisdom, care, and engineering to hold this peril under absolute control?

Abundant evidence points to the conclusion that we cannot rely on it at all.

The hazards of peaceful atomic power fall into two broad categories: the threat of violent, massive releases of radioactivity or that of slow, but deadly, seepage of harmful products into the environment.

Nuclear physicists assure us that reactors cannot explode like atomic bombs because the complex apparatus for detonating an atomic warhead is absent. This fact, however, is of little consolation when it is realized that only a *conventional* explosion, which ruptures the reactor mechanism and its containment structure, could produce havoc on a scale eclipsing any industrial accident on record or any single act of war, including the atomic destruction of Hiroshima or Nagasaki.

There are numerous ways in which such an explosion can take place in a reactor. For example, liquid sodium, which is used in some reactors as a coolant, is a devilishly tricky element that under certain circumstances burns violently on contact with air. Accidental exposure of sodium could initiate a chain of reactions: rupturing fuel assemblies, damaging components and shielding, and destroying primary and secondary emergency safeguards. If coolant is lost, as it could be in some types of reactors, fuel could melt and recongeal, forming "puddles" that could explode upon reaching a critical size. If these explosions are forceful enough, and safeguards fail, some of the fission products could be released outside the plant and into the environment in the form of a gas or a cloud of fine radioactive particles. Under not uncommon atmospheric conditions such as an "inversion," in which a layer of warm air keeps a cooler layer from rising, a blanket of radioactivity could spread insidiously over the countryside. Another possibility is that fission products could be carried out of the reactor and into a city's watershed, for all reactors are being built on lakes, rivers, or other bodies of water for cooling purposes.

What would be the toll of such a calamity?

In 1957 the Atomic Energy Commission issued a study (designated Wash.—740), largely prepared by the Brookhaven National Laboratory, that attempted to assess the probabilities of such "incidents" and the potential consequences. Some of its findings were stupefying: From the explosion of a 100–200 megawatt reactor, as many as 3,400 people could be killed, 43,000 injured, and as much as 7 billion dollars of property damage done. People could be killed at distances up to 15 miles and injured up to 45. Land contamination could extend for far greater distances: agricultural quarantines might prevail over an area of 150,000 square miles, more than the combined areas of Pennsylvania, New York, and New Jersey.

The awful significance of these figures is difficult to comprehend. By way of comparison, we might look at one of the worst industrial accidents of modern times: the Texas City disaster of 1947 when a ship loaded with ammonium nitrate fertilizer exploded, virtually leveling the city, killing 561 people, and causing an estimated $67 million worth of damage. Appalling as this catastrophe was, however, it does not begin to approach the potential havoc that would be wreaked by a nuclear explosion occurring in one of the plants now being constructed close to several American cities.

The scientists and engineers who produced the Brookhaven Report optimistically ventured to give high odds against such an occurrence, asserting that the structures, systems, and safeguards of atomic plants were so engineered as to render it practically incredible. At the same time, though, the report was replete with such statements as:

> The cumulative effect of radiation on physical and chemical properties of materials, after long periods of time, is largely unknown.

> Much remains to be learned about the characteristics and behavior of nuclear systems.

> It is important to recognize that the magnitudes of many of the crucial factors in this study are not quantitatively established, either by theoretical and experimental data or adequate experience.

Even if the report had been founded on more substantial understanding of natural and technical processes, many of the grounds on which the Brookhaven team based its conclusions are shaky at best.

For one thing, all of us are familiar with technological disasters that have occurred against fantastically high odds: the sinking of the "unsinkable"

**Nuclear Plant Locations**

■ In operation
▲ Under construction
● Planned

*Titanic,* or the November 9, 1965, "blackout" of the northeastern United States, for example. The latter happening illustrates how an "incredible" event can occur in the electric utility field, most experts agreeing that the chain of circumstances that brought it about was so improbable that the odds against it defy calculation.

Congressional testimony given in 1967 by Dr. David Okrent, a former chairman of the AEC's Advisory Committee on Reactor Safeguards, demonstrated that fate is not always a respecter of enormously adverse odds. "We do have on record cases where, for example, an applicant, appearing before an atomic safety and licensing board, stated that a mathematical impossibility had occurred; namely, one tornado took out five separate power lines to a reactor. If one calculated strictly on the basis of probability and multiplied the probability for one line five times, you get a very small number indeed," said Dr. Okrent, "but it happened."

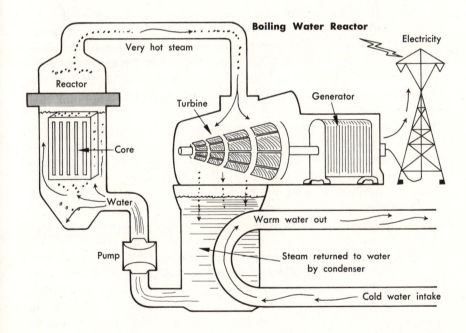

A disturbing number of reactor accidents have occurred—with sheer luck playing an important part in averting catastrophe—that seem to have been the product of incredible coincidences. On October 10, 1957, for instance, the Number One Pile (reactor) at the Windscale Works in England malfunctioned, spewing fission products over so much territory that authorities had to seize all milk and growing foodstuffs in a 400-square-mile area around the plant. A British report on the incident stated that *all* of the reactor's containment features had failed. And, closer to home, a meltdown of fuel in the Fermi reactor in Lagoona Beach, Michigan, in October, 1966, came within an ace of turning into a nuclear "runaway." An explosive release of radioactive materials was averted, but the failures of Fermi's safeguards made the event, in the words of Sheldon Novick in *Scientist and Citizen*, "a bit worse than the 'maximum credible accident.' "

The atomic industry has attempted to design components and safeguards so that failure of one vital system in a plant will not affect another, resulting in a "house of cards" collapse. However, two highly regarded authorities, Theos J. Thompson and J. G. Beckerley, in a book on reactor safety advise us not to place too much faith in claims of independent safeguards: "A structure as complex as a reactor and involving as many phenomena is likely to have relatively few completely independent components." Many manufacturers and utility operators have resisted the idea of producing "redundant safeguards" on the grounds of excessive cost.

Investigations of reactor breakdowns usually disclose a number of small, seemingly unrelated failures, which snowballed into one big one. A design flaw or a human error, a component failure here, an instrumentation failure there—all may coincide to contribute to the total event. Thompson and Beckerley, examining several atomic plant accidents, pinpointed 13 different contributing causes in three of the accidents that had occurred up to the time of their 1964 study.

Among the many factors contributing to reactor incidents, the human element is the most difficult to quantify. And perhaps for that reason, it has been largely overlooked in the AEC's assessments of reactor safety. Yet, a private researcher of nuclear accidents, Dr. Donald Oken, M.D., Associate Director of the Psychosomatic and Psychiatric Institute of Michael Reese Hospital in Chicago, reported: "A review of reports of past criticality and reactor incidents and discussions held with some of the health personnel in charge reveal a number of striking peculiarities in the behavior of many of those involved—in which they almost literally asked for trouble."

AEC annuals are full of reports of human negligence: 3,844 pounds of uranium hexafluoride lost owing to an error in opening a cylinder; a $220,000 fire in a reactor because of accidental tripping of valves by electricians during previous maintenance work; numerous vehicular accidents involving transport of nuclear materials. None of these accidents led to disaster, but who will warrant that, with the projected proliferation of power plants and satellite industries in the coming decade, a moment's misjudgment will not trigger a nightmare? Perhaps worse, the likelihood of sabotage has scarcely been weighed, despite a number of incidents and threats.

It should be apparent that if men are to build safe, successful reactors, the whole level of industrial workmanship, engineering, inspection, and quality control must be raised well above prevailing levels. The more sophisticated the

technology, the more precise the correspondence between the subtlest grada-
tions of care or negligence and that technology's success or failure. When
meters, grams, and seconds are no longer good enough, and specifications
call for millimeters, milligrams, and milliseconds, the demands made on men,
material, and machinery are accordingly intensified. Minute lapses that might
be tolerable in a conventional industrial procedure will wreck the more exacting one. And when the technology is not only exacting but hazardous in the
extreme, then a trivial oversight, a minor defect, a moment's inattention may
spell doom.

While there is little doubt that American technology is the most refined
on earth, there is ample reason to believe that it has more than met its match
in the seemingly insurmountable problems posed by the peaceful atom. Societies of professional engineers, and others concerned with establishing technical and safety criteria for the nuclear industry, have described between
2,800 and 5,000 technical standards that are necessary for a typical reactor
power plant in such areas as materials, testing, design, electrical gear, instrumentation, plant equipment, and processes. Yet, due to the rapidity with
which the nuclear industry has developed, as of March, 1967, only about 100
of these had been passed on and approved for use.

It is not surprising, then, to learn that serious technical difficulties are
turning up in reactor after reactor. At the Big Rock Point Nuclear Plant, a
relatively small reactor near Charlevoix, Michigan, control rods were found
sticking in position, studs failing or cracked, screws jostled out of place and
into key mechanisms, a valve malfunctioning for more than a dozen reasons,
foreign material lodging in critical moving parts, and welds cracked on every
one of sixteen screws holding two components in place. A reactor at Humboldt
Bay in California manifested cracks in the tubes containing fuel: in order to
keep costs down, stainless steel had been used instead of a more reliable
alloy. The Oyster Creek plant in New Jersey showed cracks in 123 of 137
fuel tubes, and welding defects at every point where tubes and control-rod
housings were joined around the reactor's vessel. Reactors in Wisconsin,
Minnesota, Connecticut, Puerto Rico, New York, and elsewhere have experienced innumerable operating difficulties, and some, such as the $55 million
Hallam plant in Nebraska have been forced to shut down for good, owing to
plant malfunction.

Chilling parallels can be drawn between failures in nuclear utility technology and in the nuclear submarine program. In October, 1962, Vice Ad-
miral Hyman G. Rickover, Director of AEC's Division of Naval Reactors, took
the atomic industry to task in a speech in New York City: "It is not well enough
understood that conventional components of advanced systems must necessarily meet higher standards. Yet it should be obvious that failures that would
be trivial if they occurred in a conventional application will have serious consequences in a nuclear plant because here radioactivity is involved. . . ."

Rickover went on to cite defective welds, forging materials substituted without authorization, violations of official specifications, poor inspection techniques, small and seemingly "unimportant" parts left out of components, faulty brazing of wires, and more. "I assure you," he declared, "I am not exaggerating the situation; in fact, I have understated it. For every case I have given, I could cite a dozen more."

The following April, the U.S. atomic submarine *Thresher*, while undergoing a deep test dive some 200 miles off the Cape Cod coast, went down with 112 naval personnel and 17 civilians and never came up again. Subsequent investigation revealed that the sub suffered from many of the same ailments described in Rickover's speech. "It is extremely unfortunate," said Senator John O. Pastore, chairman of the joint congressional committee that held hearings on the disaster, "that this tragedy had to occur to bring a number of unsatisfactory conditions into the open." We must now ask if the same will one day be said about a power plant near one of our large cities.

If a major catastrophe did occur there is good reason to believe that the consequences would be far worse than even the dismaying toll suggested by the 1957 Brookhaven Report, for a number of developments since then have made the threat considerably more formidable.

The Brookhaven Report's accident statistics, for instance, pertained to a reactor of between 100 and 200 megawatts. But while the 15 reactors currently operating in the United States average about 186 megawatts, the 87 plants going up or planned for the next decade are many times that size. Thirty-one under construction average about 726 megawatts; 42 in the planning stage average 832; 14 more, planned but without reactors ordered, will average 904. Some, such as those slated for Illinois, California, Alabama, and New York anticipate capacities of more than 1,000 megawatts. Con Edison has just announced it intends to build four units of 1,000 megawatts each on Long Island Sound near New Rochelle in teeming Westchester County —four nuclear reactors, each with a capacity five to ten times that of the reactor described in the Brookhaven Report.

These facilities will accordingly contain more uranium fuel, and because it is costly to replace spent fuel assemblies (this delicate and dangerous process can take six weeks or longer), the new reactors are designed to operate without fuel replacement far beyond the six months posited in the Brookhaven Report. As a result, the buildup of toxic fission products in tomorrow's reactors will be far greater than at present, and an accident occurring close to the end of the "fuel cycle" in such a plant could release fantastic amounts of radioactive material.

Most serious of all, perhaps, is that tomorrow's reactors are now slated for location in close proximity to population concentrations. While the Brookhaven Report had its hypothetical reactor situated about 30 miles from a major city, many of tomorrow's atomic plants will be much closer. Although the AEC has drafted "guidelines" for siting reactors, the Commission has failed to make utilities adhere to them. In 1967, Clifford K. Beck, AEC's Deputy Director of Regulation, admitted to the Joint Committee on Atomic Energy that nuclear plants in Connecticut, California, New York, and other locations "have been approved with lower distances than our general guides would have indicated when they were approved."

Also, we must remember that while a reactor may not be near the legal boundaries of a metropolis, it may lie close to a population center. Thus, while Con Edison's Indian Point plant is 24 miles from New York City (two more plants are now being built there), it is within 10 miles of an estimated population of 155,510. It need only be recalled that the Brookhaven Report foresaw people being killed by a major radioactive release at distances up to 15 miles to realize the significance of these figures.

In a recent study of nuclear plant siting made by W. K. Davis and J. E. Robb of San Francisco's Bechtel Corporation, the locations of 42 nuclear power plants (some proposed, some now operable) were examined with respect to population centers inhabited by 25,000 residents or more. Their findings are unnerving: only two plants in operation or planned are more than 30 miles from a population center. Of the rest, 14 are between 20 and 27 miles away, 15 between 10 and 16 miles, and 11 between 1 and 9 miles.

Is it necessary to build atomic plants so big and so close? The answer has to do with economics. The larger a facility is, the lower the unit cost of construction and operation and the cheaper the electricity. The longer the fuel cycle, the fewer the expensive shutdowns while spent fuel assemblies are replaced. The closer the plant is to the consumer, the lower the cost of rights of way, power lines, and other transmission equipment.

On a few occasions an aroused public has successfully opposed the situation of plants near population centers. When the Pacific Gas and Electric Company persisted in trying to build a reactor squarely over earthquake faults in an area of known seismic activity—the site was Bodega Head, north of San Francisco—a courageous conservation group forced the company to back down. It has been suggested, though, that the group might not have won had not the Alaskan earthquake of 1964, occurring while the fight was going on, underscored the recklessness of the utility's scheme.

Announcement by Con Edison at the end of 1962 of its proposal to build a large nuclear plant in Ravenswood, Queens, close to the center of New York City brought a storm of frightened and angry protest. Although the utility's chairman noted, "We are confident that a nuclear plant can be built in Long Island City, or in Times Square for that matter, without hazard to our own employees working in the plant or to the community," David E. Lilienthal, the former head of the AEC, had a contrary opinion, declaring he "would not dream of living in Queens if a huge nuclear plant were located there." Outraged citizens and a number of noted scientists prevailed.

For the most part, however, the battle has been a losing one. Con Edison, for example, after its defeat in the Ravenswood fight, has just announced an interest in building a reactor on Welfare Island, literally a stone's throw from midtown Manhattan. Also, New York's Governor Nelson Rockefeller has gone on record advocating an $8 billion electric power expansion program based extensively on nuclear energy. The state legislature approved of the program, and in 1968, voted to bolster the plan with state subsidies.

Some of the deepest concern about the size and location of atomic plants has been expressed by members of the AEC themselves. "The actual experience with reactors in general is still quite limited," said Harold Price, AEC's Director of Regulation, in 1967 congressional hearings, "and with large reactors of the type now being considered, it is nonexistent. Therefore, because there would be a large number of people close by and because of lack of experience, it is . . . a matter of judgment and prudence at present to locate reactors where the protection of distance will be present."

Price's statement is mild compared to that made in the same hearings by Nunzio J. Palladino, Chairman of the AEC's Advisory Committee on Reactor Safeguards for 1967, and Dr. David Okrent, former Chairman for 1966: "The ACRS believes that placing large nuclear reactors close to population centers will require considerable further improvements in safety, and that *none of the large power reactors now under construction is considered suitable for location in metropolitan areas* [Curtis and Hogan's italics]."

The threat of a nuclear plant catastrophe constitutes only half of the double jeopardy in which atomic power has placed us. For even if no such calamity occurs, the gradual exhaustion of what one scientist terms our environmental "radiation budget," due to unavoidable releases of radioactivity during normal operation of nuclear facilities, poses an equal and possibly more insidious threat to all living things on earth.

Most of the fission products created in a reactor are trapped. Contaminated solids, liquids, and gases are isolated, allowed to decay for a short period of time, then concentrated and shipped in drums to storage areas. These are called "high-level wastes." But technology for retaining all radioactive contaminants is either unperfected or costly, and much material of low-level radioactivity is routinely released into the air or water at the reactor site. These releases are undertaken in such a way, we are told, as to insure dispersion or dilution sufficient to prevent any predictable human exposure above harmful levels. Thus, when atomic power advocates are asked about the dangers of contaminating the environment, they imply that the relatively small amounts of radioactive materials released under "planned" conditions are harmless.

This view is a myth.

In the first place, many waste radionuclides take an extraordinarily long time to decay. The half-life (the time it takes for half of an element's atoms to disintegrate through fission) of strontium-90, for instance, is more than 27 years. Thus, even though certain long-lived isotopes are widely dispersed in air or diluted in water, their radioactivity does not cease. It remains, and over a period of time accumulates. It is therefore not pertinent to talk about the safety of any single release of "hot" effluents into the environment. At issue, rather, is their duration and cumulative radioactivity.

Further, many radioactive elements taken into the body tend to build up in specific tissues and organs to which those isotopes are attracted, increasing by many times the exposure dosage in those local areas of the body. Iodine-131, for instance, seeks the thyroid gland; strontium-90 collects in the bones; cesium-137 accumulates in muscle. Many isotopes have long half-lives, some measurable in decades.

Two more factors controvert the view that carefully monitored releases of low-level radioactivity into the environment are not pernicious. First, there is apparently no radiation threshold below which harm is impossible. Any dose, however small, will take its toll of cell material, and that damage is irreversible. Second, it may take decades for organic damage, or generations for genetic damage, to manifest itself. In 1955, for example, two British doctors reported a case of skin cancer—ultimately fatal—that had taken forty-nine years to develop following fluoroscopic irradiation of a patient.

Still another problem has received inadequate attention. Man is by no means the only creature in whom radioactive isotopes concentrate. The dietary needs of all plant and animal life dictate intake of specific elements. These concentrate even in the lowest and most basic forms of life. They are then passed up food chains, from grass to cattle to milk to man, for example. As they progress up these chains, the concentrations often increase, sometimes by hundreds of thousands of times. And if these elements are radioactive. . . .

Take zinc-65, produced in a reactor when atomic particles interact with zinc in certain components. Scrutiny of the wildlife in a pond receiving runoff from the Savannah River Plant near Aiken, South Carolina, disclosed that while the water in that pond contained only infinitesimal traces of radioactive zinc-65, the algae that lived on the water had concentrated the isotope by nearly 6,000 times. The bones of bluegills, an omnivorous fish that feeds both on algae and on algae-eating fish, showed concentrations more than 8,200 times higher than the amount found in the water. Study of the Columbia River, on which the Hanford, Washington, reactor is located, revealed that while the radioactivity of the water was relatively insignificant: 1. the radioactivity of the river plankton was 2,000 times greater; 2. the radioactivity of the fish and ducks feeding on the plankton was 15,000 and 40,000 times greater, respectively; 3. the radioactivity of young swallows fed on insects caught by their parents near the river was 500,000 times greater; and 4. the radioactivity of the egg yolks of water birds was more than a million times greater.

Here then are clear illustrations of the ways in which almost undetectable traces of radioactivity in air, water, or soil may be progressively concentrated, so that by the time it ends up on a man's plate or in his glass it is a tidy package of poison.

That nuclear facilities are producing dangerous buildups of radioisotopes in our environment can be amply documented. University of Nevada investigators, seeking a cause for concentrations of iodine-131 in cattle thyroids in wide areas of the western United States, concluded that "the principal known source of I-131 that could contribute to this level is exhaust gases from nuclear reactors and associated fuel-processing plants."

In his keynote address to the Health Physics Society Symposium at Atlanta, Georgia, early in 1968, AEC Commissioner Wilfred E. Johnson admitted that the release into the atmosphere of tritium and noble gases such as krypton-85 would present a potential problem in the future, and that, as yet, scientists had not devised a way of solving it. Krypton-85, although inert, has a 10-year half-life and tends to dissolve in fatty tissue, meaning fairly even distribution throughout the human body. Krypton-85 is particularly difficult to filter out of reactor discharges, and the accumulation of this element alone may exhaust as much as two-thirds of the "average" human's "radiation budget" for the coming century, based on the standards established by the National Committee on Radiation Protection and Measurement.

That "low-level" waste is a grossly deceptive term is obvious. In his book *Living with the Atom*, author Ritchie Calder in 1962 described an "audit" of environmental radiation that he and his colleagues, meeting at a symposium in Chicago, drew up to assess then current and future amounts of radioactivity released into atmosphere and water. Speculations covered the period 1955–1965, and because atomic power plants were few and small during that time, the figures are more significant in relation to the future. Tallying "planned releases" of radiation from such sources as commercial and test reactors, nuclear ships, uranium mills, plutonium factories, and fuel-reprocessing plants, Calder's group came to a most disquieting conclusion: "By the time we had added up all the curies which might predictably be released, by all those peaceful uses, into the environment, it came to about 13 million curies per annum." A "curie" is a standard unit of radioactivity whose lethality can be appreciated from the fact that one trillionth of one curie of radioactive gas per cubic meter of air in a uranium mine is ten times higher than the official maximum permissible dose.

Calder's figures did not include fallout due to bomb testing and similar experiments, nor did they take into account possible reactor or nuclear transportation accidents. Above all, they did not include possible escape of stored high-level radioactive wastes, the implications of which were awesome to contemplate: "what kept nagging us was the question of waste disposal and of the remaining radioactivity which must not get loose. We were told that the dangerous waste, which is kept in storage, amounted to 10,000 million curies. If you wanted to play 'the numbers game' as an irresponsible exercise, you could divide this by the population of the world and find that it is over 3 curies for every individual."

Exactly what does Calder mean by "the question of waste disposal"?

It has been estimated that a ton of spent fuel in reprocessing will produce from forty to several hundred gallons of waste. This substance is a violently lethal mixture of short- and long-lived isotopes. It would take five cubic miles of water to dilute the waste from just one ton of fuel to a safe concentration. Or, if we permitted it to decay naturally until it reached the safe level—and the word "safe" is used advisedly—just one of the isotopes, strontium-90, would still be damaging to life 1,000 years from now, when it will have only one seventeen-billionth of its current potency.

There is no known way to reduce the toxicity of these isotopes; they must decay naturally, meaning *virtually perpetual containment.* Unfortunately, mankind has exhibited little skill in perpetual creations, and procedures for handling radioactive wastes leave everything to be desired. Formerly dumped in the ocean, the most common practice today is to store the concentrates in large steel tanks shielded by earth and concrete. This method has been employed for some twenty years, and about 80 million gallons of waste are now in storage in about 200 tanks. This "liquor" generates so much heat it boils by itself for years. Most of the inventory in these caldrons is waste from weapons production, but within thirty years, the accumulation from commercial nuclear power will soar if we embark upon the expansion program now being promoted by the AEC. Dr. Donald R. Chadwick, chief of the Division of Health of the U.S. Public Health Service, estimated in 1963 that the accumulated volume of waste material would come to two billion gallons by 1995.

It is not just the volume that fills one with sickening apprehension but the techniques of disposing of this material. David Lilienthal put his finger on the crux of the matter when he stated: "These huge quantities of radioactive wastes must somehow be removed from the reactors, must—without mishap— be put into containers that will never rupture; then these vast quantities of poisonous stuff must be moved either to a burial ground or to reprocessing and concentration plants, handled again, and disposed of, by burial or otherwise, with a risk of human error at every step." Nor can it be stressed strongly enough that we are not discussing a brief danger period of days, months, or years. We are talking of periods "longer," in the words of AEC Commissioner Wilfred E. Johnson, "than the history of most governments that the world has seen."

Yet already there are many instances of the failure of storage facilities. An article in an AEC publication has cited nine cases of tank failure out of

183 tanks located in Washington, South Carolina, and Idaho. And a passage in the AEC's authorizing legislation for 1968 called for funding of $2,500,000 for the replacement of failed and failing tanks in Richland, Washington. "There is no assurance," concluded the passage, "that the need for new waste storage tanks can be forestalled." If this is the case after twenty years of storage experience, it is beyond belief that this burden will be borne without some storage failures for centuries in the future. Remember too, that these waste-holding "tank farms" are vulnerable to natural catastrophes such as earthquakes, and to man-made ones such as sabotage.

Efforts are of course being made toward effective handling of the waste problems, but many technical barriers must still be overcome. It is unlikely they will all be overcome by the end of the century, when waste tanks will boil with 6 billion curies of strontium-90, 5.3 billion curies of cesium-137, 6.07 billion curies of prometheum-147, 10.1 billion curies of cesium-144, and millions of curies of other isotopes. The amount of strontium-90 alone is 30 times more than would be released by the nuclear war envisioned in a 1959 congressional hearing.

The burden that radioactive wastes place on future generations is cruel and may prove intolerable. Physicist Joel A. Snow stated it well when he wrote in *Scientist and Citizen:* "Over periods of hundreds of years it is impossible to ensure that society will remain responsive to the problems created by the legacy of nuclear waste which we have left behind."

"Legacy" is indeed a gracious way of describing the reality of this situation, for at the very least we are saddling our children and their descendants with perpetual custodianship of our atomic refuse, and at worst may be dooming them to the same agonizing afflictions and deaths suffered by those who survived Hiroshima. Radiation has been positively linked to cancer, leukemia, brain damage, infant mortality, cataracts, sterility, genetic defects and mutations, and general shortening of life.

The implications for the survival of mankind can be glimpsed by considering just one of these effects, the genetic. In a 1960 article, James F. Crow, Professor of Genetics at the University of Wisconsin School of Medicine and president of the Genetics Society of America, stated that for every roentgen of slow radiation—the kind we can expect to receive in increasing doses from peacetime nuclear activity—about five mutations will result per 100 million genes exposed, meaning that "after a number of generations of exposure to one roentgen per generation, about one in 8,000 . . . in each generation would have severe genetic defects attributable to the radiation."

The Atomic Energy Commission is aware of the many objections that have been raised to the atomic power program: why does it continue to encourage it? Unfortunately, the Commission must perform two conflicting roles. On the one hand, it is responsible for regulating the atomic power industry. But on the other, it has been charged by Congress to promote the use of nuclear energy by the utility industry. Because of its involvement in the highest priorities of national security, enormous power and legislative advantages have been vested in the AEC, enabling it to fulfill its role as promoter with almost unhampered success—while its effectiveness as regulator has gradually atrophied. The Commission consistently denies claims that atomic power is heading for troubled waters, optimistically reassuring critics that these plants are safe, clean neighbors.

The fact that there is no foundation for this optimism is emphasized by the insurance situation on atomic facilities. Despite the AEC's own assertion that as much as $7 billion in property damage could result from an atomic power plant catastrophe, the insurance industry, working through two pools, will put up no more than $74 million, or about one per cent, to indemnify equipment manufacturers and utility operators against damage suits from the public. The federal government will add up to $486 million more, but this still leaves more than $6 billion in property damages to be picked up by victims of a Brookhaven-sized accident. And no insurance company—not even Lloyds of London—will issue property insurance to individuals against radiation damage. If there is so little risk in atomic power plants, why is insurance so inadequate?

The knowledge that man must henceforth live in constant dread of a major nuclear plant accident is disturbing enough. But we must recognize that even if such calamities are averted, the slow saturation of our environment with radioactive wastes will nevertheless be raising the odds that you or your heirs will fall victim to one of a multitude of afflictions. There is no "threshold" exposure below which we can feel safe.

We have little time to reflect on our alternatives, for the moment must soon come when no reversal will be possible. Dr. L. P. Hatch of Brookhaven National Laboratory vividly made this point when he told the Joint Committee on Atomic Energy: "If we were to go on for 50 years in the atomic power industry, and find that we had reached an impasse, that we had been doing the wrong thing with the wastes and we would like to reconsider the disposal methods, it would be entirely too late, because the problem would exist and nothing could be done to change that fact for the next, say, 600 or a thousand years." To which might be added a sobering thought stated by Dr. David Price of the U.S. Public Health Service: "We all live under the haunting fear that something may corrupt the environment to the point where man joins the dinosaurs as an obsolete form of life. And what makes these thoughts all the more disturbing is the knowledge that our fate could perhaps be sealed twenty or more years before the development of symptoms."

What must be done to avert the perils of the peaceful atom? A number of plans have been put forward for stricter regulation of activities in the nuclear utility field, such as limiting the size of reactors or their proximity to population concentrations or building more safeguards. As sensible as these proposals appear on the surface, they fail to recognize a number of important realities: first, that such arrangements would probably be opposed by utility operators and the government due to their prohibitively high costs. Since our government seems to be committed to making atomic power plants competitive with conventionally fueled plants, and because businesses are in business for profit, it is hardly likely they would buy these answers. Second, the technical problems involved in containment of radioactivity have not been successfully overcome, and there is little likelihood they will be resolved in time to prevent immense and irrevocable harm to our environment. Third, the nature of business enterprise is unfortunately such that *perfect* policing of the atomic power industry is unachievable. As we have seen in the cases of other forms of pollution, the public spirit of men seeking profit from industrial processes does not always rise as high as the welfare of society requires. It is unwise to hope that stricter regulation would do the job.

What, then, is the answer? The only course may be to turn boldly away from atomic energy as a major source of electricity production, abandoning it as this nation has abandoned other costly but unsuccessful technological enterprises.

There is no doubt that, with this nation's demand for electricity doubling every decade, new power sources are urgently needed. Nor is there doubt that our conventional fuel reserves—coal, oil, and natural gas—are rapidly being consumed. Sufficient high-grade fossil fuel reserves exist, however, to carry us to the end of this century; and new techniques for recovering these fuels from secondary sources such as oil shale could extend the time even longer. Furthermore, advances in pollution abatement technology and revolutionary new techniques, now in development, for burning conventional fuels with high efficiency, could carry us well into the next century with the fossil fuels we have. This abundance, and potential abundance, gives us at least several decades to survey possible alternatives to atomic power, select the most promising, and develop them on an appropriate scale as alternatives to nuclear power. Solar energy, tidal power, heat from the earth's core, and even garbage and solid-waste incineration have to some degree been demonstrated as promising means of electricity generation. If we subsidized research and development of those fields as liberally as we have done atomic energy, some of them would undoubtedly prove to be what atomic energy once promised, without its deadly drawbacks.

Aside from the positive prospect of profitability in these new approaches, industry will have another powerful incentive for turning to them; namely, that atomic energy is proving to be quite the opposite of the cheap, everlasting resource envisioned at the outset of the atomic age. The prices of reactors and components and costs of construction and operation have soared in the last few years, greatly damaging nuclear power's position as a competitor with conventional fuels. If insurance premiums and other indirect subsidies are brought into line with realistic estimates of what it takes to make atomic energy both safe and economical, the atom might prove to be the most *expensive* form of energy yet devised—not the cheapest. In addition, because of our wasteful fuel policies, evidence indicates that sources of low-cost uranium will be exhausted before the turn of the century. Fuel-producing breeder reactors, in which the nuclear establishment has invested such high hopes for the creation of vast, new fuel supplies, have proven a distinct technological disappointment. Even if the problems plaguing this effort were overcome in the next ten or twenty years, it may still be too late to recoup the losses of nuclear fuel reserves brought about by prodigious mismanagement.

The proposal to abandon or severely curtail the use of atomic energy is clearly a difficult one to imagine. We have only to realize, however, that by pursuing our current civilian nuclear power program, we are jeopardizing every other industry in the country; in that light, this proposal becomes the only practical alternative. In short, the entire national community stands to benefit from the abandonment of a policy which seems to be leading us toward both environmental and economic disaster.

Man's incomplete understanding of many technological principles and natural forces is not necessarily to his discredit. Indeed, that he has erected empires despite his limited knowledge is to his glory. But that he pits this ignorance and uncertainty, and the fragile yet lethal technology he has woven out of them, against the uncertainties of nature, science, and human behavior —this may well be to his everlasting sorrow.

# ADDRESS BY GOOLEY MacDOWELL TO THE HASBEENS CLUB OF CHICAGO

*Saul Bellow*

*Saul Bellow was born in Lachine, Quebec, in 1915 and was raised in Chicago where he presently teaches. He has been publishing novels since 1947 and has been for some time considered one of the best American writers. From Dangling Man (1947) to Mr. Sammler's Planet (1969), Bellow's main theme has been distraction. Modern man is confused, adrift in a sea of complexities, and much in need of what Thoreau called simplification. Gooley MacDowell is an early example of the distracted person; yet he is also one of Bellow's most compelling creations. Though only indirectly related to the ecological crisis, Gooley does represent the typical American involved in a life-and-death struggle of action and reaction, and he has something to say about the political and social nature of the crisis. Perhaps in the strictest sense, mental pollution is not an ecological matter, but to the degree that the mind reflects its environment it belongs in our understanding of the subject.*

Fellows: There have been rumors running concerning me, some funny and some two-sided, and some old ones, getting boring. I am not always against bores; I used to argue that they should be sacred, like strangers and lunatics in the simpler societies, because they keep the mysteries at the right depth; but I realize that this position overlooks great dangers. However, there's the rumor again that I'm going to marry, which is nothing but a joke on my appearance and personal neglect of years. There's a rumor that *People's Gas*

SOURCE: Saul Bellow, "Address by Gooley MacDowell to the Hasbeens Club of Chicago." Reprinted by permission from *The Hudson Review*, vol. IV, no. 2, Summer 1951. Copyright © 1951 by The Hudson Review, Inc.

wants to make me personnel director—I'm used to that one, too. And then there's this new thing, that the Nobel committee wants to start a prize for general, secular intelligence or leading laymanship, and that I'm top candidate from the Middle-west. This is some satire that people think I'm beginning to lose my edge, and with perhaps a hum of sadness to it that I'm a little calcified and the blood supply to my head is not so swift as once—when, say, I first failed in business, first got married, first became a Unitarian. But it is not a well-planned joke. It overlooks my loss to the Hasbeens, should I get that famous money; and, also, that having the reverse Midas touch, all the gold I ever came in contact with having turned into something less universally desired, I wouldn't bring the Nobels anything but peril, wearing their medal.

But this joke about my falling off aims right, though physical deterioration is not what I believe to be back of it. Only, what I have been saying in bits and drabbles has waked that suspicion, so that now I will try to report my changed views or new doubts.

It's true, fellows, that I used to be very intelligent, and now don't know but what I should give it up. I seen too many hard times, and these days I've been starting to reconsider to what end I hoped to be so smart.

Kindness sent me forward and I got praise from the first. What is there better for learning? My mother put up my exercise books in ribbons. My long dress teachers, in the mud and shack days of this town, gave me encouragement and I sprung my elbow waving to answer questions; and later on, with intelligence tests and so forth, I was up amongst the brightest apes, and have no doubt but that I would have conquered a box of Milky Ways from Professor I. Q. had the times been more advanced. But I did my best. I set much store by correspondence schools, where there were cerebellar doctors in the picture; the more confident in them if they had log-cabins drawn in behind them, and ploughs and lonesome prairie, and now were qualified to rise up to a point of information in the thick of intellectual Boston about Lycurgus or Pitt the Younger, or which birds fly longest in migration. It was so, in the hard-collar and wisely bearded photos of these men in the back pages of magazines where they advertised, that you could see the close weave of brain back of their eyes, like the reward of gazing into radio tubes—pardon, germanium crystal diodes of thought-machines—to examine the secret of their potency. And furthermore, when I was a young bachelor I was put wise to Herbert Spencer, the great practical man, with snap-purse lips and head economized of all hair and wrinkles, cleared as if for the smoother work of brain. I was in favor of all he proposed, even, for a moment, of his plan for the painless execution of criminals. Who remembers that piece of kindness? A centrifuge wheel: bind your murderer to it, start him spinning slowly and then accelerate until he dies by anemia of the brain. True capital punishment, humaner than rope, and also more noble or Copernican that society should punish into the head and by a law of motion that also rules in the stars. Yes, yes, in all the departments of life it was intelligence connected to human advancement. You might not have thought well of big money, but nevertheless there was the capacity of the Wall Street man and his major brain, displacing so much black water of finance with ticking calculations, making head into Europe and China. And as the wizard with flashing synapses rose on the world, the engineers went forward—Mr. Goethals, Mr. Eiffel, the makers of industrial bread, Mr. Sullivan raised the Monadnock Building here, Mr. Roosevelt, the first, sat in a white suit at the controls of a steam shovel in the Panama Canal, voices came out of the telephone, the country and earth filled with light forced up from Menlo Park, the front of life coruscated, the Dying Gaul was promoted to be Rodin's Thinker, a higher immolation, with his fist to his forehead—as all this happened, what was it to be wondered at that clever young fellows were wetting a finger to turn pages and set themselves ahead? As, today, the International Business Machine Company in its showroom on Michigan Boulevard prints THINK! on its walls with identical inspiration.

Now I see some of my friends putting out a foxy jaw of skepticism about me, and I know they have in reserve the memory of my little pamphlets on T. H. Huxley and my Haldeman-Julius Kansas bluebook *Evolution and Mind* with foreword by Clarence Darrow; or else my long association with the Yellow Kid whose financial genius was of such interest to the State Prosecutor and courts of Illinois, a man of great intellectual refinement. And also the fact that for years I have had the same seat, more formal than the stock exchange, under the portrait of M. le duc D'Aumale wearing his noble knickers, this clean and rational French gentleman, in the science reading-room of the Crerar, or our homespun Bodleian. They will want to know why I have it in for thought, suddenly. To them I say—Archie, Boggs, you boys—there's an interesting explanation coming, that I'm not preparing to say I'm p.o. with reason and brain.

How is any man going to account for having closed up in his head, above his teeth and palate and below his hair, what there is? This folding! This isthmus! This finding! That baroque pearl of an inmost thing! You think I'm not in awe of it? Why, here's this earth, chained down of mountains where I sway, in its silk buffets of atmosphere; here's the rooster breaking into peaks of day, while the clucks are his base, the pecking and the fostering of the eggs: the relation of the brain and other organs is what I mean. And isn't it maybe the curiosity over these internal discoveries that leads us to have captive animals—eagles in the park, canaries at home? We keep pets within and without. Imprisoned power. The heart in its cage of bone. I had goldfish myself, in the days of my victrola shop, and loved the bulge-eyed blackies so swift after flies.

This, about pets and mascots, is not so far from the subject, inasmuch as we are in the habit of regarding and considering ourselves no matter which direction we look. Is that something to surprise us? It's in the book of rules among the first and commonest of human things, and it doesn't have to be as disagreeable as it's become. But no debate, either, that it has become so, and tiresome, a complication from what was a process, and thereby makes mathematics and physics a comfort. For there it seems surest that you're not talking about yourself. In fact they probably wouldn't have developed so far, these sciences, in a less selfish age. Because this kind of consciousness can be like sharing a bed with a pal who is running a high fever, and at last you would rather shake down on the hard floor of fact, where no humanity is, in the fair cool space of vectors and the topless tent of laws, open to the universe. That's for removal from the burning *me*. And why should it be so? Fellows, there's a question!

What, Mr. President, and you others, do you think I'm turning into a Jack Cade, enraged with people who don't sign themselves with crosses, or a Know Nothing, and out to tear down the flag of *cogito*; or that pretty soon I'm going to break out with old arguments—Bishop Berkeley rocks, Paley watches? Even the latest about Percy Lowell's planet that should have been there by mathematics but didn't show up for its appointment ever? Not I. At last even Dr. Skinner's little pigeons who play ping-pong or pick out Annie Laurie on the piano can be justified in my sight, though my willingness is slow here and would be even if they played the Moonlight Sonata, for I think of Noah's messengers coming up from Ararat with little leaves. But, *grosso modo*, I give my okay because of the sublimity of the greater enterprise. Only it's in another direction I'm looking today, of thought in the singular and personal, and what you manage by it.

Around our heads we have a dome of thought as thick as atmosphere to breathe. And what's about? One thought leads to another as breath leads to breath. I find it barren just to breathe or only to have thoughts. By pulling into universal consciousness, can [one] explain everything from Democritus to Bikini? But a person can no longer keep up, and plenty are dying of good ideas. We have them in the millions, in compilations, from the *Zend Avesta* to now, all on file with the best advice for any and all human occasions. What a load you can buy for a buck, in anthologies, out of Augustine, Pascal, Aristotle, Nicholas of Cusa, super-brain Goethe, and it's a confusion for us. Look at us, deafened, hampered, obstructed, impeded, impaired and bowel-glutted with wise counsel and good precept, and the more plentiful our ideas the worse our headaches. So we ask, will some good creature pull out the plug and ease our disgusted hearts a little? We are not free to use it, that is why the advice is a loaded burden. What's the laugh on Polonius, if not just in that? And, besides, is my heart going to authorize me in this day to play chivalry and Roland, or Roman senator or good bishop? And if a good idea isn't also a law of the body, then what is it to us? There was the Mahatma in his baken-earth backyard one day, in Africa, striving with his wife that she didn't carry the chamber-pots of strangers with a willing enough pleasure. Now if you're looking for the pattern of the foundation of new great things, this man and wife bicker is what to start with; and when my soul feels its privations, and the cut-end branches of inclinations that were not allowed, I also say a knowledge to be true must be confirmed or arrive with a happiness, and we're little thin on that side, too.

Now I awoke this fine summer morning with a fly bustling in the hairs of my foot, and went to spit in the sink of my hall-bedroom in the house belonging to my brother's widow, and heard the steam of the tailor pressing below, and had a gander at the say-nothing tragedian clouds upstairs of me. Then the question rose in me, "Gooley, what is the score, you poor-man's Socrates? What is the pay-off of your devoted studies?" And I saw that I could not stand any more self-improvement originating in thought. And just a note here is necessary, that as you get better by the correction of intellect you may lose your nature, and have less and less to say because what your nature prompts your betterness turns down, so you become silent and are otherwise in danger of becoming better than you can afford. As, also, there is an acquaintance with thought whose widening cannot be stopped because the question of what it is that you yourself believe only forces you faster after what has been said, and soon that saying is all, and the record everything.

While as to being dry on the chaos waves by the management and steering of wit, haven't we taken care of that in our charter, declaring nobody can trust knowing the angles to obtain salvation? Intelligent? What for, you subject-man and personal ant, when you're so much at the mercy and soldered up inside determinations? Intelligence is the brillo of the brave! and slaveys when they scour with it shine up a disappointed face; and then what a moment on the bottom of the pot. And can you, ambitious to think the deepest penetration into manhood, tell me what effectively rules in your life and how much you are, at sixty, still your mother's and your father's child, not unfastened from that close happiness, swaying from it still with a lifted upper body like the tit-footed caterpillar? And how much great and vast sense there is to be waked up in your last years from that almost lifelong state? (Not to lose love nor yet be prevented from clear thought, is what I intend to say.) Or also what things happened, seeming neither intrinsic nor even called-for: what have these done to you, what have they made of you? And I must ask, when I see these many, ignorant of their others, and fearful, and faithless of happiness: how wise and clever must you become to be well, and what kind of specialist to stand even average among so many hard griefs and difficult battlements, when even to be decent needs heroism? What education is misery, that now is the main teacher, or the word of intelligence to human cookery and soap-manufacture? Why, then, I feel compelled to admit it seems only natural for people to confine themselves in the far corner from the largest power of their minds, which may come and whiff after them like the elephant after its peanut, and lose them in its huge digestion.

Nevertheless, I'm still in doubt and not ready yet to draw all my hopes out of the old deposited account. Except I see there are feelings of being that go beyond and beyond all I ever knew of thought, and a massiver existence of man that comes to question even in me, who was so easily hammerlocked in the first trial-grips of my life. And have the sense, as well, that there is always a furthest creature that wears various lives or forms as a garment, and the life of thought as one of the greatest of these.

# THE DEATH OF PAN

## D. H. Lawrence

*To say that David Herbert Lawrence (1885–1930) was well aware of the necessity for an eco-systematic attitude to life is an understatement. He recognized that the chaos of the twentieth century was paradoxically the result of industrial society's appetite for simplification. Anyone who reduces life to a single mode of being is for Lawrence unconsciously dedicated to the pursuit of destruction. He recognized that what the industrial world did for reasons of economy was, in fact, a perversion of what Thoreau meant by the word simplify. As a consequence, the true Lawrentian hero is one who denies the effort of the world to reduce him in any way. Much of this thought is implicit in the following selection, but the reader will be well rewarded for his efforts if he pursues the idea into Lawrence's novels, especially Women in Love (1920).*

At the beginning of the Christian era, voices were heard off the coasts of Greece, out to sea, on the Mediterranean, wailing: "Pan is dead! Great Pan is dead!"

The father of fauns and nymphs, satyrs and dryads and naiads was dead, with only the voices in the air to lament him. Humanity hardly noticed.

But who was he, really? Down the long lanes and overgrown ridings of history we catch odd glimpses of a lurking rustic god with a goat's white lightning in his eyes. A sort of fugitive, hidden among leaves, and laughing with the uncanny derision of one who feels himself defeated by something lesser than himself.

SOURCE: From *Phoenix: The Posthumous Papers of D. H. Lawrence*, edited by Edward D. McDonald. Copyright 1936 by Frieda Lawrence, copyright © renewed 1964 by The Estate of Frieda Lawrence Ravagli. All rights reserved. Reprinted by permission of The Viking Press, Inc.

An outlaw, even in the early days of the gods. A sort of Ishmael among the bushes.

Yet always his lingering title: The Great God Pan. As if he was, or had been, the greatest.

Lurking among the leafy recesses, he was almost more demon than god. To be feared, not loved or approached. A man who should see Pan by daylight fell dead, as if blasted by lightning.

Yet you might dimly see him in the night, a dark body within the darkness. And then, it was a vision filling the limbs and the trunk of a man with power, as with new, strong-mounting sap. The Pan-power! You went on your way in the darkness secretly and subtly elated with blind energy, and you could cast a spell, by your mere presence, on women and on men. But particularly on women.

In the woods and the remote places ran the children of Pan, all the nymphs and fauns of the forest and the spring and the river and the rocks. These, too, it was dangerous to see by day. The man who looked up to see the white arms of a nymph flash as she darted behind the thick wild laurels away from him followed helplessly. He was a nympholept. Fascinated by the swift limbs and the wild, fresh sides of the nymph, he followed for ever, for ever, in the endless monotony of his desire. Unless came some wise being who could absolve him from the spell.

But the nymphs, running among the trees and curling to sleep under the bushes, made the myrtles blossom more gaily, and the spring bubble up with greater urge, and the birds splash with a strength of life. And the lithe flanks of the faun gave life to the oak-groves, the vast trees hummed with energy. And the wheat sprouted like green rain returning out of the ground, in the little fields, and the vine hung its black drops in abundance, urging a secret.

Gradually men moved into cities. And they loved the display of people better than the display of a tree. They liked the glory they got of overpowering one another in war. And, above all, they loved the vainglory of their own words, the pomp of argument and the vanity of ideas.

So Pan became old and grey-bearded and goat-legged, and his passion was degraded with the lust of senility. His power to blast and to brighten dwindled. His nymphs became coarse and vulgar.

Till at last the old Pan died, and was turned into the devil of the Christians. The old god Pan became the Christian devil, with the cloven hoofs and the horns, the tail, and the laugh of derision. Old Nick, the Old Gentleman who is responsible for all our wickednesses, but especially our sensual excesses —this is all that is left of the Great God Pan.

It is strange. It is a most strange ending for a god with such a name. Pan! All! That which is everything has goat's feet and a tail! With a black face!

This really is curious.

Yet this was all that remained of Pan, except that he acquired brimstone and hell-fire, for many, many centuries. The nymphs turned into the nasty-smelling witches of a Walpurgis night, and the fauns that danced became sorcerers riding the air, or fairies no bigger than your thumb.

But Pan keeps on being reborn, in all kinds of strange shapes. There he was, at the Renaissance. And in the eighteenth century he had quite a vogue. He gave rise to an "ism," and there were many pantheists, Wordsworth one of the first. They worshipped Nature in her sweet-and-pure aspect, her Lucy Gray aspect.

"Oft have I heard of Lucy Gray," the school-child began to recite, on examination-day.

"So have I," interrupted the bored inspector.

Lucy Gray, alas, was the form that William Wordsworth thought fit to give to the Great God Pan.

And then he crossed over to the young United States: I mean Pan did. Suddenly he gets a new name. He becomes the Oversoul, the Allness of everything. To this new Lucifer Gray of a Pan Whitman sings the famous *Song of Myself*: "I am All, and All is Me." That is: "I am Pan, and Pan is me."

The old goat-legged gentleman from Greece thoughtfully strokes his beard, and answers: "All A is B, but all B is not A." Aristotle did not live for nothing. All Walt is Pan, but all Pan is not Walt.

This, even to Whitman, is incontrovertible. So the new American pantheism collapses.

Then the poets dress up a few fauns and nymphs, to let them run riskily —oh, would there were any risk—in their private "grounds." But, alas, these tame guinea pigs soon became boring. Change the game.

We still *pretend* to believe that there is One mysterious Something-or-other back of Everything, ordaining all things for the ultimate good of humanity. It wasn't back of the Germans in 1914, of course, and whether it's back of the bolshevists is still a grave question. But still, it's back of *us,* so that's all right.

Alas, poor Pan! Is this what you've come to? Legless, hornless, faceless, even smileless, you are less than everything or anything, except a lie.

And yet here, in America, the oldest of all, old Pan is still alive. When Pan was greatest, he was not even Pan. He was nameless and unconceived, mentally. Just as a small baby new from the womb may say Mama! Dada! whereas in the womb it said nothing; so humanity, in the womb of Pan, said nought. But when humanity was born into a separate idea of itself, it said Pan.

In the days before man got too much separated off from the universe, he *was* Pan, along with all the rest.

As a tree still is. A strong-willed, powerful thing-in-itself, reaching up and reaching down. With a powerful will of its own it thrusts green hands and huge limbs at the light above, and sends huge legs and gripping toes down, down between the earth and rocks, to the earth's middle.

Here, on this little ranch under the Rocky Mountains, a big pine tree rises like a guardian spirit in front of the cabin where we live. Long, long ago the Indians blazed it. And the lightning, or the storm, has cut off its crest. Yet its column is always there, alive and changeless, alive and changing. The tree has its own aura of life. And in winter the snow slips off it, and in June it sprinkles down its little catkin-like pollen-tips, and it hisses in the wind, and it makes a silence within a silence. It is a great tree, under which the house is built. And the tree is within the allness of Pan. At night, when the lamplight shines out of the window, the great trunk dimly shows, in the near darkness, like an Egyptian column, supporting some powerful mystery in the over-branching darkness. By day, it is just a tree.

It is just a tree. The chipmunks skelter a little way up it, the little black-and-white birds, tree-creepers, walk quick as mice on its rough perpendicular, tapping; the bluejays throng on its branches, high up, at dawn, and in the afternoon you hear the faintest rustle of many little wild doves alighting in its upper remoteness. It is a tree, which is still Pan.

And we live beneath it, without noticing. Yet sometimes, when one suddenly looks far up and sees those wild doves there, or when one glances quickly at the inhuman-human hammering of a woodpecker, one realizes that the tree is asserting itself as much as I am. It gives out life, as I give out life. Our two lives meet and cross one another, unknowingly: the tree's life penetrates my life, and my life the tree's. We cannot live near one another, as we do, without affecting one another.

The tree gathers up earth-power from the dark bowels of the earth, and a roaming sky-glitter from above. And all unto itself, which is a tree, woody, enormous, slow but unyielding with life, bristling with acquisitive energy, obscurely radiating some of its great strength.

It vibrates its presence into my soul, and I am with Pan. I think no man could live near a pine tree and remain quite suave and supple and compliant. Something fierce and bristling is communicated. The piny sweetness is rousing and defiant, like turpentine, the noise of the needles is keen with œons of sharpness. In the volleys of wind from the western desert, the tree hisses and resists. It does not lean eastward at all. It resists with a vast force of resistance, from within itself, and its column is a ribbed, magnificent assertion.

I have become conscious of the tree, and of its interpenetration into my life. Long ago, the Indians must have been even more acutely conscious of it, when they blazed it to leave their mark on it.

I am conscious that it helps to change me, vitally. I am even conscious that shivers of energy cross my living plasm, from the tree, and I become a degree more like unto the tree, more bristling and turpentiney, in Pan. And the tree gets a certain shade and alertness of my life, within itself.

Of course, if I like to cut myself off, and say it is all bunk, a tree is merely so much lumber not yet sawn, then in a great measure I shall be cut off. So much depends on one's attitude. One can shut many, many doors of receptivity in oneself; or one can open many doors that are shut.

I prefer to open my doors to the coming of the tree. Its raw earth-power and its raw sky-power; its resinous erectness and resistance, its sharpness of hissing needles and relentlessness of roots, all that goes to the primitive savageness of a pine tree, goes also to the strength of man.

Give me of your power, then, oh tree! And I will give you of mine.

And this is what men must have said, more naïvely, less sophisticatedly, in the days when all was Pan. It is what, in a way, the aboriginal Indians still say, and still *mean*, intensely: especially when they dance the sacred dance, with the tree; or with the spruce twigs tied above their elbows.

Give me your power, oh tree, to help me in my life. And I will give you my power: even symbolized in a rag torn from my clothing.

This is the oldest Pan.

Or again, I say: "Oh you, you big tree, standing so strong and swallowing juice from the earth's inner body, warmth from the sky, beware of me. Beware of me, because I am strongest. I am going to cut you down and take your life and make you into beams for my house, and into a fire. Prepare to deliver up your life to me."

Is this any less true then when the lumberman glances at a pine tree, sees if it will cut good lumber, dabs a mark or a number upon it, and goes his way absolutely without further thought or feeling? Is he truer to life? Is it truer to life to insulate oneself entirely from the influence of the tree's life, and to walk about in an inanimate forest of standing lumber, marketable in St. Louis, Mo.? Or is it truer to life to know, with a pantheistic sensuality, that the tree has its own life, its own assertive existence, its own living relatedness to me: that my life is added to, or militated against, by the tree's life?

Which is really truer?

Which is truer, to live among the living, or to run on wheels?

And who can sit with the Indians around a big campfire of logs, in the mountains at night, when a man rises and turns his breast and his curiously smiling bronze face away from the blaze, and stands voluptuously warming his thighs and buttocks and loins, his back to the fire, faintly smiling the inscrutable Pan-smile into the dark trees surrounding, without hearing him say, in the Pan-voice: "Aha! Tree! Aha! Tree! Who has triumphed now? I drank the heat of your blood into my face and breast, and now I am drinking it into my loins and buttocks and legs, oh tree! I am drinking your heat right through me, oh tree! Fire is life, and I take your life for mine. I am drinking it up, oh tree, even into my buttocks. Aha! Tree! I am warm! I am strong! I am happy, tree, in this cold night in the mountains!"

And the old man, glancing up and seeing the flames flapping in flamy rags at the dark smoke, in the upper fire-hurry towards the stars and the dark spaces between the stars, sits stonily and inscrutably: yet one knows that he is saying: "Go back, oh fire! Go back like honey! Go back, honey of life, to where you came from, before you were hidden in the tree. The trees climb into the sky and steal the honey of the sun, like bears stealing from a hollow tree-trunk. But when the tree falls and is put on to the fire, the honey flames and goes straight back to where it came from. And the smell of burning pine is as the smell of honey."

So the old man says, with his lightless Indian eyes. But he is careful never to utter one word of the mystery. Speech is the death of Pan, who can but laugh and sound the reed-flute.

Is it better, I ask you, to cross the room and turn on the heat at the radiator, glancing at the thermometer and saying: "We're just a bit below the level, in here"? Then to go back to the newspaper!

What can a man do with his life but live it? And what does life consist in, save a vivid relatedness between the man and the living universe that surrounds him? Yet man insulates himself more and more into mechanism, and repudiates everything but the machine and the contrivance of which he himself is master, god in the machine.

Morning comes, and white ash lies in the fire-hollow, and the old man looks at it broodingly.

"The fire is gone," he says in the Pan silence, that is so full of unutterable things. "Look! there is no more tree. We drank his warmth, and he is gone. He is way, way off in the sky, his smoke is in the blueness, with the sweet smell of a pine-wood fire, and his yellow flame is in the sun. It is morning, with the ashes of night. There is no more tree. Tree is gone. But perhaps there is fire among the ashes. I shall blow it, and it will be alive. There is always fire, between the tree that goes and the tree that stays. One day I shall go—"

So they cook their meat, and rise, and go in silence.

There is a big rock towering up above the trees, a cliff. And silently a man glances at it. You hear him say, without speech:

"Oh, you big rock! If a man fall down from you, he dies. Don't let me fall down from you. Oh, you big pale rock, you are so still, you know lots of things. You know a lot. Help me, then, with your stillness. I go to find deer. Help me find deer."

And the man slips aside, and secretly lays a twig, or a pebble, some little object in a niche of the rock, as a pact between him and the rock. The rock will give him some of its radiant-cold stillness and enduring presence, and he makes a symbolic return, of gratitude.

Is it foolish? Would it have been better to invent a gun, to shoot his game from a great distance, so that he need not approach it with any of that living stealth and preparedness with which one live thing approaches another? Is it better to have a machine in one's hands, and so avoid the life-contact: the trouble! the pains! Is it better to see the rock as a mere nothing, not worth noticing because it has no value, and you can't eat it as you can a deer?

But the old hunter steals on, in the stillness of the eternal Pan, which is so full of soundless sounds. And in his soul he is saying: "Deer! Oh, you thin-legged deer! I am coming! Where are you, with your feet like little stones bounding down a hill? I know you. Yes, I know you. But you don't know me. You don't know where I am, and you don't know me, anyhow. But I know you. I am thinking of you. I shall get you. I've got to get you. I got to; so it will be. —I shall get you, and shoot an arrow right in you."

In this state of abstraction, and subtle, hunter's communion with the quarry—a weird psychic connexion between hunter and hunted—the man creeps into the mountains.

And even a white man who is a born hunter must fall into this state. Gun or no gun! He projects his deepest, most primitive hunter's consciousness abroad, and finds his game, not by accident, nor even chiefly by looking for signs, but primarily by a psychic attraction, a sort of telepathy: the hunter's telepathy. Then when he finds his quarry, he aims with a pure, spellbound volition. If there is no flaw in his abstracted huntsman's *will,* he cannot miss. Arrow or bullet, it flies like a movement of pure will, straight to the spot. And the deer, once she has let her quivering alertness be overmastered or stilled by the hunter's subtle, hypnotic, *following* spell, she cannot escape.

This is Pan, the Pan-mystery, the Pan-power. What can men who sit at home in their studies, and drink hot milk and have lamb's-wool slippers on their feet, and write anthropology, what *can* they possibly know about men, the men of Pan?

Among the creatures of Pan there is an eternal struggle for life, between lives. Man, defenceless, rapacious man, has needed the qualities of every living thing, at one time or other. The hard, silent abidingness of rock, the surging resistance of a tree, the still evasion of a puma, the dogged earth-knowledge of the bear, the light alertness of the deer, the sky-prowling vision of the eagle: turn by turn man has needed the power of every living thing. Tree, stone, or hill, river, or little stream, or waterfall, or salmon in the fall— man can be master and complete in himself, only by assuming the living powers of each of them, as the occasion requires.

He used to make himself master by a great effort of will, and sensitive, intuitive cunning, and immense labour of body.

Then he discovered the "idea." He found that all things were related by certain *laws.* The moment man learned to abstract, he began to make engines that would do the work of his body. So, instead of concentrating upon his quarry, or upon the living things which made his universe, he concentrated upon the engines or instruments which should intervene between him and the living universe, and give him mastery.

This was the death of the great Pan. The idea and the engine came between man and all things, like a death. The old connexion, the old Allness, was severed, and can never be ideally restored. Great Pan is dead.

Yet what do we live for, except to live? Man has lived to conquer the phenomenal universe. To a great extent he has succeeded. With all the mechanism of the human world, man is to a great extent master of all life, and of most phenomena.

And what then? Once you have conquered a thing, you have lost it. Its real relation to you collapses.

A conquered world is no good to man. He sits stupefied with boredom upon his conquest.

We need the universe to live again, so that we can live with it. A conquered universe, a dead Pan, leaves us nothing to live with.

You have to abandon the conquest, before Pan will live again. You have to live to live, not to conquer. What's the good of conquering even the North Pole, if after the conquest you've nothing left but an inert fact? Better leave it a mystery.

It was better to be a hunter in the woods of Pan, than it is to be a clerk in a city store. The hunter hungered, laboured, suffered tortures of fatigue. But at least he lived in a ceaseless living relation to his surrounding universe.

At evening, when the deer was killed, he went home to the tents, and threw down the deer-meat on the swept place before the tent of his women. And the women came out to greet him softly, with a sort of reverence, as he stood before the meat, the life-stuff. He came back spent, yet full of power, bringing the life-stuff. And the children looked with black eyes at the meat, and at that wonder-being, the man, the bringer of meat.

Perhaps the children of the store-clerk look at their father with a *tiny* bit of the same mystery. And perhaps the clerk feels a fragment of the old glorification, when he hands his wife the paper dollars.

But about the tents the women move silently. Then when the cooking-fire dies low, the man crouches in silence and toasts meat on a stick, while the dogs lurk round like shadows and the children watch avidly. The man eats as the sun goes down. And as the glitter departs, he says: "Lo, the sun is going, and I stay. All goes, but still I stay. Power of deer-meat is in my belly, power of sun is in my body. I am tired, but it is with power. There the small moon gives her first sharp sign. So! So! I watch her. I will give her something; she is very sharp and bright, and I do not know her power. Lo! I will give the woman something for this moon, which troubles me above the sunset, and has power. Lo! how very curved and sharp she is! Lo! how she troubles me!"

Thus, always aware, always watchful, subtly poising himself in the world of Pan, among the power of the living universe, he sustains his life and is sustained. There is no boredom, because *everything* is alive and active, and danger is inherent in all movement. The contact between all things is keen and wary: for wariness is also a sort of reverence, or respect. And nothing, in the world of Pan, may be taken for granted.

So when the fire is extinguished, and the moon sinks, the man says to the woman: "Oh, woman, be very soft, be very soft and deep towards me, with the deep silence. Oh, woman, do not speak and stir and wound me with the sharp horns of yourself. Let me come into the deep, soft places, the dark, soft places deep as between the stars. Oh, let me lose there the weariness of the day: let me come in the power of the night. Oh, do not speak to me, nor break the deep night of my silence and my power. Be softer than dust, and darker than any flower. Oh, woman, wonderful is the craft of your softness, the distance of your dark depths. Oh, open silently the deep that has no end, and do not turn the horns of the moon against me."

This is the might of Pan, and the power of Pan.

And still, in America, among the Indians, the oldest Pan is alive. But here, also, dying fast.

It is useless to glorify the savage. For he will kill Pan with his own hands, for the sake of a motor-car. And a bored savage, for whom Pan is dead, is the stupefied image of all boredom.

And we cannot return to the primitive life, to live in tepees and hunt with bows and arrows.

Yet live we must. And once life has been conquered, it is pretty difficult to live. What are we going to do, with a conquered universe? The Pan relationship, which the world of man once had with all the world, was better than anything man has now. The savage, today, if you give him the chance, will become more mechanical and unliving than any civilized man. But civilized man, having conquered the universe, may as well leave off bossing it. Because when all is said and done, life itself consists in a live relatedness between man and his universe: sun, moon, stars, earth, trees, flowers, birds, animals, men, everything—and not in a "conquest" of anything by anything. Even the conquest of the air makes the world smaller, tighter, and more airless.

And whether we are a store-clerk or a bus-conductor, we can still choose between the living universe of Pan, and the mechanical conquered universe of modern humanity. The machine has no windows. But even the most mechanized human being has only got his windows nailed up, or bricked in.

# ANOTHER WEEK ON THE CONCORD AND MERRIMACK RIVERS

*Raymond Mungo*

God help us,
refugees in winter dress,
Skating home on thin ice
from the Apocalypse.
    —Verandah Porche

To one who habitually endeavors to contemplate the true state of things, the political state can hardly be said to have any existence whatever. It is unreal, incredible, and insignificant to him, and for him to endeavor to extract the truth from such lean material is like making sugar from linen rags, when sugar-cane may be had. Generally speaking, the political news, whether domestic or foreign, might be writen to-day for the next ten years with sufficient accuracy. Most revolutions in society have not power to interest, still less alarm us; but tell me that our rivers are drying up, or the genus pine dying out in the country, and I might attend.
    —Henry D. Thoreau, *A Week on the Concord and Merrimack Rivers*

*Raymond Mungo tried to remain in the movement. He tried as a radical journalist for the Boston University News. After graduation, he tried by working for the Liberation News Service, an organization that sent dispatches to student and underground papers around the country. Now he and his friends live in an isolated farm in Vermont. In the following article, why he chose to "drop out" becomes painfully clear.*

SOURCE: From the book *Total Loss Farm: A Year in the Life* by Raymond Mungo. Copyright © 1970 by Raymond Mungo. Reprinted by permission of E. P. Dutton & Co., Inc., publishers.

## Friday: Portsmouth, N.H.

The farm in Vermont had fooled us, just as we hoped it would when we moved there in early '68; it had tricked even battle-scarred former youth militants into seeing the world as bright clusters of Day-Glo orange and red forest, rolling open meadows, sparkling brooks and streams. I had lived in industrial, eastern New England all my life, though, as well as worse places like New York and Washington, D.C., so I might have known better. But Vermont had blurred my memory, and when we finally left the farm for Portsmouth, I was all Thoreau and Frost, October up North, ain't life grand, all fresh and eager to begin rowing up the Concord and Merrimack rivers in the vanished footsteps of old Henry D. himself. Verandah Porche, queen of the Bay State Poets for Peace, packed the failing '59 VW and we went tearing down the mountain, kicking up good earth from the dirt road and barely slowing down for the 18th-century graveyard and all manner of wild animals now madly racing for shelter against the sharp winds of autumn in these hills. The frost was on the pumpkin, it was our second autumn together, and warm vibrations made the yellow farmhouse fairly glow in the dying daylight as we pointed east, over the Connecticut River, heading for our rendezvous with what *he* called "the placid current of our dreams." Knockout October day in 1969 in Vermont. All the trees had dropped acid.

The idea had come to me in a dream. It was one of those nights after Steve brought the Sunshine (wotta drug) when I'd wake up and sit bolt upright, alarmed at a sudden capacity, or *power,* I had acquired, to *see* far. I could see eternity in the vast darkness outside my window and inside my head, and I remembered feeling that way when but an infant. In my dream I was floating silently downstream in a birchbark canoe, speechless me watching vistas of bright New England autumn open up with each bend, slipping unnoticed between crimson mountains, blessing the warm sun by day and sleeping on beds of fresh leaves under a canary harvest moon by night. I was on the road to no special place, but no interstate highway with Savarinettes and Sunoco for this kid; in my dream, I was on a natural highway through the planet, the everlovin' me-sustainin' planet that never lets you down. Said Henry: "I have not yet put my foot through it."

It was the farm that had allowed me the luxury of this vision, for the farm had given me the insulation from America which the peace movement promised but cruelly denied. When we lived in Boston, Chicago, San Francisco, Washington (you name it, we lived there; some of us still live there), we dreamed of a New Age born of violent insurrection. We danced on the graves of war dead in Vietnam, every corpse was ammunition for Our Side; we set up a countergovernment down there in Washington, had marches, rallies and meetings; tried to fight fire with fire. Then Johnson resigned, yes, and the universities began to fall, the best and oldest ones first, and by God every 13-year-old in the suburbs was smoking dope and our numbers multiplying into the millions. But I woke up in the spring of 1968 and said, "This is not what I had in mind," because the movement had become my enemy; the movement was not flowers and doves and spontaneity, but another vicious system, the seed of a heartless bureaucracy, a minority Party vying for power rather than peace. It was then that we put away the schedule for revolution, gathered together our dear ones and all our resources, and set off to Vermont in search of the New Age.

The New Age we were looking for proved to be very old indeed, and I've often wondered aloud at my luck for being 23 years old in a time and place in which only the past offers hope and inspiration; the future offers only artifice and blight. I travel now in a society of friends who heat their houses with hand-cut wood and eliminate in outhouses, who cut pine shingles with draw-knives and haul maple sugar sap on sleds, who weed potatoes with their university-trained hands, pushing long hair out of their way and thus marking their foreheads with beautiful penitent dust. We till the soil to atone for our fathers' destruction of it. We smell. We live far from the market-places in America by our own volition, and the powerful men left behind are happy to have us out of their way. They do not yet realize that their heirs will refuse to inhabit their hollow cities, will find them poisonous and lethal, will run back to the Stone Age if necessary for survival and peace.

Yet this canoe trip had to be made because there was adventure out there. We expected to find the Concord and Merrimack rivers polluted but still beautiful, and to witness firsthand the startling juxtaposition of old New England, land and water and mountains, and new America, factories and highways and dams; and to thus educate ourselves further in the works of God and man. We pushed on relentlessly, top speed 50 mph, in our eggshell Volkswagen (Hitler's manifestly correct conception of the common man's car), 100 miles to the sea. The week following, the week we'd spend in our canoe, was the very week when our countrymen would celebrate Columbus Day (anniversary of the European discovery of Americans), the New York Mets in the World (American) Series, and the National Moratorium to demand an "early end to the war." Since we mourn the ruthless extinction of the natives, have outgrown baseball, and long ago commenced our own total Moratorium on constructive participation in this society, our presence and support was irrelevant to all of these national pastimes. We hoped only to paddle silently through the world, searching for traces of what has been lost.

Portsmouth was in an uproar.

\* \* \*

George and Martha Dodge are the parents of the revolution as well as of seven sons, all of whom have now come home to Portsmouth, one of the oldest ports on the Atlantic side and of some importance to the United States

Armed Forces. Gus, as he is nicknamed, is a respected physician in the city; Martha was a Nichols and still fries her own October donuts. Both are descendants of the oldest New England families, both old-fashioned, hospitable, warm, full of common sense, both admirers of Eldridge Cleaver and passionately involved, almost wracked, in attempts to right some of the American wrongs. In short, they are good candidates for an old homestead in Vermont, and yet themselves the most attractive natural resource left in Portsmouth. Another feature of the town is its extraordinary number (and quality) of 17th-century and 18th-century houses, built with virgin lumber which has yet to begin rotting or even chipping, but many of these houses are being stupidly and arbitrarily destroyed. (More about this in a moment.) Their sons, youngest first, are Peter, 14, who claims he can drive a motorcycle; Hovey, 16, who puts together electronic systems, including piecemeal stereo systems capable of blasting out "Goddamn the Pusher Man" and other hits from *Easy Rider* at astonishing volume and fidelity; Frank, 19, an accomplished cellist; Mark, 22, a soulful painter; Laurie, 25, a New Age carpenter; David, 27, a man of many pursuits who at the moment is restoring his house on South Street; and Buzzy, the oldest but no particular age, who can do anything.

It was Buzzy we had come to get, for Buzzy was our Native Indian Guide to the Concord and Merrimack rivers, and Buzzy could do anything. Had not Buzzy camped out at 60° below zero in Alaska? Wasn't it Buzzy who ran the rapids of the Pemigewasset? Didn't Buzzy fix the freezer with a clothespin or something? Buzzy can build a fire out of wet pine, sleep in a hollow log, make a shed into a mansion, or scale a snow-peaked mountain. If you are thinking of some perilous undertaking, my friend, my advice is to take Buzzy along. He is gifted with a calm and intelligent temperament, and a general all-around competence which is nothing less than astounding, particularly to half-freaked former militants trying gamely to live the life and discover what the planet is made up of. Mind you, Buzzy is no more remarkable than the rest of the Dodges, each in his or her fashion, but I haven't paper enough to go into the whole family (maybe some other time, over a fire, when we are alone), and the Dodges aren't seeking the publicity anyway. Buzzy and his wife, the former Erika Schmidt-Corvoisier, had been tripping around in Spain for a year or so there, but they were back in Portsmouth, now restoring from scratch an old house on Hanover Street and living in it with neither insulation in the walls nor a furnace in the basement, but with fireplaces older than anyone could remember. We apologized to Erika, Verandah and I, but we needed Buzzy for an historic river trip. She understood.

We went over to the main house, the Dodge Commune I called it, where the canoe was waiting for us, stored in the garage alongside children's bicycles, rakes, spare parts, nuts-and-bolts jar, the accumulation of seven sons' childhoods in Portsmouth by the sea. Our old friend Laurie, who lived with us in Vermont before the inexplicable magnet of Portsmouth drew him away, took us aside for a long walk through the Desolation Row of fine old buildings scheduled for demolition by Portsmouth Urban Renewal, and he showed us these houses from a carpenter's careful perspective. We touched the beams 14 inches thick, the planks wider than an arm-span, and gingerly stepped over broken glass where vandals had wrecked and robbed after the tenants of these buildings were forced to leave. There had been no protest over the demolition of the 17th-century in Portsmouth, not more than a whimper really, and I felt my long-dormant sense of outrage beginning to rekindle, and I knew I had to split. For outrage leads to action, and action leads . . . where? Usually into a morass. It was strange, though, my outrage reborn not over some plan for future progressive society, but over concern for preservation of ancient hoary stuff from way back. That kind of stuff, I had always thought, is for Historical Society ladies. But when the whole world becomes one McDonald's Hamburger stand after another, you too will cry out for even a scrap of integrity.

Back at the Dodge Manse, everything was in healthy chaos as the entire family readied for a trip to Martha's mother's farm in Sturbridge, Mass. The driveway was lined with vehicles which showed the scars of their years of heavy use. Laurie's red pickup was chosen to carry the canoe, first to Sturbridge, then back to Boston (Cambridge), from which it would be driven on a friendly Volvo station wagon to Concord, Mass., where the river trip would begin. A lecturer from the University of New Hampshire appeared while Buzzy was going through the intense gymnastics required to fasten the canoe to the truck, and he stood by urging Buzzy to come talk to his class about ecology and such while Buzzy said, "Yeah, sure" and "If I can find the time," all the while spinning Indian knots of every esoteric variety and the rest of the family carrying luggage (Hovey with his complete stereo system) for the two-day visit to the ancestral farm, and Dr. Gus looking angrily for the missing Peter. Laurie danced in his boots as he painted OCTOBER 15 in big black letters on the sides of the upturned canoe; good advertising for the Moratorium, he said, and even if the Moratorium showed signs of being a schmucky liberal thing, it was the best we had.

Porche and I hadn't counted on a Dodge excursion, and we found ourselves with two days to kill as Saturday dawned. To stay in Portsmouth with the Dodges all gone would have been too depressing, we agreed, so we repacked camping gear and artifacts of outdoor living into aforementioned VW, and decided to wait it out in old college hangouts, blast from the past, in Cambridge.

We split the map south along the green line designated as the Atlantic, uncomplicated by route numbers and little Esso markers, went to hole up in of all places, Cambridge.

## Saturday: Danvers, Mass.

Interstate Route 95, like many another road in Massachusetts, is forever incomplete. The signs bravely contend that this Detour is merely Temporary, but the same Detour has existed since I was born and reared in these parts, and I cannot be convinced that Interstate Route 95 will ever be finished. For the time being, motorists (who probably deserve it) are required to get off 95—which is supposed to be a north-south road from Boston to Canada—at Topsfield, Mass., and take the last 15 miles into Boston on Route 1, assaulted on all sides by gas stations, boogie restaurants (Mr. Boogie, Boogie King, McDonald's Boogie, Boogie Delight, etc.), furniture outlets, pseudo-Native junk and tourist emporiums with names like Trading Post and Wampum Shop, drive-in movies; anything goes down on old Route 1.

In the course of effortlessly rolling south on the paved planet, however, we finally entered Danvers on old Route 1 and something in my head exploded. I had not been in Danvers for six or seven years, but it was the town where I had done four years once, as a student at St. John's Preparatory School, hereinafter called The Prep. I remembered that October was always spectacularly beautiful there, though I could recall little else good about The Prep. But here was I in Danvers once again and who knew how many years would pass before I'd be back? We pulled off the highway at the candy outlet and headed up Route 62, past the State Mental Hospital, where the old lady got on our bus every morning in 1960 and asked us "Is your name John? Is it Peter?"

I have had some experience with mental hospitals in Massachusetts, though never, through chance or whatever, as an inmate; but I could never tell you in mere words the horror and despair they enclose in vain striving to rid the air in the Commonwealth of beserk and helpless vibrations. The Commonwealth as a whole is full of nightmares, universities, and museums aside or perhaps inclusive; all the authorities out of control, people getting screwed right and left, but whaddya gonna do kid all the politishun's crooked everybody knows.

I have also had some experience with Roman Catholicism, which is still alive in Massachusetts, where it too is out of control. They say there is no more virulent anti-Catholic than a former one; I'm living proof of the bigotry that comes of rebellion to indoctrination. Recent news reports, which one can never trust, indicate that it is slowly dying of its own anachronisms but I have seen too many millions of crucifix-kissing 8-year-olds to be satisfied with gradual progress. Listen, the Stephen Daedalus withdrawal, which began for me at puberty, ain't no joke, and it is thoroughly avoidable.

"What kind of a school is this?" Verandah asked as we pulled uphill to the lofty and expensive spires of The Prep, where some kind of Parents' Weekend was evidently in progress. Well, it's a school which houses the dead. The dead me resides there still. It's a school where terror of God is a tool, where violence between teacher and student was common, where sex was reserved for toilet stalls. This school was nowhere to go on a fine Saturday afternoon, yet there we were and all the parents looking shocked and the dolled-up Brothers looking confused.

Once a Catlick, always a Catlick, that's what really scared me I suppose. These guys and the Good Sisters before them really did a job on me, and I feared getting out of the friendly VW would expose me to the germs of the old disease, that I'd meet some former teacher who'd ask me had I been keeping the faith, and I'd mumble something indistinct because I couldn't summon the audacity to rant and scream, "You mothers are going to pay for all this!" as would have been appropriate. Actually they might never have asked me about the faith for my opinions are well-understood at The Prep, or so an old school friend, now a Marine lieutenant in Vietnam, once told me; he said I had the status of a What-Went-Wrong case there. But we did get out of the car and we walked around the neatly mowed grass and under the glowing October trees now dull in comparison to Vermont and overshadowed, as they were landscaped to be, by the crosses and towers of the old school. I didn't meet any old teachers and I never introduced myself to anybody. I did try to talk to one old friend but the secretary informed me that some of the Brothers had gone out with some of the Boys and my friend was not around. Out with the Boys, all of whom we encountered looked strictly 1955, short hair and ties, and I knew the poor bastards would mostly miss the boat, just as most of my classmates did. Missing the boat is just about the worst thing that can happen to a young man in America today, for where is he if he's still on the other side? In the company of the constipated, that's where. The best way he can phrase his situation is in terms of reform—he is reforming the churches, schools, corporations, by belonging to them. "Well, that's fine if it's the best you can do," say the self-righteous freaks from their tight-knit brotherhood of hair and leisure and though my opinions belong to the latter class, my heart goes out to the lonely ones who missed that goddamn boat and will never see another chance.

We all got on the boat late in life, I understand, and perhaps our children if we can overcome our fear of bringing them into the world, perhaps they will be afloat from the first. That sort of apparent progress would be gratifying, and I have met wonderful Acid Children in my time. This interlude is over. We raced back to the car and resumed our path to Cambridge, watching the signs carefully to avoid the big big hex in this, the traditional Season of the Witch in New England.

## Sunday: Cambridge, Mass.

I was reading the *Boston Sunday Globe* Financial Section for lack of other employment or reading matter when I came upon a new account of the spectacular success of a chain of artsie-fartsie shops called Cambridge Spice and Tea Exporters (or something close to that). These shops sell ornaments for the home, bamboo ding-dongs to hang over the window, incense, colorful but useless items of all sorts; and the proprietor was there quoted to the effect that the word "Cambridge" on the shops gives them a magic quality that brings in the bread right quick. And of course! Funny I never realized it before, but Cambridge is the home base, one of the centers, at least, of useless conceits for the affluent American, including the longhaired variety. Harvard University, if I may say so, could vanish tomorrow (in fact it *may*) with no appreciable loss to the physical or intellectual health of the nation. Those who wished to study Catullus would continue to do so; and those whose lives are considerably less earnest would doubtless find some other occupation, perhaps more rewarding than hanging out in the Yard. The great irony of Cambridge is that, despite its vaunted status as a center of the arts, education, technology, and political wisdom, it is in reality a Bore. It stultifies, rather than encourages, productive thought and employment, by throwing up countless insuperable obstacles to peace of mind and simple locomotion from one place to another. Why, if all the creative energy expended in Cambridge on paying telephone bills, signing documents, finding a cab, buying a milkshake, bitching at the landlord and shoplifting from Harvard Coop could be channeled into writing, playing, loving, and working, the results would probably be stupendous. At the moment, it is simply a marketplace of fatuous ideas and implements for those who seek to amuse themselves while Babylon falls around them. Thoreau on Boston:

> I see a great many barrels and fig-drums—piles of wood for umbrella-sticks,—blocks of granite and ice,—great heaps of goods, and the means of packing and conveying them,—much wrapping-paper and twine,—many crates and hogsheads and trucks,—and that is Boston. The more barrels, the more Boston. The museums and scientific societies and libraries are accidental. They gather around the sands to save carting. The wharf-rats and customhouse officers, and broken-down poets, seeking a fortune amid the barrels. (*Cape Cod*)

There are some useful items to be purchased in Cambridge (and Boston) but they are hard to find; it is next to impossible, for example, to find oak beams for building, spare uncut wood for burning, or fresh vegetables for eating, and certainly not worth the trouble you would encounter in getting them—unless you are inextricably bound to the city. Similarly, there are many men and women worth meeting there, people whose lives are neither devoted to poisoning the environment and water nor to idle and dispassionate amusements, but I have found such folks increasingly difficult to locate. You remember so-and-so who used to live on Brookline Street? Split to the Coast. Somebody else is in Nova Scotia, many are in Vermont, a few have taken to caves in Crete, a whole group went way up in British Columbia. Perhaps I am speaking only of a limited generational attitude, surely the managers of Cambridge corporations and deans of the schools have not all split too, but I know that there is more to it than youth and mobility, more than that youthful restlessness which George Apley was sent to Europe to work off. No, there's a definite panic on the hip scene in Cambridge, people going to uncommonly arduous lengths (debt, sacrifice, the prospect of cold toes and brown rice forever) to get away while there's still time. Although we grew up, intellectually and emotionally at least, in Cambridge, and once made the big scene there in scores of apartments and houses, V and I now could find only one friendly place to lay our heads and weary bums, and that was at Peter Simon's. We went looking for Peter's head of wild curly red hair, he looks like a freaked-out Howdy Doody really, a photographer, sure that when we found it there'd be new Beatles and Band music, orange juice in the frig, place to take your shoes off; and so there was.

We had brought along for the canoe trip the kind of things that made sense: sleeping bags, tarp, tools, cooking utensils, potatoes and other vegetables we'd grown in the summer, several gallons of honest-to-God Vermont water (no bacterial content) in the event the waters of the Merrimack should be beyond boiling. We couldn't bake bread on the river banks, surely, as we do at home, and, sensing that Henry's advice on buying bread from farmers just didn't apply these days, I went to a local sooper-dooper and acquired two loaves of Yah-Yah Bread at 20 cents the loaf and, almost as an afterthought, got a jar of Skippy Peanut Butter for about 40 cents. The Yah-Yah Bread was packed in a psychedoolic magenta plastic with cartoons of hipsters (one boy, one girl) on the outside and Avalon Ballroom lettering, the kind you must twist your head to read, so it did catch my eye. And I have liked Skippy since I discovered (1) peanuts will not grow very well in Vermont, (2) the jar can be used as a measuring cup (but only when it's empty), (3) the Skippy heiress is 22 and some kind of pill freak who busts up cocktail parties in New York. I noticed that the Skippy contained no BHA or BHT but that the Yah-Yah Bread did; these chemicals are often called "preservatives," and although I can't responsibly suggest they will kill *you*, they do contain the element which makes most commercial foods taste *dead*. We have found that an astonishingly wide variety of food items contain BHA or BHT or both, so I can only conclude that most of my countrymen subsist on the stuff. They are hooked. The sole advantage of preservatives to the consumer, it seems, is that he can now save money by buying day-old or month-old baked goods and be certain that they will taste like cold putty no matter their birthday. We did spend a goodly part of the harvest season giving away all the fruits and vegetables we couldn't use to city people (old friends and family) who freaked on what a tomato, or a peach, really is. The middle-aged and elderly ones remembered; the young ones learned. One and all reflected on how sinister and subtle the Dead Food craze came on, how you didn't notice it taking over until it was too late. The old Victory Garden thing may be in for a revival, friends, but I suspect it will reach only a marginal part of the population, the others will be too busy at the shop or office, dump DDT or other chemical killers on their crop, or be afraid to eat an ear of corn that's white, a tomato with a hole in it, a carrot with dirt on it. Tough luck for them what think it's Easier to go to the sooper-dooper and get those nice *clean* apples wrapped in cellophane, uniform in size and shining like mirrors, the kind I have never seen growing on any tree. How about *you*?

We escaped the supermarket, thus, without being tempted by the Meat, Poultry, or Vegetable departments, not to mention the paperbacks and plants. And we then did what everybody does in Cambridge, which amounted to what Bob Dylan called Too Much of Nothing. We waited for the morning to come, the daybreak which would put us on the rivers in our canoe at last; we got stoned and listened to the Beatles; we got bored and went out to spend some money, finally choosing a hip movie-house on Massachusetts Avenue and killing some hours with old Orson Welles. We did not get raped, mugged, or robbed as it turned out. We heard the noises and smelled the smells, drank the water and breathed the air. It was altogether quite a risky adventure. Our guides, Plucky Peter and his lady-friend Nancy, who is *only 17*, could not have been more hospitable and reassuring; in fact, they agreed to accompany us on parts of the river trip, grateful for some excuse to cut boring college classes they said. And Nancy even cooked a fine meal out of some farm vegetables on a stove which produced instant heat from gasoline which comes from under the street!

The canoe arrived after dark, good old Buzzy with it spinning yarns of rapids and dams, islands to camp on (the name Merrimack meant to the natives "river of many islands"), wild animules, the likelihood of rain. He and Verandah went to sleep early, I stayed up nervously watching commercials on television (including one post-midnight Stoned Voice urging kids not to smoke dope *because it's illegal*), went to sleep on the floor and dreamed of wild muskrats and other creatures of the past.

## Monday: Concord, Mass.

Monday dawned quietly even in the Hub of the Universe for Monday this time around was a Holiday, the day after Columbus Day. I guessed that those who had gone off on three-day weekends had not yet returned, and the others were all sleeping late; because here it was Monday morning, and

Central Square was not putrid with humanity, just a few winos hanging around and no policemen for traffic. The canoe advertising OCTOBER 15 was loaded onto Peter's Volvo while I hurry-hid my VW in the neighbor Harry's backyard; Harry was not around anyway, Harry had split to Vermont, but I left Harry a note explaining that since his backyard was full of garbage anyway, it might as well have my VW. Our canoe was 18 feet long and three feet wide, bright orange, and aluminum. Buzzy had fashioned wheels for it out of a block of wood and two old tricycle wheels, not unlike Thoreau's contraption I thought; they gave the canoe a faithful if bumpy ride around dams and such.

We took Route 2 past the shopping plazas and biochemical warfare factories out to Concord. There are two sets of signs in Concord, one leading to Walden, the other to the Concord Reformatory. The former is a state park with rules and regulations posted on the trees, the latter a prison for boys with a fancy-pants highway sign in front: "Welcome to Concord, Home of Emerson, Thoreau and the Alcotts." The Reformatory, a vast grey dungeon, is complemented by a farm where, I am told, the Boys learn vital agricultural skills. And not a few other tricks. The Brothers and the Boys. Pity the Boy who grows up in Massachusetts, if she has as many greystone towers to enclose him as it seems.

We stopped for advice, which way to the Concord River please, at a gas station. The man there obliged us, but all the while acting like we were wasting his valuable time. There were no other customers. The spot he led us to proved to be a park, full of monuments and walkways, grass mowed as with a Gillette Techmatic but a lovely spot notwithstanding. As we readied the canoe for embarkment, a uniformed gent approached us grimly, and I was sure there'd be some Commonwealth law against canoes but no, he merely wanted to admire the rig and satisfy his curiosity. It is quite legal to launch your boat in Concord still, though they have placed speed limit signs on the bridges ("River Speed Limit: 10 MPH ENFORCED."), and so we rolled ours to what looked like a good place and waited for a moment, very like the moment you take before diving off a high covered bridge into a gurgling fresh-water pond in July. Peter took funny-face pictures while a small band of strollers, tourists, or townspeople who can tell the difference, leaf-peepers we called them because they took Kodak Brownie shots of this or that red tree, gathered about to watch and wave. There was no obvious animosity between us this bright morning, for unlike the gas station, the place itself was beautiful, we were together, and it was a great day for a boat trip. Something in all men smiles on the idea of a cruise up the planet. We knew we'd be heading downstream, or north, to the mouth of the Merrimack, but the river itself had no easily discernible current; rather it looked from the shore like a quiet and friendly scar on the earth, made of such stuff you could put your foot through. Buzzy knew by some mysterious instinct which way was north, but I argued the point for a while. Then, as we were climbing into our silent craft, a noisy crowd of Canadian honkers drifted into view overhead, flying V-formation (V for victory, Vietnam, Verandah, Vermont) due south, and I declare even the tired holiday crowd broke into smiles. Canadian geese over Concord, it's enough to make you believe in God.

> The Musketaquid, or Grass-ground River, though probably as old as the Nile or Euphrates, did not begin to have a place in civilized history until the fame of its grassy meadows and its fish attracted settlers out of England in 1635, when it received the other but kindred name of CONCORD from the first plantation on its banks, which appears to have been commenced in a spirit of peace and harmony. It will be Grass-ground River as long as grass grows and water runs here; it will be Concord River only while men lead peaceable lives on its banks. To an extinct race it was grass-ground, where they hunted and fished; and it is still perennial grass-ground to Concord farmers, who own the Great Meadows, and get the hay from year to year.

Of course, get the hay! But the Great Meadows are mostly woods now, called the Great Meadows National Wildlife Refuge according to the brightly painted signs posted here and there on the banks, obviously intended for the information of those who would ride the river in boats. And as we paddled along, we did meet other boats, speedboats mostly with vroom-vroom motors and gaseous fumes who circled our canoe and laughed as it rocked in the unnatural waves of their passing. And one old couple, strictly Monet, paddling a tiny wisp of a canoe. Despite everything, though, the land *did* goddamn it open in a great vista, rising up on both sides to support scampering squirrels and the like, and while it lasted, the National Wildlife Refuge seemed to me a worthy piece of territory.

Buzzy tired of the paddles before Porche and I did, and over our protests, elected to turn on his pint-sized outboard, which went bap-bap rather than vroom-vroom, and moved the canoe no faster than the paddles but with less effort on our part, of course. I used this respite from work to survey the terrain with the close eye of loving ignorance, and I watched the Wildlife Refuge become plain old Concord and a pastel ranch house come into view. Everything moved so slowly, it was like a super-down drug, and we were spared no details of this modern American prefab architecture—and, beyond it, the rising towers of yon civilization. Fishermen began to appear, at first alone and then in groups; and though we dutifully inquired of each what he had caught that morning, we never found a man with so much as a catfish to show for his efforts. Clearly, I thought, it is Columbus Day (or the day after) and these people are fishing for old times' sake and not in hopes of actually catching something. The last group of fishers were segregated—a half-dozen white people on one side of the Concord, and as many black people on the other. The river was narrow and shallow enough at that point to walk across, so I guessed that these people wanted it that way, preferred at least to do their fruitless casting among friends. Soon enough, several hours later, we were in Carlisle, at the Carlisle Bridge, and I'd become concerned that the river still showed no sign of a current. It was just about standing still and we the only moving things in the landscape. Verandah trailed her fingers through the water from the bow. From my perch in the center, I remarked, "It's pretty but it's dead."

"Maybe we're dead" was all she said.

*   *   *

From Carlisle, where we met Peter Simon and enjoyed a Skippy and Yah-Yah lunch, we went on to Billerica with high hopes of making Lowell that day and thus getting over the New Hampshire border the next. For reasons obviously unassociated with fact, I expected the scenery, colors, and water in New Hampshire to be superior to those in Massachusetts, and we reassured ourselves that, bad as the Concord was now becoming, we were at least taking the worst medicine first. The entrance to Billerica by water resembles the old MGM views of distant forts in the wild West; for the first sign of the approaching town is an American flag flapping in the breeze like somebody's long-johns on the line, and planted on top of a hideous red-brick mill with a mammoth black smokestack. No smoke today, though, for it was a holiday remember (and we do need constant reminders on days like Columbus Day and Washington's Birthday, so difficult has it become for us to relate to them), and the only sign of life was a wilted elderly watchman, who sat behind the factory gates merely watching cars go by. The mill, called North Billerica Company (presumably manufacturers of North Billericas), was built on a dam which we didn't notice until we very nearly went over it, and seemed to be rooted in the water itself. That is, the sides of the buildings extended below the riverline, making the banks absolutely inaccessible except through the mill-yard itself, for several hundred yards. And the watch-man, clearly, was the old Keeper of the Locks whom Henry had charmed into letting him pass on the Sabbath. Thus did this kind man unchain the gates of the North Billerica Company and lead us through to the safe side of the dam —where, for the first time, we paddled through water actually being used, before our very eyes, as an open sewer. Worse yet, we recognized that the scuz and sludge pouring forth from the mill through 6-inch drainpipes would follow us downstream, that it was in fact better to navigate on dead but quiet waters than on water teeming with Elimination, at times even belching out gaseous bubbles, and smelling like fresh bait for tsetse flies and vultures. From North Billerica to the end of our journey, we would see only two other craft on these waters, one a crude raft bearing three boys (more or less 10-years-old) and a smiling dog, straight Huck Finn stuff but the kids said not a word as we passed them by, and the other a hardware-store rubber bathtub floating two 13- or 14-year-old boys who were headed for Concord

Reformatory, you could just tell. This latter pair were reincarnations of the Bad Boys I'd known back in Lawrence, which is on the Merrimack, boys whom I had joined in some Bad adventures on the river until I finally couldn't make their grade.

Boys will find charm in junk, as every red-blooded smalltown scoutmaster knows; boys will hang around burned-out houses, old railroad yards, town dumps, the backs of breweries, and find there unlimited access to toys for the body and mind. We met these two as our canoe bumped to a stop against huge rocks surrounding a factory which had burned to the ground, only the smokestack erect, nobody else around, as we hauled our gear out of the boat onto broken glass and pieces of brick and charred timbers that fell through when you stepped on them, in this unspeakably North Vietnamese place, Dresden in Billerica, this corner of Massachusetts which could be the scene after World War III. The boys informed us in a heavy local accent—Oh yah, Oh yah—that we'd pass "three rapids and a dam" before the Concord emptied into the Merrimack in Lowell, then left us alone again. Buzzy ran the first set of rapids alone while Verandah and I hauled knapsacks and sleeping bags, paddles and outboard, through the wreckage to the place where the river deepened. It was then that we began to notice the trees, even the trees in this place were palsied and skinny, their colors muted. An old stump I was using for support caved in on me. And to venture anywhere near the trees or brush meant to be covered with clinging brown dead burrs, pickies I call them, that fall into your socks and irritate your skin. We were grateful to get back in the canoe and leave that nightmare once-and-past factory behind. It was the worst place I had ever lived, a place where nothing could be salvaged, not even a piece of wire or useful stick on the ground.

A mile or so downstream, just south of Lowell we reckoned, the second set of rapids began and the canoe quickly became trapped between and on top of rocks which shared the water now with old tires, a refrigerator, a washing machine, wrecked cars and trucks, metal hoops, and bobbing clumps of feces. Verandah had to get out in mid-stream to lighten our load, and she disappeared into the pickies. A little later, I too got out as the canoe turned sideways, broadstream, and Buzzy warned in a calm and dejected tone, "We are going to capsize very soon, we will capsize if we don't get out of here." Boots tied about my neck and dungarees rolled up over my knees, there's me slipping on rocks slimy with who wants to to know what, making for a bank which appears impossible to scale. I lost sight of Buzzy as the canoe bounced and careened downstream, but caught Verandah in my free eye, silently waiting for me at the top of the rise. The current, which was imperceptible before, was ferocious now, as the shallow water rushed downhill over rocks left there by the Billerica dams; and more than once I felt myself falling over them, too, breaking bones I thought in my mad rush to the sea on a road greased with shit. The bank, when I reached it, was knee-deep in garbage of all kinds—metal, paper, and glass. Rolls of toilet paper had been strung like Christmas tinsel on the brittle limbs of the trees, and cardboard containers by the hundreds, flattened by snow and made soggy by rain, had formed layers of mush. I was the creature from the black lagoon, or a soul in purgatory, stretching forth his hand for a lift out of my slime from the mysterious beautiful lady Up There.

When the ascent was made and breaths caught, we discovered ourselves in a railroad yard whose tracks were varnished amber with rust and freight cars left there, open, to suffer all weather and never move again. Union Pacific, they proudly announced. Nobody was around. I fancied myself a television reporter for some new galaxy, bringing the folks back home a documentary of the continent on Earth that died: "it all began to break down, folks, in fourteen hundred and ninety-two, when Columbus sailed the ocean blue. To an extinct race, it was grass-ground river."

We walked the graveled planet now for maybe a half-mile, shouting for Buzzy from time to time. We found him at the end of the third set of rapids with gloom all over his face. The canoe took a bad leap, he said, the outboard was lost somewhere under water too black to reveal it, the Yah-Yah bread soaked to the consistency of liquid BHT, all the bedding and clothing and food dripping wet.

And the sun was setting somewhere but we could not see it.

And the air was turning colder though we couldn't say why.

And the land was impossible to camp on, it would be a bed of broken glass and rusty nails.

Clearly we could only push on to Lowell, where Peter Simon had been waiting hours for us no doubt, and push we did until we floated into the heart of that town after dark, almost bumping the edge of a vast dam in our blindness, then groping and paddling back and forth across this Concord to find a bank which was neither solid vertical concrete nor sealed off by a high chain-link fence, operating by the light of *The Lowell Sun* neon billboard and finally hauling OCTOBER 15 from the water behind a taxicab garage and wheeling it through the crowded center of town, wondering where we could safely be alive.

Lowell is a sister-city to Lawrence and Haverhill, all three being one-river towns born of the "industrial revolution" and very close in spirit to those almost-charming images of factory-towns in British literature from Blake to the Beatles. Ethnic neighborhoods remain and national churches (mostly Catholic) thrive there still—the Greeks still fiercely chauvinistic, the French Canadians still hard drinkers, the Italians still fond of block parties in honor of the Three Saints. There is a strikingly 19th-century downtown area but, despite the energetic promotion of the oldest merchants in town, it is slowly corroding as it loses ground to the highway shopping plazas. Life there is sooty, and even the young people look hard and wrinkled. Though it is only a stone's throw from cultured, boring Boston, it may as well be a thousand miles away for all the intellectual influence it has absorbed. We didn't know it at the moment we were strolling down Central Street with the canoe between us, but Peter Simon had earlier fled the city, terrified at the fierce looks and obscene catcalls which his long hair had provoked. I was not afraid, though, for I knew that the natives, while resenting our freedom, were yet too pacified and dulled by their daily lives to risk energetic hostility on us. Strangers may securely enough walk the streets of Lowell, Lawrence, or Haverhill, for the locals will kill only each other. Arriving in Lowell was for me a grand homecoming.

Kerouac came back to Lowell after all those years making scenes, and that has scared me crazy since I've known who Kerouac was. "If all else fails," I thought, "we could always go see Kerouac, maybe he'd put us up." Came back to Lowell even though nobody goes anywhere from there, he must have come back to die, that's the only thing makes sense by the gee. Stopped writing he did, just sat there in crummy Lowell with beer and television, and *The Lowell Sun* at four in the afternoon, delivered by the local altar boy at Saint Ann's, or Sacred Heart, or Saint Pat's. Was he an altar boy, chief Boy Scout, candidate for the priesthood, did he win a Ladies Sodality scholarship to The Prep? God, Kerouac, did you have a paper route too and hit all the bars on Christmas Eve? Christ, Kerouac, you're blowing my mind living in Lowell, will you never go back to Big Sur? Kerouac, listen: Frost came from Lawrence, too, hey from my neighborhood in South Lawrence, but he *got out* man and he didn't come back. Robert Frost! And didn't Jack Kennedy make him poet laureate or something? Kerouac, see: Leonard *Bernstein* came from here, but *he* got out! Everybody from Lowell and Lawrence had half a break in this world *split*. You stay here, you're as good as dead baby.

I wanted to go get Kerouac and put him in the middle of the canoe with his bottle and take him north to New Hampshire. Instead, I went looking for my younger brother Rick but his Greek landlord said he moved to a street that don't *exist* any more in Lowell. "Your brother hiza nice boy, I tell him 'two things, Rick: don't smoke no marihoony, geta you degree!' " I used to think a degree was the only ticket out of Lawrence and Lowell, too, but here I am on Central Street with my canoe!

We were befriended by a corpulent *Boston Record-American* reporter (Hearst sheet, cheesecake and crime mostly), who put us and the canoe on the back of his truck which he normally uses for carting secondhand furniture; man's got to make a living. He was also, he said, a member of the Lowell police force and found out about us from the police radio's moment-by-moment broadcast report of our progress through the city. He called the cops on a street-side phone and arranged for us to sleep on the Boulevard river bank, past all the dams and fetid canals of Lowell, and there we took our rest at last. The Boulevard traffic passed several yards from our heads at 60 and 70 miles per hour, and some local teenagers drove their jalopy up to our encampment with bright lights on at 2 or 3 A.M. The bank was littered with broken beer bottles, but I slept soundly nonetheless. We had no food now, so I got up in the night and walked up the Boulevard to where I knew an all-night pizza stand existed; and, in the process, bumped into a parked car with two kids fucking noisily in the back seat. Of course, I thought, Lowell is the last place on the planet where kids still ball in dad's car because there is no place to go, there are no private apartments for kids or independent kid-societies. Walking back with coffee-cake and hamburgers, I noticed dozens of parked cars just off the road, a road without sidewalks, where nobody but me had walked for a long time. And just before I got back to our encampment, I met an old man with whiskey on his breath who looked me straight in the eye and said, "Going to Lawrence?"

Around midnight, a group of married couples arrived with Dunkin Donuts for us to eat; they had heard of the legendary canoe, it was all over town, they wanted to see if it was really so. One man used to fish for salmon in the Merrimack, but he "wouldn't piss in it now." His wife blamed the rich people who own the mills, they are the ones she said who have destroyed the water. All who came to talk with us that night said how many years had passed since they last saw a real boat seriously navigating up the Merrimack River. "Are you sure," one woman asked in a harsh voice, "nobody's makin ya do it?"

## Tuesday: Lowell, Mass.

Culture-hero Steve McQueen has said, "I would rather wake up in the middle of nowhere than in any city on Earth." Naturally, I second that. Morning in Lowell cannot properly be called "sunrise," for it is the General Electric plant and not the burning star which first appears on the horizon. Our *Record-American* reporter friend returned to take pictures of us for his newspaper but we waited around a long time hoping for Peter Simon to arrive in the magic Volvo which could both fetch new groceries and go searching for the lost outboard. We would be paddling upstream now, and in the face of a stiff wind, so the motor might have proved useful in a pinch. But there was no Peter, no coffee, no breakfast and no hope, so we shoved off at 9 A.M., with only the *Boston Record-American* for witness. We had camped, it turned out, next to a row of garbage cans on which somebody in Lowell (maybe someday she'll come come come along) had painted peace signs and slogans like "Smile on your brother" and "Let's clean up Big Muddy." It was a noble but pathetic gesture, this youthful assumption that the Dirt in the Merrimack was nothing worse than Mud, and that it could be cleaned up if only each of us smiled more. As the rows of factories proved beyond doubt, and there is something hard and undeniable in this, Lowell would cease to be Lowell if it did not pollute the Merrimack River. Lowell and its sister-cities create shoes, textiles, and paper for you and me—who, as literate people, do not live on the Merrimack River anyway. The industries in Lowell pay their employees very poorly indeed, yet their profits cannot be what they used to be, for the shops are slowly and one-by-one closing down. We paddled furiously against the wind to get the hell out, aiming ourselves toward Tyngsboro by noon and Nashua, New Hampshire, by nightfall.

The Merrimack is substantially wider and deeper than the Concord, a real river and not just a stream, so for the first time I felt that flush of anxiety which comes after knowing you are too far-out to swim back in the event of trouble. It was back-aching work but we could manage about two miles per hour, which seemed to me fast enough for any sensible voyage. I set myself little targets, such as the big drive-in movie screen on Route 113, and over-took each one in my stride. I enjoy slow progress and gradual change in my own life as much as I deplore it in social trends; but I am sufficiently tuned-in to the century to realize that we men never really get *anywhere*. It's always more of the same, so to speak—birth, life, death, walking abroad in a shower of your days, how soon having Time the subtle thief of youth stealing on his wing your three-and-twentieth year, etc. etc. Life does move exquisitely slow, all the crap in newspapers about "revolutionary developments" aside, and we do tend to end up where we started. The absurdity of our situation, too, lay in the fact that we could have gotten from Lowell to Tyngsboro in three minutes rather than three hours, but there was no reason to go to Tyngsboro *anyway* as none of us believed it would be the idyllic spot Henry described; thus we never felt we were *wasting time*.

Two or three miles up from Lowell, as we paddled through water abso-lutely white with swirling pools of some awful chemical substance, we heard Peter Simon's voice calling as from afar. He and Nancy were on the opposite bank, trapped in the Volvo by a pack of ravenous house-dogs, yet overjoyed to have found us again. We paddled over to them and mutually decided on a spot just up a piece to disembark and confer. Verandah and Nancy stayed behind to cook a breakfast of oatmeal and eggs while we menfolk took off in the car to look for that outboard, got a flat on the Boulevard, got soaking wet, got in trouble with an elderly French Canadian lady who objected to Buzzy's using her backyard as an approach to the river until I calmed her in the best Lawrence-Lowell half-Canuck accent I could muster from memories of my grandmother. In all, got nothing accomplished and returned to the breakfast site close to noon, Peter swearing it was gonna rain and Buzzy just swearing. The outboard had cost B his last 60 dollars, and was purchased especially for this trip; moreover, he was beginning to feel sick in the stomach, and wondered just what poisons we might be picking up from the fair Merrimack.

I wanted Pierre to join us at that point, abandon his car and get on the boat. Fancying myself Kesey and all of us Merry Pranksters, I said, "Peter, you must be On the Boat or Off the Boat." But Tuesday was a Mets day, Peter said, and though he would follow us upstream and generally watch out, he must stay close to the car radio to keep tabs on Tom Seaver and so-and-so's stealing third. It meant nothing to me, but since Peter thought it was important, who was I to belittle it? Some people get their energy off Kesey and Kerouac and Thoreau, others off Seaver and Swoboda; stocks and bonds, movies and periodicals, movements and rallies, rivers and oceans, balls and strikes; you name it, somebody lives on it. Friends of mine have been addicted to such dangerous drugs as television, bourbon, and *The New York Times,* daily *and* Sunday. I myself have been addicted to Pall Mall cigarettes for years, and have more than once gone hungry to support my habit; I am also a black coffee freak, and have been known to drink 15 to 20 cups in a day. Everything in me which responds to reason prays for the imminent day when mass-produced and commercially distributed goods will simply stop coming, all the bright red Pall Mall trucks will break down in North Carolina and all the Colombian coffee boats will rot in their harbors. Then we, poor weaklings, will have at least a chance to aspire to that personal independence which we all so desperately need. We will be addicted to making do for ourselves, each of us will be President of the United States and responsible for the social welfare of the whole world, we will rise to our godheads at the same time we stoop to gather scrap wood for the fire. We will be able to afford, then, to offer and accept a little help from our friends.

So Peter was hooked on the Mets and there seemed no solution but to plan the rest of the trip *around* this handicap. Peter had to break camp early, drive to towns for newspaper reports of the previous day's game, leave the canoe to its own progress while he sought out television stores where the American Series would be coming across display color sets, return to us radiant with news of the latest victories. The Mets were *winners* at least, that's more than I could say for Pall Malls—which I consumed, though moderately, throughout the journey.

These pathetic addictions came together in Tyngsboro in an odd fashion. When we arrived at the bridge there, Peter was nowhere to be found, off watching the Mets; and we three were out of cigarettes and of course carrying no money. Buzzy, to the rescue, found a selection of old two-cent and nickel soda pop bottles imbedded in the silt bank, and cashed them in for a pack of smokes at the variety store conveniently located on top of the bank. All else we found there was a single half-rotted sunfish, five inches or so, washed ashore.

We were always looking for "a nice little island" on which to camp. The only one we found that day was King's Island, which is now a golf course with buildings, garages, a bar, and a bridge to the highway. Three lady golfers, the kind with jewel-encrusted sunglasses roped to their necks on aluminum chains, spied us from the ninth hole, and one chirped, "Well isn't that *adorable*." A painted sign on the bank read "Watch out for golf balls," and the river around the island had obviously become a God-made water trap for the wives of the Lowell-Nashua managerial class. It became evident near here, too, that many of the houses along the banks had eliminated the need for septic tanks by flushing directly into the river through underground pipes.

Both Buzzy and Verandah being now sick at their centers, and the prospect of sleeping in industrial Nashua too bleak to consider, we elected after much procrastination to drive around that city altogether, and thus ended up resuming the trip and camping out in Bow Junction, New Hampshire, birthplace of Mary Baker Eddy. Peter parked the Volvo on what we assumed to be a lonely access road and we paddled to what looked like a stretch of serious forest, arriving there just in time to spread out a few tarps and start a fire before dark fell. Stumbling about in the night in search of a place to Eliminate, I discovered that the woods were only 30 to 40 feet wide, bordered by the river on one side and a real, if dirt, road on the other; and they were only a quarter-mile long, bordered by immense machines of one kind or other on either end. The access road was studded with houses suburban-style, whose lights shined brightly at us and were reflected in the water, and the traffic on it sounded high-speed. We had been once more cruelly tricked. Sirens filled the air and our heads. Brakes screeched and a metallic thud bounced off our ears. The quiet but persistent rumble of technology charged the atmosphere, never letting up; it was the trembling of the earth which you, friend, can hear tonight if you but focus your attention on it. The earth is crying, what can I do to help it? Give it a Demerol?

## Wednesday: Bow Junction, N.H.

I love man-kind but I hate the institutions of the dead unkind. Men execute nothing so faithfully as the wills of the dead, to the last codicil and letter. *They* rule this world, and the living are but their executors.

—Thoreau, *A Week on the Concord and Merrimack Rivers*

When we were babes in college and thought ourselves the only people in America smart enough to be unilaterally opposed to the United States' presence in Vietnam, we'd sit around the Protestant house at B.U., though we were none of us Protestants, and say, "This war won't end until every mother who loses a son, every wife who loses a husband, knows that their men died *in vain*." As long as the families of the 42,000 dead in fruitless combat could congratulate themselves on giving a boy to a good cause, more deaths would be unavoidable, we analyzed. It seemed the very will of the dead that America continue its genocidal assault on the East, the voices of those Southside Bad Boys crying out "Get him back, Emile!" to the runty kid from Sacre Coeur Parish. I'm not sure when this attitude began to corrode, sometimes I flatter myself with the thought that I did my part to bring it about (though a fat lot of good it has done over *there*); but I see with my own eyes that the wife of a dead Marine in Manchester, New Hampshire, on the Merrimack, refuses to have her husband's coffin draped in the Stars and Stripes. There is great mourning in New Hampshire over a group of six men who come back in boxes; five are buried with all attendant military honors, the sixth with Bob Dylan and angry rhetoric. In Manchester, New Hampshire, the most reactionary town in all of New England. So the will of the dead *now* is that we take revenge on the government, on Lyndon Johnson (remember that stinker?) and Richard Nixon and Lew Hershey, McNamara, Rusk, Rostow, Clifford, Laird, Westmoreland, Abrams, as if these men together and alone caused it to happen, and not the entire lot of us. The American people, in taking revenge on the gooks, have all but destroyed the paradisial terrain and refined culture of Vietnam; now they will turn on themselves and do the same at home. What is ambiguously called "the system" will crumble and fall, it is all too clear. The economy, military effectiveness, control and discipline of the young, none of these is looking too good for "the system." What will replace it? Does it matter?

After Marshall Bloom's suicide, I was exhorted by some old friends to come back to Washington, where my personal adventure with Bloom began, and rip up a cloud in the streets; have a reunion with my former allies in the movement. I declined. Just as I have avoided Chicago, Berkeley, New York City, even Woodstock, where all the heavy scenes have been going down, I shall absent myself from Washington on November 15. For I am choosing to refuse to execute the wills of the dead. Marshall had asked me, in his note, to be an executor of sorts, distributing his personal things from a second-floor closet at the farm to his friends around the country and on the farms; but I can't even do that, at least I haven't been able to yet.

*The New York Times* seized on Marshall's death to print a five-column headline, "Suicide Puzzles Friends of Founder of Radical News Service," and an article which mocked his conviction that activists will move to rural areas because "the city burns people out." *The Times* suggested that the last laugh was on Marshall and his friends, for while the citified branch of our Liberation News Service was still churning out propaganda from Claremont Avenue in New York, we were running vacuum-cleaner hoses from exhaust pipes into vent windows and expiring of despair. And it is true winter is here, Michael's toe was broken by a cow, Richard is in the county hospital with an esoteric fever, John's VW was turned-over up on Route 91, Peter's father died last week in Pennsylvania, Pepper is in Rochester waiting for hers to go, the freezer broke down and much of the harvest moved to another house until we can fix it, no storm windows for lack of money and howling winds out-side. But it has nothing to do with the city versus the country, it has only to do with the strange twists in our lives which yet excite the attention of the newspapers who display our photographs and write our biographies as professional hippies and postrevolutionaries; and it has to do with Marshall himself, and there will never be another.

Marshall's death was the logical extension of the Concord and Merri-mack rivers trip; indeed, it followed hard on the heels of the boating. Sensitive as he was, he no doubt saw the opportunity to embellish the awfulest October in history and couldn't pass it up; get all the bad shit out of the way, he must have thought, before the new decade begins. What bad angel, thus, has elected to sit over our chimney? When your crop don't fail and your house don't burn down, your best friend will leave you stranded and helpless. Winter will come and snow you in, yet you can't move back to the city despite it because any natural hardship is better than an unnatural life. Every winter the hospitals in Vermont declare dead old men who just one evening neglected to light their stoves.

And here I had the chicken house one-quarter shingled, too, when it happened, and after that Saturday it rained day and night for six days. Everybody stared at each other, each was broken down in his unique way. Nothing got accomplished and yet there was nowhere to go.

Death generates death, then, though we know in our remaining animal instincts that organic material makes carrots grow. It will be a long winter with ghosts behind the walls, and what wise men could be certain that we will make it to the spring? Spring, or life, is always a surprise and a gift, not something we have earned any firm right to.

> It will be long ere the marshes resume,
> It will be long ere the earliest bird,
> So close all the windows and not hear the wind,
> But see all wind-stirred.
>
> —Frost

So the army of corpses, some freshly laid in the ground and others now grown cold and bony, led the people of my country to create a Moratorium, which was nothing more or less than a Memorial Day of the new regime. Didn't Ho Chi Minh have generals? Thus will the Provisional Revolutionary Government have its holidays, and the time of Vietnam will be marked in history books in Skokie, Illinois, as an era of great plague and disaster in the nation. And monuments raised to the great men who "gave their lives" in the service of destroying the old. Marshall wasn't like that, he searched for the life in things, but found it unsatisfactory in the end. He was always taking us down with him, demanding a group involvement in his pain, and he has done it again; and all in the course of living like crazy and kicking up, as John said, a lot of shit for 25 years old.

Was he serious about it or is this just Super-Burn? Will he show up in the cucumber patch next July, and will we say "Marshall, you son of a *bitch*"? Or will this empty numb half-heartedness go on forever, and will we always be sailing the River Styx in our canoe, surveying the damage? Spring is right around the coroner.

\*     \*     \*

From Bow Junction all the way to Plymouth, further north than Thoreau ever managed to get, we jumped from canoe to Volvo as sections of the river gave out underneath us, became too foul to navigate, turned into a bed of high sharp rocks, and trickled weakly through dams and obstructions thrown up by cities like Concord and Manchester, the latter being as one and all recognize the worst city on the planet. We drove to Plymouth at last, determined to find some water worth paddling through, and believing that the Pemigewasset, which runs through that town and becomes the Merrimack just north of Manchester, would still be relatively unspoiled. But in the course of the afternoon's rowing from there down to Ashland, we encountered more rapids alongside a sandbar which, when we sank into it, proved to be quick-sand mixed with shit and putrefaction impossible to describe. And we passed a yellow machine engaged in pushing trees into the water and despoiling the air with vast clouds of exhaust, so that even the atmosphere was no longer enjoyable and the sky invisible.

We also discovered that Route 93, which runs from Boston (via a long Detour, of course) up through Lawrence and north, follows the course of the Merrimack exactly, so that no camping spot or island left on the river can be free from the vroom-vroom noises of hell-for-leather diesel trucks and all-night passenger cars tooling up and down the planet bringing people their Pall Malls and Kentucky Bourbon, DDT and mass-produced foam rubber parlor chairs, and a million other things. And these monsters unkindly refused to declare Moratorium since they are not people anyway and thus insensitive to the needs of the living or the demands of the dead. Peter left his car, though, at a place in Ashland or Bridgewater where two bridges crossed the Pemigewasset, one for the railroad and the other for traffic, and we found ourselves all together as night fell on a forest glen in which all the trees were marked with surveyor's identifying paint, signifying that they were scheduled to be bulldozed in the near future. We made the last wood fire that place will know.

The stars were out despite everything and I gave them my thorough uneducated scrutiny (I have never been able to find the Big Dipper, though I can immediately recognize the Northern Lights when they come around in March) as I thought and thought about the war. For the first time I could remember, I felt not the slightest indignity at being punished for an evil I did not create or support. "You get what you pay for," as the fat Texans say. We lived off the destructive energy in Vietnam *even though* we *were opposed to it,* and now our efforts to find and encourage life are of doubtful promise at best. But we're still alive and trying, and I suppose you are too. Do you suppose it is too late?

Shall we go out and rebuild this thing together? That was on my mind. Will we be able to start anew without nature, with only mankind, to support us? Dresden in ashes was yet potentially a prosperous center for the manufacture of Volkswagens, what will come out of an Atlantic Ocean which casts death and waste on the beaches as well as foam and salt? For lack of anything more overwhelming to tackle, I am willing to try it. At least most of the time. Do you have the strength to join?

## Thursday: Ashland, N.H.

Breakfast was hearty and the coffee was strong, so this kid was raring once again to go, though by now with no illusions of having a pleasant or honestly working experience. He longed for his dog, Barf Barf, and thanked whatever stars put him in Vermont for the fact that Mr. B. didn't have to drink *this* water. He wondered what the point was in further subjecting his body and soul to such a diseased and hopeless piece of the earth, but pushed these reservations aside to climb into the bow for more of Buzzy's dead-serious lessons in steering. He was not prepared to discover, a full mile from the camping-place, that Peter Simon had lost his wallet somewhere among those doomed trees, with money, driver's license, and Bank Americard; and to eat up a large part of the day in searching the banks for the exact spot in Ashland where we'd camped and then finding the missing papers. God forbid that we should wander the rivers and forests of the planet without our papers in order! Why, friends of mine have been incarcerated for weeks simply for lacking the right papers while passing through Cheyenne, Wyoming. As much as we might philosophically contend that we are free creatures on God's earth, we do not question when a brother says, "Turn around, bowman, for my driver's license and Bank Americard."

Great confusion now ensued as we considered which way to go: north to the White Mountain National Forest, south to Concord again, east to Portsmouth, west to Vermont? It hardly made any difference, we'd so badly botched up Thoreau's itinerary by then, and so much of the original waters were now inaccessible to living creatures. The question was resolved by paddling back to where we had left Peter's car so he could drive to Plymouth and watch the Mets win their Series. Somebody hit a homer and somebody else got hit by a pitch. I imagined our party in a Camel ad (we'd paddle a mile etc.) and loudly said, "You other guys, *start walkin'*." We fooled around in Bridgewater, Ashland, and Holderness until we found a small tributary which led us to a stand of virgin pine holding out majestically in full view of an abandoned homestead and a railroad trestle. Buzzy guessed that the pine was on too great a pitch to be of use to 1930 American lumbering equipment, but in this nuclear age, we knew, it would not long go on rising. I hugged one of the trees and could not hardly stretch my arms around it.

With all the time lost in wallets and such, darkness seemed to fall inordinately early, but of course we were approaching the solstice with every day and might have expected as much. While I was in the cities, I lived by night and slept all day, for the streets of town were always more bearable under thin cover of grey; their lights made it easy to walk, and all the enclosed spaces were brightly lit with fraudulent sunshine, so I had the *impression* that I was alive. In the woods, though, nightfall is literally the end of the day. The degree to which you may perform outdoor chores depends on variables like the temperature, the moon, and the stars. You *must* make your hay while the sun shines. It terrifies me at times, so ill-adjusted am I to progress, to think that these very terms (names for the planets and stars) are just about obsolete in the day-to-day language of working people in Manchester and Lowell, professional people in L.A. and Paris, even Greenhouse Farmers in Pennsylvania.

The waning hours of afternoon also brought rain; and, disgusted, we set out in the car for . . . somewhere. The conversation in the back seat was in the quiet tones you can imagine defeated football players using after the big game. Buzzy spoke of real rivers he had sailed, most of them outside the United States; and I protested mildly that Vermont was still OK, then wondered how long it would take for my words to be ready-to-eat. The general talk rested on the subject of expatriation, the hows and wheres of it I mean. I imagined a family in Greenland taking in Verandah and me as "refugees from America"; it would not be an extraordinary scene in history. National boundaries mean nothing in the New Age, of course, and all we know of American history would make us anxious to leave, were the genuine natives not so thoroughly destroyed and the prospect of finding an untarnished culture and geography so dismal. Besides, our leaving would be the same as our staying, just the shifting of bodies from one spot to another on the big checkerboard, and the land never noticing.

And that's where the story ends I suppose, with the land, though the trip ended in Dr. Gus Dodge's house in Portsmouth, with Peter getting injected with gamma globulin as protection against Merrimack hepatitis. (As a Merrimack native, I am immune.) The land at the farm, at this writing, is alive and well if soaked with rain. It stretches out as far as my eyes can see, forming exquisite perspectives on all sides and limited only by the open sky which protects it. It generates new life at a furious pace, such that our main problem is keeping the forest from reclaiming itself; trees, saplings, grass, hay, vegetables, spices, flowers, and weeds crop up in riotous confusion, making oxygen and protein for deer, muskrats, coons, owls, porcupines, skunks, bobcats, snakes, goats, cows, horses, honeybees, rabbits, mice, cats, dogs, and people and a million other fine fellows and gals great and small. "Live off the land," our fathers said, and so we do. They didn't tell us to live in groups, they preferred the lonely family circle, so we have rejected that part. They didn't care enough about living off the rivers, oceans, and skies. We'll eat no meat or fish, it is clear. We'll burn no oil or gases in our houses, and finally in our cars. We'll bury our organic waste as deep as we can. We'll try to stay alive, for what else can we do? Friend, we are barking up the right trees.

*　　*　　*

Henry concludes in his Week:

>My friend is not of some other race or family of men, but flesh of my flesh, bone of my bone. He is my real brother. I see his nature groping yonder so like mine. We do not live far apart. Have not the fates associated us in many ways? It says in the Vishnu Purana: "Seven paces together is sufficient for the friendship of the virtuous, but thou and I have dwelt together." Is it of no significance that we have so long partaken of the same loaf, drank at the same fountain, breathed the same air summer and winter, felt the same heat and cold; that the same fruits have been pleased to refresh us both, and we have never had a thought of different fiber the one from the other!

>As surely as the sunset in my latest November shall translate me to the ethereal world, and remind me of the ruddy morning of youth; as surely as the last strain of music which falls on my decaying ear shall make age to be forgotten, or, in short, the manifold influences of nature survive during the term of our natural life, so surely my Friend shall forever be my Friend, and reflect a ray of God to me, and time shall foster and adorn and consecrate our Friendship, no less than the ruins of temples. As I love nature, as I love singing birds, and gleaming stubble, and flowing rivers, and morning and evening, and summer and winter, I love thee, my Friend.

—November, 1969.

# IV TOWARD
## A FUTURE

# INTRODUCTION

Perhaps more than other cultivated nations, America has been suspicious of sophistication, preferring to rely heavily on its anti-intellectualist approach to life and its problems. As a people it has been particularly impressed by the way in which learning has a habit of forestalling action—a realization that paradoxically may provide a partial explanation of the enormous technological advantage it has over the rest of the world. However, because it is a nation of utopian impulse, convinced that its actions are intrinsically good, the sophistication that does exist is spirited almost to the point of being apocalyptic; and indeed, there is in America a vision that is entirely prophetic. Those who subscribe to this world vision picture America in the historic landscape of other young and adventurous nations from the past. They see its achievements and its failures as part of the normal picture repeated time and time again in the legend of civilization. In their Christian sense of indulgence, they argue that America is one segment, perhaps the culminating segment, of a historical process that has been up to now cyclical in nature. As Crane Brinton has observed, the favorite parallel has been that of the United States and the Holy Roman Empire. Of course, not all prophetic visions of America have been filled with a sense of impending doom. Even if they are doom oriented, however, there is the lingering quality of American thought, which is best understood as it appears in the story of the phoenix—that mythological bird that is born again from the ashes of its own corpse. Sophistication has bred in the mind of the American intellectual a vision of life that is cyclical, repetitive, rhythmic, or evolutionary, and thus, by implication, removed from the office of active participation. As sure as there is darkness today, there will be a great dawning on the morrow.

There are those, too, who do not read the history of man as a spiraling cycle of providential accidents, who have little faith in the next messiah, and who do not naively assume that in the wings is a Newton, an Einstein, a cancer-eating drug, biding its time before stepping on stage. These people are not prepared to think of ecological disaster as one more arbitrary beginning in the development of the race called man. They feel that as we are the enemy, we must also be the questing knight, and that genius is generally the product of enormous work done by a great many people. It is in these beliefs that they explore the terrain for ways out of the present dilemma.

In the following section you will find a representation of such thought. Some of the ideas may be too utopian; others will be found to be unduly modest. They are offered as beginnings, serious beginnings rife with immense challenge for the reader.

# THE LAND: FIFTEEN YEARS AFTER

## Louis Bromfield

*In the late 1930's Louis Bromfield, the Pulitzer Prize winning novel-ist, gave up fiction and the life of European society to sink roots in the valley of his birth in Ohio. The land was ravaged and abandoned as a result of mismanagement and greed. Within fifteen years he created a miracle, namely Malabar Farm where, using ecologically sensitive agricultural practices, he restored what was equivalent to 2,000 years of topsoils in a little over a decade. The valley became one of the most productive in the country. He was temporarily influential with some agriculturalists, but this changed with the development of the chemically sanitized "agri-business" of the 1960's. We must return to his models, as his techniques will be needed to restore landscapes. "The Land: Fifteen Years After" is the introductory chapter to his last book.*

Fifteen years have passed since the snowy winter day when I turned the corner of the road into Pleasant Valley and said to myself, "This is the place." I had come back after twenty-five years of living in the world to my own country, to the valley I had known as a boy. I knew the country in the marrow of my bones; I knew it even in the recurring dreams which happened in strange countries here and there over half the world. I knew the marshes and the hills, the thick, hardwood forests, the wide fields and the beautiful hills behind which lay one lovely small valley after another, each a new, a rich, mysterious self-contained world on its own.

SOURCE: "The Land: Fifteen Years After" from *From My Experience* by Louis Bromfield. Copy-right © 1955 by Louis Bromfield. Reprinted by permission of Harper & Row, Publishers, Inc.

I was sick of the troubles, the follies and the squabbles of the Europe which I had known and loved for so long. I wanted peace and I wanted roots for the rest of my life.[1]

When I saw the valley again after twenty-five years it was under deep snow and the farms I bought were under deep snow. I have recounted the whole story in detail in *Pleasant Valley*, while the first impressions were still fresh in my mind, and I am glad that I did this for I can read it all and know now how little I understood the changes that occurred while I was away and how little I foresaw or understood what lay ahead and how nearly all the values of my life were to be changed and enriched. What I saw then, I saw through a haze of nostalgia, with homesick eyes. What I was seeking in part at least was something that was already on its way out of American life.

On that first snowy evening when I knocked at the door of the farmhouse which stood where the Big House now stands, I was, like many a man on the verge of middle age, knocking at the door of my long-gone boyhood. Tired and a little sick in spirit, I wanted to go back and, like many a foolish person, perhaps like all of us, I thought or hoped that going back was possible. I was sick even of writing novels and stories, although they had brought me considerable fortune and fame in nearly every country in the world. All fiction, save perhaps such books as *War and Peace*, which is more history than fiction, seemed to me at last to be without consequence and even trivial in contrast to all that was going on in the world about me. I knew and partly understood—better than most, I think—what was going on and what was ahead, certainly far better than the great majority of Americans, because I had lived for nearly a generation at the very midst of the turmoil and the decay which ended finally with the humiliation of Munich and the Second World War.

Most of the fellow countrymen to whom I talked on my return seemed almost childish in their naïveté and their lack of understanding concerning the significance of what was happening in Europe and in the world—just as many of them today seem childish in their refusal to face a world which is utterly changed, a world in which Soviet Russia and Red China and the awakening of Asia and the decadence of Europe and the end of the colonial empires are all simple facts which cannot be wished nor laughed off nor evaded. Peace and decency can only come by and through recognition of such facts and a recognition above all that the old world which many of us perhaps found agreeable enough is not coming back.

Those of us who lived in Europe and Asia between the first two world wars knew and understood pretty well both Nazism and Fascism and we knew

[1] I cannot say that with regard to the troubles, the follies, and the squabbles I found much change or relief. The record of my own country in these times with its politics, its meddling in the affairs of other nations, its spasmodic Utopianism, its militarism, its saber rattling, its attempts to dominate the world and dictate the policies of other nations, has been no record in which to take pride or to justify a sense of superiority in any American.

too all about Communism in all its manifestations, and knew that there was nothing to choose between the doctrines; one was merely a perversion of an unnatural political and economic philosophy and vice versa. They were both derived from the philosophy of a sick and psychopathic German called Marx. The forces behind them were actually the forces of a world revolution which manifested itself in countless ways not the least of which was the slow death of the already obsolete colonial empires. These doctrines were created not merely by a philosophy, wishful thinking or even fanaticism: they were created by immense economic and political stresses and strains which involved such things as overpopulation, shortages of foods and raw materials, abysmal living standards, restricted markets. In short, Nazism and Communism were merely the political manifestations of profound economic ills. They did not in themselves cause war and revolution. They were basically the outward political manifestations of much more real and deep-rooted evils. The guilt for the war could not be fixed on any one nation or group of nations. Every nation in one degree or another shared responsibility. The responsibility still remains. We in this country have solved nothing nor made any great contribution toward creating and maintaining peace. The Utopians and the Militarists have indeed merely accomplished the contrary.

Fifteen years after I hoped bleakly to escape from all the evils I knew so well at first hand, I have discovered bleakly that there is nothing superior about my own people and that they do not have any special wisdom or vision. We have merely been more fortunate than other peoples. We are generous because we can afford to be generous. We are perhaps open-hearted because we are still a young people, but we still understand very little about the evils of the world or how they can be cured or at least modified. We lack almost entirely the capacity of putting ourselves in the place of other peoples and the knowledge of the average citizen concerning the life and the circumstances of other nations and peoples is primitive, frequently enough even among those who occupy high places in our government.

When I came back to the valley on that snowy morning fifteen years ago, I was trying to escape the evils of the past and the weariness of Europe. Somehow in the misty recesses of my mind there existed a happy image of this valley in which I had spent a happy and complete boyhood—the image of a valley shut away, immaculate and inviolate, self-sustained and complete and peaceful. It was an image in which the sun was always shining. It was not of course an image born of the mind and the intellect but one born of the emotions. At that time I did not fully understand what it was that moved me so powerfully. I have only come to understand it through the gradual corrosion of disillusionment and pessimism and since I have found something which I did not then know existed, something which I did not even know I was seeking . . . something which it is difficult or impossible to describe save that it is a combination perhaps of reality and truth and of values which were unsuspected.

All this is closely related to the earth, the sky, to animals and growing plants and trees and my fellow men. All this is of course of immense importance to me personally; it is important to others only as a bit, a fragment, of human experience, that element which perhaps more than any other sets men apart from animals who, so far as we know, are not capable of reflection or philosophy.

And so in the light of all this, the writing of fiction, unless it was merely a story to divert a tired world or provide relaxation for it, came presently to seem silly. It still does, no matter how pompous, how pretentious, how self-important, how cult-ridden the writer or the product. In this age fiction writing is simply a way of making a living and for my money not a very satisfactory or even self-respecting one. There are better and more satisfying things to do. One degree sillier are the writings of those who write importantly about novels. Once when a person said to me, "Oh, I never read novels!" I was inclined to regard him with snobbish condescension as a Philistine. Now I am not so sure.

So when the door opened in the Anson house more than fifteen years ago and the familiar childhood smell of a farm kitchen came to me, I was aware of a sudden delight as if in reality I had stepped from the recurrent weariness and disillusionment back into the realities of my boyhood. The smell was one in which there was blended the odor of woodsmoke, of apple butter, of roasting pork sausage, of pancakes, of spices. Mrs. Anson, workworn and no longer young, her arthritic hands wrapped in her apron, stood in the doorway with her back to the gaslight of the big kitchen walled with hand-cut stone from the low cliffs behind the house. With a puzzled smile lighted by some ancient memory and recognition, she invited me in, and brought me hot coffee and spiced cookies. When I told her my name she remembered me dimly as a small boy who used to camp on the adjoining Douglass Place which belonged to her cousins, and who fished in Switzer's Run that flowed through the valley below the house.

When I asked her if the place was for sale, she answered that she believed not but that we might talk to her husband who was out in the barn milking. And so she put on a shawl and we walked in the blue winter twilight across the squeaking snow to the barn where old Mr. Anson sat on a three-legged stool drawing milk from a Guernsey cow. And again at the barn door the smell of the warm stable, the granary, the steaming manure came to me in the form of the perfumes of Araby. For the moment I again renounced all the world in which I had lived most of my life and escaped from it back into the past . . . into the world of horses-and-buggies, of muddy roads, of church suppers, of everything that was the America of my childhood. I who had been everywhere and known the world and "all the answers" wanted only to come back into this world.

The Ansons could not make up their minds whether they wanted to sell until they talked to their children, and so after another cup of coffee in the big stone kitchen I set off down the winding road past the cottage where Ceeley Rose had poisoned her parents and brothers, across the little bridge and finally to the highway. It was a *blue* winter night with that peculiar quality of blue in the sky and in the air itself which one finds in our part of the world and only a few other places, places like Northern France and the Castilian plain about Madrid. The air seems to become luminous like the unreal blue of skies on the cyclorama of a theater. I was happy. I was hopeful that I had really escaped and had come home.

That night the whole valley was covered by snow and the little creek fringed with ice. What lay beneath the snow I could not see. I was only to discover it when the spring came with a rush of green and wild flowers.

When the snow was gone, I discovered that the valley of my childhood was no longer there. Something had happened to it. It had been ravaged by time and by the cruel and careless treatment of the land. As a small boy I had never noticed that these once small, lovely, rich valleys throughout our countryside had already begun to change, growing a little more gullied and bare with each year, or that the pastures grew thinner and more weedy and the ears of corn a little smaller each season. When the snow was gone, I began to understand what had happened to it . . . perhaps because for so long I had lived in the rich green country of Northern France where the land is loved and where it is respected and cherished almost with avarice, perhaps out of grim necessity, but none the less respected and even held precious. It was clear that no such thing had happened in the valley. Some of the farms which lay below and around the Anson Place no longer raised crops at all. They had been rented out to year-by-year tenants or to neighbors who took everything off them until they would no longer grow anything. The houses were occupied by industrial workers who spent their days in the city factories.

Except where starving sheep on wasted farms were turned loose in the woods to find meager living as best they could off the seedlings and the sun-starved vegetation, the forests and the marshes were still the same. Here and there a woodlot had been brutally murdered by some fly-by-night timber speculator, the trees sold perhaps to raise money for interest on the mortgage of some dying farm; but otherwise the growth of the wild grasses, the trees, the briars provided evidence of the original and fundamental richness of the whole countryside. Wherever there was desolation and sterility it had been created by man, by ignorance, by greed and by a strange belief inherent in early generations of American farmers that their land owed them a living.

When I look back now, the vague and visionary idea I had in returning home seems ludicrous and a little pathetic. Somehow the picture I saw of the future was one in which vaguely there were blended the carefree happiness of my boyhood and the life in a great house in the countryside of England which, in the great days, was perhaps the best and most civilized life man has ever known. If in the dream there was any other element it was that of security; I wanted a place which, again vaguely, would be like the medieval fortress-manor of France where a whole community once found security and self-sufficiency. In a troubled world I wanted a place which, if necessary, could withstand a siege and where, if necessary, one could get out the rifle and shotgun for defense.

Today, fifteen years later, we at Malabar have not achieved these romantic dreams nor have I won the escape into the boyhood past which brought about the decision to return. A return to the past can never be accomplished and the sense of fortified isolation and security is no longer possible in the world of automobiles, of radios, of telephones and airplanes. One must live with one's times and those who understand this and make the proper adjustments and concessions and compromises are the happy ones. In the end I did not find at all what I was seeking on that snowy night; I found something much better . . . a whole new life, and a useful life and one in which I have been able to make a contribution which may not be forgotten overnight and with the first funeral wreath, like most of the writing of our day, but one which will go on and on. And I managed to find and to create, not the unreal almost fictional life for which I hoped, but a tangible world of great and insistent reality, made up of such things as houses, and ponds, fertile soils, a beautiful and rich landscape and the friendship and perhaps the respect of my fellow men and fellow farmers. The people who come to the Big House are not the fashionable, the rich, the famous, the wits, the intellectuals (although there is a sprinkling of all these) but plain people and farmers and cattlemen from all parts of the world.

Perhaps it will turn out that I have left behind some contributions not only to the science of agriculture, which is the only profession in the world which encompasses *all* sciences and *all* the laws of the universe, but to the realm of human philosophy as well. None of this could I have done within the shallow world of a writer living as most writers live. Without implying in any way a comparison or any conceit, I am sure that Tolstoy understood all this on his estate at Yasna Polyana, Voltaire at Ferney, Virgil on his Tuscan farm; indeed most writers since the time of Hesiod who felt sooner or later the illusion and the futility of fancy words and sought some sturdier and earthier satisfaction. Flaubert was neither a happy man nor a complete one, nor was Turgenev or Henry James. One has only to read their letters or their diaries and sometimes their stories and poems to discover a shadowy sense of impotence and inadequacy.

The *complete* man is a rarity. Leonardo was one and Michelangelo and Shakespeare and Balzac. They lived; they brawled; they had roots; they were immoral: they had vices as well as virtues; they were totally lacking in preciosity and the pale, moldy qualities of the poseur or the seeker after publicity and sensation. The *complete* man is a happy man, even in misery and tragedy, because he has always an inner awareness that he has lived a *complete* existence, in vice and virtue, in success and failure, in satisfaction and disappointment, in distinction and vulgarity. Not only is he complete; he is much more, he *is* a man.

The older I grow the more I become aware wistfully of that goal of completeness. It is not something that can be attained by wishing or even by plotting and determination. The man who sets out deliberately to be a complete man defeats himself, for from the beginning he is of necessity self-conscious, contriving and calculating. He becomes the fake, the poseur, the phony. Some attempt to turn their own inadequacies into a defense by affecting a sense of snobbery or superiority. In this sense a writer like Henry James is pathetic. So are many writers of our own times with their lacy preciosities, their affectations, their pomp and pretensions, their fundamental shallowness and decadence.

But enough of all that. If such thoughts have any importance, it is an importance which primarily touches only the individual and the individual only in relation to the satisfactions which he may win from a life which is all too short if one considers the glories, the complexities, the mysteries, the fascination, of the world and the universe in which we live. Certainly one of the happiest of men is the good farmer who lives close to the storm and the forest, the drought and the hail, who knows and understands well his kinfolk the beasts and the birds, whose whole life is determined by the realities, whose sense of beauty and poetry is born of the earth, whose satisfactions, whether in love or the production of a broad rich field, are direct and fundamental, vigorous, simple, profound and deeply satisfying. Even the act of begetting his offspring has about it a vigor, a force, a directness in which there is at once the violence of the storm and the gentleness of a young willow against the spring sky.

I was born of farming and land-owning people and have never been for long away from a base with roots in the earth, despite the fact that the fortune and circumstances of life have from time to time brought me into the most worldly and sophisticated and fashionable of circles in a score of countries and capitals and frequently among the great, the famous and those, in the world of politics, who have made the history of our times. Still I know of no intellectual satisfaction greater than that of talking to a good intelligent farmer or livestock breeder who, instinctively perhaps, knows what many less fortunate men endeavor most of their lives in vain to learn from books, or the satisfaction of seeing a whole landscape, a whole small world change from a half-desert into a rich ordered green valley inhabited by happy people, secure and prosperous, who each day create and add a little more to the world in which they live, who each season see their valley grow richer and more beautiful, who are aware alike of the beauty of the deer coming down to the ponds in the evening and the mystery and magnificence of a prize-winning potato or stalk of celery, who recognize alike the beauty of a field with a rich crop in which there are no "poor spots" and the beauty of a fine sow and her litter.

These are, it seems to me, among the people who belong, the fortunate ones who know and have always known whither they were bound, from the first hour of consciousness and memory to the peace of falling asleep for the last time never to waken again, to fall asleep in that tranquility born only of satisfaction and fullness and completion. In a sense they know the whole peace of the eternal creator who has built and left behind him achievements in stone, in thought, in good black earth, in a painting upon a wall or some discovery which has helped forward a little on the long and difficult path his fellow men and those who come after him. It is these individuals who belong and who need not trouble themselves about an after life, for at the end there is no terror of what is to follow and no reluctance to fall asleep forever, since it is the direct and the natural thing for them to do. They are at once and in essence humble and simple people, no matter what fame they achieve or what admiration they have won from their fellow men. Because they belong and so have found their place and their adjustment in the intricate and complex pattern of the universe they find even the tragedies and the suffering of their own lives only a part of a general vast pattern. They are, perhaps, more sharply aware of the significance of tragedy and suffering than less limited and egotistical folk; that very awareness and significance blunt the sharp edge of suffering and presently the pain wears away, leaving only peace and even perhaps a strange sense of beauty and richness.

The mysteries of the human mind are certainly fascinating but more limited and, for me at least, less important and less interesting than the cosmic mysteries which take place within a cubic foot of rich productive soil, for essentially these "mysteries" of the human mind are merely a part of an infinitely greater and more intricate and complex mystery which utterly baffles all of us, even the wisest and most learned. One of the great errors of our time, and one which has brought us in our time much misery, is the attribution of an overweening and disproportionate importance to man and his mind. Man himself, as a physical machine, as a mechanistic functional and living organism, is indeed marvelous as is every part of the universe; but his ego and self-importance, in our time, are given a distorted, decadent and tragicomic importance.

Man is merely a part of the universe, and not a very great part, which happens to be fortunate principally in having evolved such traits and powers as consciousness, reflection, logic and thought. The wise and happy man is the one who finds himself in adjustment to this truth, who never needs, in moments of disillusionment and despair, to cut himself down to size because it has never occurred to him, in the beginning or at any time, to inflate his own importance whether through ignorance, morbidity, egotism or undergoing psychoanalysis (which is merely another name for one of the age-old manifestations of brooding impotence and frustration of the incomplete man).

It is only later in life, in the midst of what is still a somewhat turbulent and certainly a varied existence, that any full understanding and satisfaction of this sense of *belonging*, of being a small and relatively unimportant part of something vast but infinitely friendly, has begun to come to me. It is only now that I have come to understand that from earliest childhood, this passion to *belong*, to lose one's self in the whole pattern of life, was the strong and overwhelming force that unconsciously has directed every thought, every act, every motive of my existence.

This urge is by no means an uncommon one and is perhaps shared in one degree or another by every man or woman of real intelligence, although many fail to recognize it for what it is or to understand it. Certainly it is the profoundly motivating pattern in the life of a man like Albert Schweitzer just as it is of many a humbler thinker and teacher. It is indeed the very force of every good and real teacher; it is that element which distinguishes the creator from the desiccated savant, that distinguished Christ from Socrates, that makes all formal education seem futile and useless unless it is *understood* and *used*. Mere *learnedness* becomes insignificant and even useless when the fire is absent. This rarity of understanding is the reason why throughout the ages there has always been a cry that there are not enough teachers and why all the knowledge and all the formal training in the world produced by all the universities and colleges since the beginning of time cannot produce one more real teacher. I am afraid it is true that teachers, like good cattlemen and good farmers and good engineers and good cooks, are born and not made. So essentially are most really good people.

\*     \*     \*

And now, dear reader, if you have survived these philosophical ramblings we shall get on to the business at hand . . . the business of drying hay, of finding means for making more money for the farmer, of helping to feed hungry people, and in general the satisfactions of life in the country. But do not overlook the fact that all of these things real, finite and even perhaps materialistic, are all very much a part of the pattern in the life of at least one busy and happy man even though he is a long way from being a complete one, just as they have become in these times more and more a part of the pattern in the lives of countless middle-aged and elderly businessmen, industrial workers, engineers, clerks, retired mechanics, and have been part of the pattern in the lives of a number of Presidents from Washington and Jefferson to General Eisenhower . . . men who have discovered sooner or later that somehow farming, and especially good and intelligent farming, comes closer to providing a key to adjustment and understanding of the universe than all the algebraic formulas of Albert Einstein.

Considering the general insignificance and unimportance of man, the pleasures of agriculture are perhaps more real and gratifying than the pleasures and even the excesses of the purely mathematical mind (which are certainly not pleasures to be underestimated). If the pleasures of mathematical debauchery or orgies in physics are to be treated as limited, it is only because they are denied the great mass of humanity and because all too often they induce and create the deformities and limitations of the incomplete man. Rousseau was in many respects a fool and at times a humbug and a liar but he had something in his conception of the Natural Man, something in which even the great wise and cynical mind of Voltaire, an infinitely more intellectual and sophisticated man, found a perpetual source of envy.

The greatest creative and intellectual vice of our times, and a factor which causes increasing distress and even tragedy, is the overspecialization which man has partly chosen and which has been partly forced upon him by the shrinking of the world, by the incredible speeding up of daily life and the materialistic impact of technological development upon our daily existence. The superspecialist tends to become not only an incomplete man but a deformed one. What perhaps limits forever the superspecialist or the "pure" intellectual stewing in his own juice is the fact that in his intensity and in the sharp limitations of his narrow field, he loses frequently the power to understand or even to conceive the principle of the complete man, flesh, appetites, weakness, follies and all, and so he loses in time all understanding in the true sense of the universe or even of the small world which surrounds him.

# WHAT WE MUST DO

*John Platt*

*John Platt is the associate director of the Mental Health Research Institute in Ann Arbor, Michigan. He is well known for his The Step to Man (1966) and has for a long time been interested in strategies for survival. "What We Must Do" reflects his faith in the ability of man to organize his knowledge and his efforts in a way that would prevent crises from piling up as they have for the 1970's. In particular, Platt is interested in buying time so that the present transformation does not kill us before some solutions appear.*

There is only one crisis in the world. It is the crisis of transformation. The trouble is that it is now coming upon us as a storm of crisis problems from every direction. But if we look quantitatively at the course of our changes in this century, we can see immediately why the problems are building up so rapidly at this time, and we will see that it has now become urgent for us to mobilize all our intelligence to solve these problems if we are to keep from killing ourselves in the next few years.

The essence of the matter is that the human race is on a steeply rising "S-curve" of change. We are undergoing a great historical transition to new levels of technological power all over the world. We all know about these changes, but we do not often stop to realize how large they are in orders of magnitude, or how rapid and enormous compared to all previous changes in history. In the last century, we have increased our speeds of communication by a factor of $10^7$; our speeds of travel by $10^2$; our speeds of data handling by $10^6$; our energy resources by $10^3$; our power of weapons by $10^6$; our ability

SOURCE: John Platt, "What We Must Do," *Science*, vol. 162, 28 November 1969, pp. 1115–1121. Copyright © 1969 by the American Association for the Advancement of Science.

to control diseases by something like $10^2$; and our rate of population growth to $10^3$ times what it was a few thousand years ago.

Could anyone suppose that human relations around the world would not be affected to their very roots by such changes? Within the last 25 years, the Western world has moved into an age of jet planes, missiles and satellites, nuclear power and nuclear terror. We have acquired computers and automation, a service and leisure economy, superhighways, superagriculture, supermedicine, mass higher education, universal TV, oral contraceptives, environmental pollution, and urban crises. The rest of the world is also moving rapidly and may catch up with all these powers and problems within a very short time. It is hardly surprising that young people under 30, who have grown up familiar with these things from childhood, have developed very different expectations and concerns from the older generation that grew up in another world.

What many people do not realize is that many of these technological changes are now approaching certain natural limits. The "S-curve" is beginning to level off. We may never have faster communications or more TV or larger weapons or a higher level of danger than we have now. This means that if we could learn how to manage these new powers and problems in the next few years without killing ourselves by our obsolete structures and behavior, we might be able to create new and more effective social structures that would last for many generations. We might be able to move into that new world of abundance and diversity and well-being for all mankind which technology has now made possible.

The trouble is that we may not survive these next few years. The human race today is like a rocket on a launching pad. We have been building up to this moment of takeoff for a long time, and if we can get safely through the takeoff period, we may fly on a new and exciting course for a long time to come. But at this moment, as the powerful new engines are fired, their thrust and roar shakes and stresses every part of the ship and may cause the whole thing to blow up before we can steer it on its way. Our problem today is to harness and direct these tremendous new forces through this dangerous transition period to the new world instead of to destruction. But unless we can do this, the rapidly increasing strains and crises of the next decade may kill us all. They will make the last 20 years look like a peaceful interlude.

## The Next 10 Years

Several types of crisis may reach the point of explosion in the next 10 years: nuclear escalation, famine, participatory crises, racial crises, and what have been called the crises of administrative legitimacy. It is worth singling out two or three of these to see how imminent and dangerous they are, so that we can fully realize how very little time we have for preventing or controlling them.

Take the problem of nuclear war, for example. A few years ago, Leo Szilard estimated the "half-life" of the human race with respect to nuclear escalation as being between 10 and 20 years. His reasoning then is still valid now. As long as we continue to have no adequate stabilizing peace-keeping structures for the world, we continue to live under the daily threat not only of local wars but of nuclear escalation with overkill and megatonnage enough to destroy all life on earth. Every year or two there is a confrontation between nuclear powers—Korea, Laos, Berlin, Suez, Quemoy, Cuba, Vietnam, and the rest. MacArthur wanted to use nuclear weapons in Korea; and in the Cuban missile crisis, John Kennedy is said to have estimated the probability of a nuclear exchange as about 25 percent.

The danger is not so much that of the unexpected, such as a radar error or even a new nuclear dictator, as it is that our present systems will work exactly as planned!—from border testing, strategic gambles, threat and counter-threat, all the way up to that "second-strike capability" that is already aimed, armed, and triggered to wipe out hundreds of millions of people in a 3-hour duel!

What is the probability of this in the average incident? 10 percent? 5 percent? There is no average incident. But it is easy to see that five or ten more such confrontations in this game of "nuclear roulette" might indeed give us only a 50-50 chance of living until 1980 or 1990. This is a shorter life expectancy than people have ever had in the world before. All our medical increases in length of life are meaningless, as long as our nuclear lifetime is so short.

Many agricultural experts also think that within this next decade the great famines will begin, with deaths that may reach 100 million people in densely populated countries like India and China. Some contradict this, claiming that the remarkable new grains and new agricultural methods introduced in the last 3 years in Southeast Asia may now be able to keep the food supply ahead of population growth. But others think that the reeducation of farmers and consumers to use the new grains cannot proceed fast enough to make a difference.

But if famine does come, it is clear that it will be catastrophic. Besides the direct human suffering, it will further increase our international instabilities, with food riots, troops called out, governments falling, and international interventions that will change the whole political map of the world. It could make Vietnam look like a popgun.

In addition, the next decade is likely to see continued crises of legitimacy of all our overloaded administrations, from universities and unions to cities and national governments. Everywhere there is protest and refusal to accept the solutions handed down by some central elite. The student revolutions circle the globe. Suburbs protest as well as ghettoes, Right as well as Left. There are many new sources of collision and protest, but it is clear that the general problem is in large part structural rather than political. Our traditional methods of election and management no longer give administrations the skill and capacity they need to handle their complex new burdens and decisions. They become swollen, unresponsive—and repudiated. Every day now some distinguished administrator is pressured out of office by protesting constituents.

In spite of the violence of some of these confrontations, this may seem like a trivial problem compared to war or famine—until we realize the dangerous effects of these instabilities on the stability of the whole system. In a nuclear crisis or in any of our other crises today, administrators or negotiators may often work out some basis of agreement between conflicting groups or nations, only to find themselves rejected by their people on one or both sides, who are then left with no mechanism except to escalate their battles further.

## The Crisis of Crises

What finally makes all of our crises still more dangerous is that they are now coming on top of each other. Most administrations are able to endure or even enjoy an occasional crisis, with everyone working late together and getting a new sense of importance and unity. What they are not prepared to deal with are multiple crises, a crisis of crises all at one time. This is what happened in New York City in 1968 when the Ocean Hill–Brownsville teacher and race strike was combined with a police strike, on top of a garbage strike, on top of a longshoremen's strike, all within a few days of each other.

When something like this happens, the staff gets jumpy with smoke and coffee and alcohol, the mediators become exhausted, and the administrators find themselves running two crises behind. Every problem may escalate because those involved no longer have time to think straight. What would have happened in the Cuban missile crisis if the East Coast power blackout had occurred by accident that same day? Or if the "hot line" between Washington and Moscow had gone dead? There might have been hours of misinterpretation, and some fatally different decisions.

I think this multiplication of domestic and international crisis today will shorten that short half-life. In the continued absence of better ways of heading off these multiple crises, our half-life may no longer be 10 or 20 years, but more like 5 to 10 years, or less. We may have even less than a 50-50 chance of living until 1980.

This statement may seem uncertain and excessively dramatic. But is there any scientist who would make a much more optimistic estimate after considering all the different sources of danger and how they are increasing? The shortness of the time is due to the exponential and multiplying character of our problems and not to what particular numbers or guesses we put in. Anyone who feels more hopeful about getting past the nightmares of the 1970's has only to look beyond them to the monsters of pollution and population rising up in the 1980's and 1990's. Whether we have 10 years or more like 20 or 30, unless we systematically find new large-scale solutions, we are in the gravest danger of destroying our society, our world, and ourselves in any of a number of different ways well before the end of this century. Many futurologists who have predicted what the world will be like in the year 2000 have neglected to tell us that.

Nevertheless the real reason for trying to make rational estimates of these deadlines is not because of their shock value but because they give us at least a rough idea of how much time we may have for finding and mounting some large-scale solutions. The time is short but, as we shall see, it is not too short to give us a chance that something can be done, if we begin immediately.

From this point, there is no place to go but up. Human predictions are always conditional. The future always depends on what we do and can be made worse or better by stupid or intelligent action. To change our earlier analogy, today we are like men coming out of a coal mine who suddenly begin to hear the rock rumbling, but who have also begun to see a little square of light at the end of the tunnel. Against this background, I am an optimist—in that I want to insist that there is a square of light and that it is worth trying to get to. I think what we must do is to start running as fast as possible toward that light, working to increase the probability of our survival through the next decade by some measurable amount.

For the light at the end of the tunnel is very bright indeed. If we can only devise new mechanisms to help us survive this round of terrible crises, we have a chance of moving into a new world of incredible potentialities for all mankind. But if we cannot get through the next decade, we may never reach it.

## Task Forces for Social Research and Development

What can we do? I think that nothing less than the application of the full intelligence of our society is likely to be adequate. These problems will require the humane and constructive efforts of everyone involved. But I think they will also require something very similar to the mobilization of scientists for solving crisis problems in wartime. I believe we are going to need large numbers of scientists forming something like research teams or task forces for social research and development. We need full-time interdisciplinary teams combining men of different specialties, natural scientists, social scientists, doctors, engineers, teachers, lawyers, and many other trained and inventive minds, who can put together our stores of knowledge and powerful new ideas into improved technical methods, organizational designs, or "social inventions" that have a chance of being adopted soon enough and widely enough to be effective. Even a great mobilization of scientists may not be enough. There is no guarantee that these problems can be solved, or solved in time, no matter what we do. But for problems of this scale and urgency, this kind of focusing of our brains and knowledge may be the only chance we have.

Scientists, of course, are not the only ones who can make contributions. Millions of citizens, business and labor leaders, city and government officials, and workers in existing agencies, are already doing all they can to solve these problems. No scientific innovation will be effective without extensive advice and help from all these groups.

But it is the new science and technology that have made our problems so immense and intractable. Technology did not create human conflicts and inequities, but it has made them unendurable. And where science and technology have expanded the problems in this way, it may be only more scientific understanding and better technology that can carry us past them. The cure for the pollution of the rivers by detergents is the use of nonpolluting detergents. The cure for bad management designs is better management designs.

Also, in many of these areas, there are few people outside the research community who have the basic knowledge necessary for radically new solutions. In our great biological problems, it is the new ideas from cell biology and ecology that may be crucial. In our social-organizational problems, it may be the new theories of organization and management and behavior theory and game theory that offer the only hope. Scientific research and development groups of some kind may be the only effective mechanism by which many of these new ideas can be converted into practical invention and action.

The time scale on which such task forces would have to operate is very different from what is usual in science. In the past, most scientists have tended to work on something like a 30-year time scale, hoping that their careful studies would fit into some great intellectual synthesis that might be years away. Of course when they become politically concerned, they begin to work on something more like a 3-month time scale, collecting signatures or trying to persuade the government to start or stop some program.

But 30 years is too long, and 3 months is too short, to cope with the major crises that might destroy us in the next 10 years. Our urgent problems now are more like wartime problems, where we need to work as rapidly as is consistent with large-scale effectiveness. We need to think rather in terms of a 3-year time scale—or more broadly, a 1- to 5-year time scale. In World War II, the ten thousand scientists who were mobilized for war research knew they did not have 30 years, or even 10 years, to come up with answers. But they did have time for the new research, design, and construction that brought sonar and radar and atomic energy to operational effectiveness within 1 to 4 years. Today we need the same large-scale mobilization for innovation and action and the same sense of constructive urgency.

## Priorities: A Crisis Intensity Chart

In any such enterprise, it is most important to be clear about which problems are the real priority problems. To get this straight, it is valuable to try to separate the different problem areas according to some measures of their magnitude and urgency. A possible classification of this kind is shown in Tables 1 and 2. In these tables, I have tried to rank a number of present or potential problems or crises, vertically, according to an estimate of their order of intensity or "seriousness," and horizontally, by a rough estimate of their time to reach climactic importance. Table 1 is such a classification for the United States for the next 1 to 5 years, the next 5 to 20 years, and the next 20 to 50 years. Table 2 is a similar classification for world problems and crises.

The successive rows indicate something like order-of-magnitude differences in the intensity of the crises, as estimated by a rough product of the size of population that might be hurt or affected, multiplied by some estimated average effect in the disruption of their lives. Thus the first row corresponds to total or near-total annihilation; the second row, to great destruction or change affecting everybody; the third row, to a lower tension affecting a smaller part of the population or a smaller part of everyone's life, and so on.

Informed men might easily disagree about one row up or down in intensity, or one column left or right in the time scales, but these order-of-magnitude differences are already so great that it would be surprising to find much larger disagreements. Clearly, an important initial step in any serious problem study would be to refine such estimates.

In both tables, the one crisis that must be ranked at the top in total danger and imminence is, of course, the danger of large-scale or total annihilation by nuclear escalation or by radiological-chemical-biological-warfare (RCBW). This kind of crisis will continue through both the 1- to 5-year time period and the 5- to 20-year period as Crisis Number 1, unless and until we get a safer peace-keeping arrangement. But in the 20- to 50-year column, following the reasoning already given, I think we must simply put a big "✠" at this level, on the grounds that the peace-keeping stabilization problem will either be solved by that time or we will probably be dead.

At the second level, the 1- to 5-year period may not be a period of great destruction (except nuclear) in either the United States or the world. But the problems at this level are building up, and within the 5- to 20-year period, many scientists fear the destruction of our whole biological and ecological balance in the United States by mismanagement or pollution. Others fear political catastrophe within this period, as a result of participatory confrontations or backlash or even dictatorship, if our divisive social and structural problems are not solved before that time.

On a world scale in this period, famine and ecological catastrophe head the list of destructive problems. We will come back later to the items in the 20- to 50-year column.

The third level of crisis problems in the United States includes those that are already upon us: administrative management of communities and cities, slums, participatory democracy, and racial conflict. In the 5- to 20-year period, the problems of pollution and poverty or major failures of law and justice could escalate to this level of tension if they are not solved. The last column is left blank because secondary events and second-order effects will interfere seriously with any attempt to make longer-range predictions at these lower levels.

The items in the lower part of the tables are not intended to be exhaustive. Some are common headline problems which are included simply to

| | Grade | Estimated crisis intensity (number affected $\times$ degree of effect) | |
|---|---|---|---|
| **TABLE 1** Classification of problems and crises by estimated time and intensity (United States). | 1. | | Total annihilation |
| | 2. | $10^8$ | Great destruction or change (physical, biological, or political) |
| | 3. | $10^7$ | Widespread almost unbearable tension |
| | 4. | $10^6$ | Large-scale distress |
| | 5. | $10^5$ | Tension producing responsive change |
| | 6. | | Other problems—important, but adequately researched |
| | 7. | | Exaggerated dangers and hopes |
| | 8. | | Noncrisis problems being "overstudied" |

show how they might rank quantitatively in this kind of comparison. Anyone concerned with any of them will find it a useful exercise to estimate for himself their order of seriousness, in terms of the number of people they actually affect and the average distress they cause. Transportation problems and neighborhood ugliness, for example, are listed as grade 4 problems in the United States because they depress the lives of tens of millions for 1 or 2 hours every day. Violent crime may affect a corresponding number every year or two. These evils are not negligible, and they are worth the efforts of enormous numbers of people to cure them and to keep them cured—but on the other hand, they will not destroy our society.

The grade 5 crises are those where the hue and cry has been raised and where responsive changes of some kind are already under way. Cancer goes here, along with problems like auto safety and an adequate water supply.

| Estimated time to crisis* | | |
|---|---|---|
| 1 to 5 years | 5 to 20 years | 20 to 50 years |
| Nuclear or RCBW escalation | Nuclear or RCBW escalation | ✖ (Solved or dead) |
| (Too soon) | Participatory democracy<br>Ecological balance | Political theory and<br>    economic structure<br>Population planning<br>Patterns of living<br>Education<br>Communications<br>Integrative philosophy |
| Administrative management<br>Slums<br>Participatory democracy<br>Racial conflict | Pollution<br>Poverty<br>Law and justice | ? |
| Transportation<br>Neighborhood ugliness<br>Crime | Communications gap | ? |
| Cancer and heart<br>Smoking and drugs<br>Artificial organs<br>Accidents<br>Sonic boom<br>Water supply<br>Marine resources<br>Privacy on computers | Educational inadequacy | ? |
| Military R & D<br>New educational methods<br>Mental illness<br>Fusion power | Military R & D | |
| Mind control<br>Heart transplants<br>Definition of death | Sperm banks<br>Freezing bodies<br>Unemployment from<br>    automation | Eugenics |
| Man in space<br>Most basic science | | |

* If no major effort is made at anticipatory solution.

**TABLE 2**
Classification of
problems and crises
by estimated time and
intensity (World).

| Grade | Estimated crisis intensity (number affected $\times$ degree of effect) | |
|---|---|---|
| 1. | $10^{10}$ | Total annihilation |
| 2. | $10^9$ | Great destruction or change (physical, biological, or political) |
| 3. | $10^8$ | Widespread almost unbearable tension |
| 4. | $10^7$ | Large-scale distress |
| 5. | $10^6$ | Tension producing responsive change |
| 6. | | Other problems—important, but adequately researched |
| 7. | | Exaggerated dangers and hopes |
| 8. | | Noncrisis problems being "overstudied" |

This is not to say that we have solved the problem of cancer, but rather that good people are working on it and are making as much progress as we could expect from anyone. (At this level of social intensity, it should be kept in mind that there are also positive opportunities for research, such as the automation of clinical biochemistry or the invention of new channels of personal communication, which might affect the 20-year future as greatly as the new drugs and solid state devices of 20 years ago have begun to affect the present.)

## Where the Scientists Are

Below grade 5, three less quantitative categories are listed, where the scientists begin to outnumber the problems. Grade 6 consists of problems that many people believe to be important but that are adequately researched at the present time. Military R & D belongs in this category. Our huge military establishment creates many social problems, both of national priority and international stability, but even in its own terms, war research, which engrosses hundreds of thousands of scientists and engineers, is being taken care of

| Estimated time to crisis* | | |
| --- | --- | --- |
| 1 to 5 years | 5 to 20 years | 20 to 50 years |
| Nuclear or RCBW escalation | Nuclear or RCBW escalation | ✖ (Solved or dead) |
| (Too soon) | Famines | Economic structure |
| | Ecological balance | and political theory |
| | Development failures | Population and |
| | Local wars | ecological balance |
| | Rich-poor gap | Patterns of living |
| | | Universal education |
| | | Communications-integration |
| | ↑ | Management of world |
| | | Integrative philosophy |
| Administrative management | Poverty | |
| Need for participation | Pollution | |
| Group and racial conflict | Racial wars | |
| Poverty-rising expectations | Political rigidity | |
| Environmental degradation | Strong dictatorships | ? |
| Transportation | Housing | |
| Diseases | Education | |
| Loss of old cultures | Independence of big powers | ? |
| | Communications gap | |
| Regional organization | ? | ? |
| Water supplies | | |
| Technical development design | | |
| Intelligent monetary design | | |
| | | Eugenics |
| | | Melting of ice caps |
| Man in space | | |
| Most basic science | | |

* If no major effort is made at anticipatory solution.

generously. Likewise, fusion power is being studied at the $100-million level, though even if we had it tomorrow, it would scarcely change our rates of application of nuclear energy in generating more electric power for the world.

Grade 7 contains the exaggerated problems which are being talked about or worked on out of all proportion to their true importance, such as heart transplants, which can never affect more than a few thousands of people out of the billions in the world. It is sad to note that the symposia on "social implications of science" at many national scientific meetings are often on the problems of grade 7.

In the last category, grade 8, are two subjects which I am sorry to say I must call "overstudied," at least with respect to the real crisis problems today. The Man in Space flights to the moon and back are the most beautiful technical achievements of man, but they are not urgent except for national display, and they absorb tens of thousands of our most ingenious technical brains.

And in the "overstudied" list I have begun to think we must now put most of our basic science. This is a hard conclusion, because all of science is so important in the long run and because it is still so small compared, say, to advertising or the tobacco industry. But basic scientific thinking is a scarce resource. In a national emergency, we would suddenly find that a host of our scientific problems could be postponed for several years in favor of more urgent research. Should not our total human emergency make the same claims? Long-range science is useless unless we survive to use it. Tens of thousands of our best trained minds may now be needed for something more important than "science as usual."

The arrows at level 2 in the tables are intended to indicate that problems may escalate to a higher level of crisis in the next time period if they are not solved. The arrows toward level 2 in the last columns of both tables show the escalation of all our problems upward to some general reconstruction in the 20- to 50-year time period, if we survive. Probably no human institution will continue unchanged for another 50 years, because they will all be changed by the crises if they are not changed in advance to prevent them. There will surely be widespread rearrangements in all our ways of life everywhere, from our patterns of society to our whole philosophy of man. Will they be more humane, or less? Will the world come to resemble a diverse and open humanist democracy? Or Orwell's 1984? Or a postnuclear desert with its scientists hanged? It is our acts of commitment and leadership in the next few months and years that will decide.

## Mobilizing Scientists

It is a unique experience for us to have peacetime problems, or technical problems which are not industrial problems, on such a scale. We do not know quite where to start, and there is no mechanism yet for generating ideas systematically or paying teams to turn them into successful solutions.

But the comparison with wartime research and development may not be inappropriate. Perhaps the antisubmarine warfare work or the atomic energy project of the 1940's provide the closest parallels to what we must do in terms of the novelty, scale, and urgency of the problems, the initiative needed, and the kind of large success that has to be achieved. In the antisubmarine campaign, Blackett assembled a few scientists and other ingenious minds in his "back room," and within a few months they had worked out the "operations analysis" that made an order-of-magnitude difference in the success of the campaign. In the atomic energy work, scientists started off with extracurricular research, formed a central committee to channel their secret communications, and then studied the possible solutions for some time before they went to the government for large-scale support for the great development laboratories and production plants.

Fortunately, work on our crisis problems today would not require secrecy. Our great problems today are all beginning to be world problems, and scientists from many countries would have important insights to contribute.

Probably the first step in crisis studies now should be the organization of intense technical discussion and education groups in every laboratory. Promising lines of interest could then lead to the setting up of part-time or full-time studies and teams and coordinating committees. Administrators and boards of directors might find active crisis research important to their own organizations in many cases. Several foundations and federal agencies already have in-house research and make outside grants in many of these crisis areas, and they would be important initial sources of support.

But the step that will probably be required in a short time is the creation of whole new centers, perhaps comparable to Los Alamos or the RAND Corporation, where interdisciplinary groups can be assembled to work full-time on solutions to these crisis problems. Many different kinds of centers will eventually be necessary, including research centers, development centers, training centers, and even production centers for new sociotechnical inventions. The problems of our time—the $100-billion food problem or the $100-billion arms control problem—are no smaller than World War II in scale and importance, and it would be absurd to think that a few academic research teams or a few agency laboratories could do the job.

## Social Inventions

The thing that discourages many scientists—even social scientists—from thinking in these research-and-development terms is their failure to realize that there are such things as social inventions and that they can have large-scale effects in a surprisingly short time. A recent study with Karl Deutsch has examined some 40 of the great achievements in social science in this century, to see where they were made and by whom and how long they took to become effective. They include developments such as the following:

Keynesian economics
Opinion polls and statistical sampling
Input-output economics
Operations analysis
Information theory and feedback theory
Theory of games and economic behavior
Operant conditioning and programmed learning
Planned programming and budgeting (PPB)
Non-zero-sum game theory

Many of these have made remarkable differences within just a few years in our ability to handle social problems or management problems. The opinion poll became a national necessity within a single election period. The theory of games, published in 1946, had become and important component of American strategic thinking by RAND and the Defense Department by 1953, in spite of the limitation of the theory at that time to zero-sum games, with their dangerous bluffing and "brinkmanship." Today, within less than a decade, the PPB management technique is sweeping through every large organization.

This list is particularly interesting because it shows how much can be done outside official government agencies when inventive men put their brains together. Most of the achievements were the work of teams of two or more men, almost all of them located in intellectual centers such as Princeton or the two Cambridges.

The list might be extended by adding commercial social inventions with rapid and widespread effects, like credit cards. And sociotechnical inventions, like computers and automation or like oral contraceptives, which were in widespread use within 10 years after they were developed. In addition, there are political innovations like the New Deal, which made great changes in our economic life within 4 years, and the pay-as-you-go income tax, which transformed federal taxing power within 2 years.

On the international scene, the Peace Corps, the "hot line," the Test-Ban Treaty, the Antarctic Treaty, and the Nonproliferation Treaty were all implemented within 2 to 10 years after their initial proposal. These are only small contributions, a tiny patchwork part of the basic international stabilization system that is needed, but they show that the time to adopt new structural designs may be surprisingly short. Our cliches about "social lag" are very misleading. Over half of the major social innovations since 1940 were adopted or had widespread social effects within less than 12 years—a time as short as, or shorter than, the average time for adoption of technological innovations.

## Areas for Task Forces

Is it possible to create more of these social inventions systematically to deal with our present crisis problems? I think it is. It may be worth listing a few specific areas where new task forces might start.

1. *Peace-keeping mechanisms and feedback stabilization.* Our various nuclear treaties are a beginning. But how about a technical group that sits down and thinks about the whole range of possible and impossible stabilization and peace-keeping mechanisms? Stabilization feedback-design might be a complex modern counterpart of the "checks and balances" used in designing the constitutional structure of the United States 200 years ago. With our new knowledge today about feedbacks, group behavior, and game theory, it ought to be possible to design more complex and even more successful structures.

Some peace-keeping mechanisms that might be hard to adopt today could still be worked out and tested and publicized, awaiting a more favorable moment. Sometimes the very existence of new possibilities can change the atmosphere. Sometimes, in a crisis, men may finally be willing to try out new ways and may find some previously prepared plan of enormous help.

2. *Biotechnology.* Humanity must feed and care for the children who are already in the world, even while we try to level off the further population explosion that makes this so difficult. Some novel proposals, such as food from coal, or genetic copying of champion animals, or still simpler contraceptive methods, could possibly have large-scale effects on human welfare within 10 to 15 years. New chemical, statistical, and management methods for measuring and maintaining the ecological balance could be of very great importance.

3. *Game theory.* As we have seen, zero-sum game theory has not been too academic to be used for national strategy and policy analysis. Unfortunately, in zero-sum games, what I win, you lose, and what you win, I lose. This may be the way poker works, but it is not the way the world works. We are collectively in a non-zero-sum game in which we will all lose together in nuclear holocaust or race conflict or economic nationalism, or all win together in survival and prosperity. Some of the many variations of non-zero-sum game theory, applied to group conflict and cooperation, might show us profitable new approaches to replace our sterile and dangerous confrontation strategies.

4. *Psychological and social theories.* Many teams are needed to explore in detail and in practice how the powerful new ideas of behavior theory and the new ideas of responsive living might be used to improve family life or community and management structures. New ideas of information handling and management theory need to be turned into practical recipes for reducing the daily frustrations of small businesses, schools, hospitals, churches, and town meetings. New economic inventions are needed, such as urban development corporations. A deeper systems analysis is urgently needed to see if there is not some practical way to separate full employment from inflation. Inflation pinches the poor, increases labor-management disputes, and multiplies all our domestic conflicts and our sense of despair.

5. *Social indicators.* We need new social indicators, like the cost-of-living index, for measuring a thousand social goods and evils. Good indicators can have great "multiplier effects" in helping to maximize our welfare and minimize our ills. Engineers and physical scientists working with social scientists might come up with ingenious new methods of measuring many of these important but elusive parameters.

6. *Channels of effectiveness.* Detailed case studies of the reasons for success or failure of various social inventions could also have a large multiplier effect. Handbooks showing what channels or methods are now most effective for different small-scale and large-scale social problems would be of immense value.

The list could go on and on. In fact, each study group will have its own pet projects. Why not? Society is at least as complex as, say, an automobile with its several thousand parts. It will probably require as many research-and-development teams as the auto industry in order to explore all the inventions it needs to solve its problems. But it is clear that there are many areas of great potential crying out for brilliant minds and brilliant teams to get to work on them.

## Future Satisfactions and Present Solutions

This is an enormous program. But there is nothing impossible about mounting and financing it, if we, as concerned men, go into it with commitment and leadership. Yes, there will be a need for money and power to overcome organizational difficulties and vested interests. But it is worth remembering that the only real source of power in the world is the gap between what is and what might be. Why else do men work and save and plan? If there is some future increase in human satisfaction that we can point to and realistically anticipate, men will be willing to pay something for it and invest in it in the hope of that return. In economics, they pay with money; in politics, with their votes and time and sometimes with their jail sentences and their lives.

Social change, peaceful or turbulent, is powered by "what might be." This means that for peaceful change, to get over some impossible barrier of unresponsiveness or complexity or group conflict, what is needed is an inventive man or group—a "social entrepreneur"—who can connect the pieces and show how to turn the advantage of "what might be" into some present advantage for every participating party. To get toll roads, when highways were hopeless, a legislative-corporation mechanism was invented that turned the future need into present profits for construction workers and bondholders and continuing profitability for the state and all the drivers.

This principle of broad-payoff anticipatory design has guided many successful social plans. Regular task forces using systems analysis to find payoffs over the barriers might give us such successful solutions much more often. The new world that could lie ahead, with its blocks and malfunctions removed, would be fantastically wealthy. It seems almost certain that there must be many systematic ways for intelligence to convert that large payoff into the profitable solution of our present problems.

The only possible conclusion is a call to action. Who will commit himself to this kind of search for more ingenious and fundamental solutions? Who will begin to assemble the research teams and the funds? Who will begin to create those full-time interdisciplinary centers that will be necessary for testing detailed designs and turning them into effective applications?

The task is clear. The task is huge. The time is horribly short. In the past, we have had science for intellectual pleasure, and science for the control of nature. We have had science for war. But today, the whole human experiment may hang on the question of how fast we now press the development of science for survival.

# SURVIVAL U: PROSPECTUS FOR A REALLY RELEVANT UNIVERSITY

*John Fischer*

> It gets pretty depressing to watch what is going on in the world and realize that your education is not equipping you to do anything about it.
>
> —From a letter by a University of California senior

*Ecology is a subject that is as complex as it is vital. It would appear then that our educational institutions will have to treat it seriously in the very near future. The following proposal by John Fischer outlines some of the steps that could be taken.*

She is not a radical, and has never taken part in any demonstration. She will graduate with honors, and profound disillusionment. From listening to her —and a good many like-minded students at California and East Coast campuses—I think I am beginning to understand what they mean when they say that a liberal-arts education isn't relevant.

They meant it is incoherent. It doesn't cohere. It consists of bits and pieces which don't stick together, and have no common purpose. One of our leading Negro educators, Arthur Lewis of Princeton, recently summed it up better than I can. America is the only country, he said, where youngsters are required "to fritter away their precious years in meaningless peregrination from subject to subject . . . spending twelve weeks getting some tidbits of religion, twelve weeks learning French, twelve weeks seeing whether the history professor is stimulating, twelve weeks seeking entertainment from the economics professor, twelve weeks confirming that one is not going to be able to master calculus."

These fragments are meaningless because they are not organized around any central purpose, or vision of the world. The typical liberal-arts college has no clearly defined goals. It merely offers a smorgasbord of courses, in hopes that if a student nibbles at a few dishes from the humanities table, plus a snack of science, and a garnish of art or anthropology, he may emerge as "a cultivated man"—whatever that means. Except for a few surviving church schools, no university even pretends to have a unifying philosophy. Individual teachers may have personal ideologies—but since they are likely to range, on any given campus, from Marxism to worship of the scientific method to exaltation of the irrational (à la Norman O. Brown), they don't cohere either. They often leave a student convinced at the end of four years that any given idea is probably about as valid as any other—and that none of them has much relationship to the others, or to the decisions he is going to have to make the day after graduation.

Education was not always like that. The earliest European universities had a precise purpose: to train an elite for the service of the Church. Everything they taught was focused to that end. Thomas Aquinas had spelled it all out: what subjects had to be mastered, how each connected with every other, and what meaning they had for man and God.

Later, for a span of several centuries, Oxford and Cambridge had an equally clear function: to train administrators to run an empire. So too did Harvard and Yale at the time they were founded; their job was to produce the clergymen, lawyers, and doctors that a new country needed. In each case, the curriculum was rigidly prescribed. A student learned what he needed, to prepare himself to be a competent priest, district officer, or surgeon. He had no doubts about the relevance of his courses—and no time to fret about expanding his consciousness or currying his sensual awareness.

This is still true of our professional schools. I have yet to hear an engineering or medical student complain that his education is meaningless. Only in the liberal-arts colleges—which boast that "we are not trade schools"—do the youngsters get that feeling that they are drowning in a cloud of feathers.

For a long while some of our less complacent academics have been trying to restore coherence to American education. When Robert Hutchins was at Chicago, he tried to use the Great Books to build a comprehensible framework for the main ideas of civilized man. His experiment is still being carried on, with some modifications, at St. John's—but it has not proved irresistibly contagious. Sure, the thoughts of Plato and Machiavelli are still pertinent, so far as they go—but somehow they don't seem quite enough armor for a world beset with splitting atoms, urban guerrillas, nineteen varieties of psychotherapists, amplified guitars, napalm, computers, astronauts, and an atmosphere polluted simultaneously with auto exhaust and TV commercials.

Another strategy for linking together the bits-and-pieces has been attempted at Harvard and at a number of other universities. They require their students to take at least two years of survey courses, known variously as core studies, general education, or world civilization. These too have been something less than triumphantly successful. Most faculty members don't like to teach them, regarding them as superficial and synthetic. (And right they are, since no survey course that I know of has a strong unifying concept to give it focus.) Moreover, the senior professors shun such courses in favor of their own narrow specialties. Consequently, the core studies which are meant to place all human experience—well, at least the brightest nuggets—into One Big Picture usually end up in the perfunctory hands of resentful junior teachers. Naturally the undergraduates don't take them seriously either.

Any successful reform of American education, I am now convinced, will have to be far more revolutionary than anything yet attempted. At a minimum, it should be:

1. Founded on a single guiding concept—an idea capable of knotting together all strands of study, thus giving them both coherence and visible purpose.

2. Capable of equipping young people to do something about "what is going on in the world"—notably the things which bother them most, including war, injustice, racial conflict, and the quality of life.

Maybe it isn't possible. Perhaps knowledge is proliferating so fast, and in so many directions, that it can never again be ordered into a coherent whole, so that molecular biology, Robert Lowell's poetry, and highway engineering will seem relevant to each other and to the lives of ordinary people. Quite possibly the knowledge explosion, as Peter F. Drucker has called it, dooms us to scholarship which grows steadily more specialized, fragmented, and incomprehensible.

The Soviet experience is hardly encouraging. Russian education is built on what is meant to be a unifying ideology: Marxism-Leninism. In theory, it provides an organizing principle for all scholarly activity—whether history, literature, genetics, or military science. Its purpose is explicit: to train a Communist elite for the greater power and glory of the Soviet state, just as the medieval universities trained a priesthood to serve the Church.

Yet according to all accounts that I have seen, it doesn't work very well. Soviet intellectuals apparently are almost as restless and unhappy as our own. Increasing numbers of them are finding Marxism-Leninism too simplistic, too narrowly doctrinaire, too oppressive; the bravest are risking prison in order to pursue their own heretical visions of reality.

Is it conceivable, then, that we might hit upon another idea which could serve as the organizing principle for many fields of scholarly inquiry; which is relevant to the urgent needs of our time; and which would not, on the other hand, impose an ideological strait jacket, as both ecclesiastical and Marxist education attempted to do?

Just possibly it could be done. For the last two or three years I have been probing around among professors, college administrators, and students —and so far I have come up with only one idea which might fit the specifications. It is simply the idea of survival.

For the first time in history, the future of the human race is now in serious question. This fact is hard to believe, or even think about—yet it is the message which a growing number of scientists are trying, almost frantically, to get across to us. Listen, for example, to Professor Richard A. Falk of Princeton and of the Center for Advanced Study in the Behavioral Sciences:

> The planet and mankind are in grave danger of irreversible catastrophe. . . . Man may be skeptical about following the flight of the dodo into extinction, but the evidence points increasingly to just such a pursuit. . . . There are four interconnected threats to the planet—wars of mass destruction, overpopulation, pollution, and the depletion of resources. They have a cumulative effect. A problem in one area renders it more difficult to solve the problems in any other area. . . . The basis of all four problems is the inadequacy of the sovereign states to manage the affairs of mankind in the twentieth century.

Similar warnings could be quoted from a long list of other social scientists, biologists, and physicists, among them such distinguished thinkers as Rene Dubos, Buckminster Fuller, Loren Eiseley, George Wald, and Barry Commoner. They are not hopeless. Most of them believe that we still have a chance to bring our weapons, our population growth, and the destruction of our environment under control before it is too late. But the time is short, and so far there is no evidence that enough people are taking them seriously.

That would be the prime aim of the experimental university I'm suggesting here: to look seriously at the interlinking threats to human existence, and to learn what we can do to fight them off.

Let's call it Survival U. It will not be a multiversity, offering courses in every conceivable field. Its motto—emblazoned on a life jacket rampant—will be: "What must we do to be saved?" If a course does not help to answer that question, it will not be taught here. Students interested in musicology, junk sculpture, the Theater of the Absurd, and the literary *dicta* of Leslie Fiedler can go somewhere else.

Neither will our professors be detached, dispassionate scholars. To get hired, each will have to demonstrate an emotional commitment to our cause.

Moreover, he will be expected to be a moralist; for this generation of students, like no other in my lifetime, is hungering and thirsting after righteousness. What it wants is a moral system it can believe in—and that is what our university will try to provide. In every class it will preach the primordial ethic of survival.

The biology department, for example, will point out that it is sinful for anybody to have more than two children. It has long since become glaringly evident that unless the earth's cancerous growth of population can be halted, all other problems—poverty, war, racial strife, uninhabitable cities, and the rest—are beyond solution. So the department naturally will teach all known methods of birth control, and much of its research will be aimed at perfecting cheaper and better ones.

Its second lesson in biological morality will be: "Nobody has a right to poison the environment we live in." This maxim will be illustrated by a list of public enemies. At the top will stand the politicians, scientists, and military men—of whatever country—who make and deploy atomic weapons; for if these are ever used, even in so-called defensive systems like the ABM, the atmosphere will be so contaminated with strontium-90 and other radioactive isotopes that human survival seems most unlikely. Also on the list will be anybody who makes or tests chemical and biological weapons—or who even attempts to get rid of obsolete nerve gas, as our Army recently proposed, by dumping the stuff in the sea.

Only slightly less wicked, our biology profs will indicate, is the farmer who drenches his land with DDT. Such insecticides remain virulent indefinitely, and as they wash into the streams and oceans they poison fish, water fowl, and eventually the people who eat them. Worse yet—as John Hay noted in his recently published *In Defense of Nature*—"The original small, diluted concentrations of these chemicals tend to build up in a food chain so as to end in a concentration that may be thousands of times as strong." It is rapidly spreading throughout the globe. DDT already has been found in the tissues of Eskimos and of Antarctic penguins, so it seems probable that similar deposits are gradually building up in your body and mine. The minimum fatal dosage is still unknown.

Before he finishes this course, a student may begin to feel twinges of conscience himself. Is his motorcycle exhaust adding carbon monoxide to the smog we breathe? Is his sewage polluting the nearest river? If so, he will be reminded of two proverbs. From Jesus: "Let him who is without sin among you cast the first stone." From Pogo: "We have met the enemy and he is us."

In like fashion, our engineering students will learn not only how to build dams and highways, but where *not* to build them. Unless they understand that it is immoral to flood the Grand Canyon or destroy the Everglades with a jetport, they will never pass the final exam. Indeed, our engineering graduates will be trained to ask a key question about every contract offered them: "What will be its effect on human life?" That obviously will lead to other questions which every engineer ought to comprehend as thoroughly as his slide rule. Is this new highway really necessary? Would it be wiser to use the money for mass transit—or to decongest traffic by building a new city somewhere else? Is an offshore oil well really a good idea, in view of what happened to Santa Barbara?

Our engineering faculty also will specialize in training men for a new growth industry: garbage disposal. Americans already are spending $4.5 billion a year to collect and get rid of the garbage which we produce more profusely than any other people (more than five pounds a day for each of us). But unless we are resigned to stifling in our own trash, we are going to have to come up with at least an additional $835 million a year.[1] Any industry with a growth rate of 18 percent offers obvious attractions to a bright young man —and if he can figure out a new way to get rid of our offal, his fortune will be unlimited.

Because the old ways no longer work. Every big city in the United States is running out of dumping grounds. Burning won't do either, since the air is dangerously polluted already—and in any case, 75 percent of the incinerators in use are inadequate. For some 150 years Californians happily piled their garbage into San Francisco Bay, but they can't much longer. Dump-and-fill operations already have reduced it to half its original size, and in a few more decades it would be possible to walk dry-shod from Oakland to the Embarca-dero. Consequently San Francisco is now planning to ship garbage 375 miles to the yet-uncluttered deserts of Lassen County by special train—known locally as "The Twentieth Stenchery Limited" and "The Excess Express." The city may actually get away with this scheme, since hardly anybody lives in Lassen County except Indians, and who cares about them? But what is the answer for the metropolis that doesn't have an unspoiled desert handy?

A few ingenious notions are cropping up here and there. The Japanese are experimenting with a machine which compacts garbage, under great heat and pressure, into building blocks. A New York businessman is thinking of building a garbage mountain somewhere upstate, and equipping it with ski runs to amortize the cost. An aluminum company plans to collect and reprocess used aluminum cans—which, unlike the old-fashioned tin can, will not rust away. Our engineering department will try to Think Big along these lines. That way lies not only new careers, but salvation.

Survival U's Department of Earth Sciences will be headed—if we are lucky —by Dr. Charles F. Park, Jr., now professor of geology and mineral engineer-ing at Stanford. He knows as well as anybody how fast mankind is using up the world's supply of raw materials. In a paper written for the American

[1] According to Richard D. Vaughn, chief of the Solid Wastes Program of HEW, in his recent horror story entitled "1968 Survey of Community Solid Waste Practices."

Geographical Society he punctured one of America's most engaging (and pernicious) myths: our belief that an ever-expanding economy can keep living standards rising indefinitely.

It won't happen; because, as Dr. Park demonstrates, the tonnage of metal in the earth's crust won't last indefinitely. Already we are running short of silver, mercury, tin, and cobalt—all in growing demand by the high-technology industries. Even the commoner metals may soon be in short supply. The United States alone is consuming one ton of iron and eighteen pounds of copper every year, for each of its inhabitants. Poorer countries, struggling to industrialize, hope to raise their consumption of these two key materials to something like that level. If they should succeed—and if the globe's population doubles in the next forty years, as it will at present growth rates—then the world will have to produce, somehow, *twelve times* as much iron and copper every year as it does now. Dr. Parks sees little hope that such production levels can ever be reached, much less sustained indefinitely. The same thing, of course—doubled in spades—goes for other raw materials: timber, oil, natural gas, and water, to note only a few.

Survival U, therefore, will prepare its students to consume less. This does not necessarily mean an immediate drop in living standards—perhaps only a change in the yardstick by which we measure them. Conceivably Americans might be happier with fewer automobiles, neon signs, beer cans, supersonic jets, barbecue grills, and similar metallic fluff. But happy or not, our students had better learn how to live The Simpler Life, because that is what most of them are likely to have before they reach middle age.

To help them understand how very precious resources really are, our mathematics department will teach a new kind of bookkeeping: social accounting. It will train people to analyze budgets—both government and corporate—with an eye not merely to immediate dollar costs, but to the long-range costs to society.

By conventional bookkeeping methods, for example, the coal companies strip-mining away the hillsides of Kentucky and West Virginia show a handsome profit. Their ledgers, however, show only a fraction of the true cost of their operations. They take no account of destroyed land which can never bear another crop; of rivers poisoned by mud and seeping acid from the spoil banks; of floods which sweep over farms and towns downstream, because the ravaged slopes can no longer hold the rainfall. Although these costs are not borne by the mining farms, they are nevertheless real. They fall mostly on the taxpayers, who have to pay for disaster relief, flood-control levees, and the resettlement of Appalachian farm families forced off the land. As soon as our students (the taxpayers of tomorrow) learn to read a social balance sheet, they obviously will throw the strip miners into bankruptcy.

Another case study will analyze the proposal of the Inhuman Real Estate Corporation to build a fifty-story skyscraper in the most congested area of midtown Manhattan. If 90 percent of the office space can be rented at $12 per square foot, it looks like a sound investment, according to antique accounting methods. To uncover the true facts, however, our students will investigate the cost of moving 12,000 additional workers in and out of midtown during rush hours. The first (and least) item is $8 million worth of new city buses. When they are crammed into the already clogged avenues, the daily loss of man-hours in traffic jams may run to a couple of million more. The fumes from their diesel engines will cause an estimated 9 percent increase in New York's incidence of emphysema and lung cancer: this requires the construction of three new hospitals. To supply them, plus the new building, with water—already perilously short in the city—a new reservoir has to be built on the headwaters of the Delaware River, 140 miles away. Some of the dairy farmers pushed out of the drowned valley will move promptly into the Bronx and go on relief. The subtraction of their milk output from the city's supply leads to a price increase of two cents a quart. For a Harlem mother with seven hungry children, that is the last straw. She summons her neighbors to join her in riot, seven blocks go up in flames, and the mayor demands higher taxes to hire more police. . . .

Instead of a sound investment, Inhuman Towers now looks like criminal folly, which would be forbidden by any sensible government. Our students will keep that in mind when they walk across campus to their government class.

Its main goal will be to discover why our institutions have done so badly in their efforts (as Dr. Falk put it) "to manage the affairs of mankind in the twentieth century." This will be a compulsory course for all freshmen, taught by professors who are capable of looking critically at every political artifact, from the Constitution to the local county council. They will start by pointing out that we are living in a state of near-anarchy, because we have no government capable of dealing effectively with public problems.

Instead we have a hodgepodge of 80,000 local governments—villages, townships, counties, cities, port authorities, sewer districts, and special purpose agencies. Their authority is so limited, and their jurisdictions so confused and overlapping, that most of them are virtually impotent. The states, which in theory could put this mess into some sort of order, usually have shown little interest and less competence. When Washington is called to help out—as it increasingly has been for the last thirty-five years—it often has proved ham-handed and entangled in its own archaic bureaucracy. The end result is that nobody in authority has been able to take care of the country's mounting needs. Our welfare rolls keep growing, our air and water get dirtier, housing gets scarcer, airports jam up, road traffic clots, railways fall apart, prices rise, ghettos burn, schools turn out more losing confidence in American institutions. In their present state, they don't deserve much confidence.

The advanced student of government at Survival U will try to find out whether these institutions can be renewed and rebuilt. They will take a hard look at the few places—Jacksonville, Minnesota, Nashville, Appalachia—which are creating new forms of government. Will these work any better, and if so, how can they be duplicated elsewhere? Can the states be brought to life, or should we start thinking about an entirely different kind of arrangement? Ten regional prefectures, perhaps, to replace the fifty states? Or should we take seriously Norman Mailer's suggestion for a new kind of city-state to govern our great metropolises? (He merely called for New York City to secede from its state; but that isn't radical enough. To be truly governable, the new Republic of New York City ought to include chunks of New Jersey and Connecticut as well.) Alternatively, can we find some way to break up Megalopolis, and spread our population into smaller and more livable communities throughout the continent? Why should we keep 70 percent of our people crowded into less than 2 percent of our land area, anyway?

Looking beyond our borders, our students will be encouraged to ask even harder questions. Are nation-states actually feasible, now that they have power to destroy each other in a single afternoon? Can we agree on something else to take their place, before the balance of terror becomes unstable? What price would most people be willing to pay for a more durable kind of human organization—more taxes, giving up national flags, perhaps the sacrifice of some of our hard-won liberties?

All these courses (and everything else taught at Survival U) are really branches of a single science. Human ecology is one of the youngest disciplines, and probably the most important. It is the study of the relationship between man and his environment, both natural and technological. It teaches us to understand the consequences of our actions—how sulfur-laden fuel oil burned in England produces an acid rain that damages the forests of Scandinavia, why a well-meant farm subsidy can force millions of Negro tenants off the land and lead to Watts and Hough. A graduate who comprehends ecology will know how to look at "what is going on in the world," and he will be equipped to do something about it. Whether he ends up as a city planner, a politician, an enlightened engineer, a teacher, or a reporter, he will have had a relevant education. All of its parts will hang together in a coherent whole.

And if we can get enough such graduates, man and his environment may survive a while longer, against all the odds.

# ECOLOGY AND REVOLUTIONARY THOUGHT

*Murray Bookchin*

*Murray Bookchin refers to himself as an anarchist and what he means by that becomes very clear in the following article. In his opinion, time has altered the established meaning of the term so that today the true anarchist is a person who subscribes to a much more complex and subtle notion of how the world is to be held together. As an anarchist the only threat he poses to society is to those whose efforts to synthesize now appear to be heading us toward extinction. His notion of annihilation is affirmative. Furthermore, he strongly believes that ecology will prove the clue to social revolution. Although in his writing style one senses that he wants to be cautious about utopian thought, there is much that is truly exciting in his plan for the future.*

In almost every period since the Renaissance, the development of revolutionary thought has been heavily influenced by a branch of science, often in conjunction with a school of philosophy.

Astronomy in the time of Copernicus and Galileo helped to guide a sweeping movement of ideas from the medieval world, riddled by superstition, into one pervaded by a critical rationalism, openly naturalistic and humanistic in outlook. During the Enlightenment—the era that culminated in the Great French Revolution—this liberatory movement of ideas was reinforced by advances in mechanics and mathematics. The Victorian Era was shaken to its very foundations by evolutionary theories in biology and anthropology, by Marx's reworking of Ricardian economics, and towards its end, by Freudian psychology.

SOURCE: Murray Bookchin, "Ecology and Revolutionary Thought," *Anarchos*, no. 1, February 1968. Reprinted by permission of the author.

In our time, we have seen the assimilation of these once liberatory sciences by the established social order. Indeed, we have begun to regard science itself as an instrument of control over the thought processes and physical being of man. This distrust of science and of the scientific method is not without justification. "Many sensitive people, especially artists," observes Abraham Maslow, "are afraid that science besmirches and depresses, that it tears things apart rather than integrating them, thereby killing rather than creating." What is perhaps equally important, modern science has lost its critical edge. Largely functional or instrumental in intent, the branches of science that once tore at the chains of man are now used to perpetuate and gild them. Even philosophy has yielded to instrumentalism and tends to be little more than a body of logical contrivances, the handmaiden of the computer rather than the revolutionary.

There is one science, however, that may yet restore and even transcend the liberatory estate of the traditional sciences and philosophies. It passes rather loosely under the name of "ecology"—a term coined by Haeckel a century ago to denote "the investigation of the total relations of the animal both to its inorganic and to its organic environment." A first glance, Haeckel's definition sounds innocuous enough; and ecology, narrowly conceived as one of the biological sciences, is often reduced to a variety of biometrics in which field workers focus on food chains and statistical studies of animal populations. There is an ecology of health that would hardly offend the sensibilities of the American Medical Association and a concept of social ecology that would conform to the most well-engineered notions of the New York City Planning Commission.

Broadly conceived, however, ecology deals with the balance of nature. Inasmuch as nature includes man, the science basically deals with the harmonization of nature and man. This focus has explosive implications. The explosive implications of an ecological approach arise not only from the fact that ecology is intrinsically a critical science—in fact, critical on a scale that the most radical systems of political enconomy failed to attain—but it is also an integrative and reconstructive science. This integrative, reconstructive aspect of ecology, carried through to all its implications, leads directly into anarchic areas of social thought. For in the final analysis, it is impossible to achieve a harmonization of man and nature without creating a human community that lives in a lasting balance with its natural environment.

## The Critical Nature of Ecology

Let us examine the critical edge of ecology—a unique feature of the science in a period of general scientific docility.

Basically, this critical edge derives from the subject-matter of ecology—from its very domain. The issues with which ecology deals are imperishable in the sense that they cannot be ignored without bringing into question the viability of the planet, indeed the survival of man himself. The critical edge of ecology is due not so much to the power of human reason—a power which science hallowed during its most revolutionary periods—but to a still higher power, the sovereignty of nature over man and all his activities. It may be that man is manipulable, as the owners of the mass media argue, or that elements of nature are manipulable, as the engineers demonstrate by their dazzling achievements, but ecology clearly shows that the *totality* of the natural world—nature taken in *all* its aspects, cycles, and interrelationships—cancels out all human pretensions to mastery over the planet. The great wastelands of North Africa and the eroded hills of Greece, once areas of a thriving agriculture or a rich natural flora, are historic evidence of nature's revenge against human parasitism, be it in the form of soil exploitation or deforestation.

Yet none of these historical examples compare in weight and scope with the effects of man's despoilation—and nature's revenge—since the days of the Industrial Revolution, and especially since the end of the Second World War. Ancient examples of human parasitism were essentially local in scope; they were precisely *examples* of man's potential for destruction and nothing more. Often, they were compensated by remarkable improvements in the natural ecology of a region, as witness the European peasantry's superb re-working of the soil during centuries of cultivation and the even more superb achievements of Inca agriculturists in terracing the Andes Mountains during pre-Columbian times.

Modern man's despoilation of the environment is global in scope, like his imperialisms. It is even extra-terrestrial, as witness the disturbances of the Van Allen Belt a few years ago. Human parasitism, today, disrupts not only the atmosphere, climate, water resources, soil, flora, and fauna of a region; it upsets virtually all the basic cycles of nature and threatens to undermine the stability of the environment on a worldwide scale.

To gauge the scope of modern man's disruptive role: it has been estimated that the burning of fossil fuels (coal and oil) annually adds 600 million tons of carbon dioxide to the air, an average of about .03 percent of the total atmospheric mass—this, I may add, aside from an incalculable quantity of toxicants. Since the Industrial Revolution, the overall atmospheric mass of carbon dioxide has increased by 13 percent over earlier, more stable, levels. It could be argued on very sound theoretical grounds that this mounting blanket of carbon dioxide, by intercepting heat radiated from the earth into outer space, leads to rising atmospheric temperatures, to a more violent circulation of air, to more destructive storm patterns, and eventually, it will lead to a melting of the polar ice caps (possibly in two or three centuries), rising sea levels, and the inundation of vast land areas. Far removed as such a deluge may be, the changing proportion of carbon dioxide to other atmospheric gases is symbolic of the impact man is having on the balance of nature.

A more immediate ecological issue is man's extensive pollution of the earth's waterways. What counts, here, is not the fact that man befouls a given stream, river, or lake—a thing he has done for ages—but rather the magnitude water pollution has reached in the past two generations.

Nearly all the surface waters of the United States are polluted. Many American waterways are open cesspools that properly qualify as extensions of urban sewage systems. It would be a euphemism to describe them any longer as rivers or lakes. More significantly, large portions of groundwater are sufficiently polluted to be undrinkable, even medically hazardous, and a number of local hepatitis epidemics have been traced to polluted wells in suburban areas. In contrast to surface-water pollution, groundwater or subsurface-water pollution is immensely difficult to eliminate and tends to linger on for decades after the sources of pollution have been removed.

An article in a mass-circulation magazine appropriately describes the polluted waterways of the United States as "Our Dying Waters." This despairing, apocalyptic description of the water-pollution problem in the United States really applies to the world at large. The waters of the earth, conceived as factors in a large ecological system, are literally dying. Massive pollution is destroying the once pristine rivers and lakes of Africa, Asia, and Latin America as media of life, as well as the long-abused waterways of highly industrialized continents. Even the open sea has not been spared from extensive pollution. And I speak, here, not only of radioactive pollutants from nuclear bomb tests and power reactors, which apparently reach all the flora and fauna of the sea. It suffices to point out that the discharge of diesel-oil wastes from ships in the Atlantic has become a massive pollution problem, claiming marine life in enormous numbers every year.

Accounts of this kind can be repeated for virtually every part of the biosphere. Pages can be written on the immense losses of productive soil that occur annually in almost every continent of the earth; on the extensive loss of the tree cover in areas vulnerable to erosion; on lethal air-pollution episodes in major urban areas; on the worldwide distribution of toxic agents, such as radioactive isotopes and lead; on the chemicalization of man's immediate environment—one might say his very dinner table—with pesticide residues and food additives. Pieced together like bits of a jigsaw puzzle, these affronts to the environment form a pattern of destruction that has no precedent in man's long history on the earth.

Obviously, man would be dismissed as a highly destructive parasite, who threatens to destroy his host—the natural world—and eventually himself. In ecology, however, the word "parasite," used in this oversimplified sense, is not an answer to a question, but comprises the question itself. Ecologists know that a destructive parasitism of this kind usually reflects a disruption of an ecological situation; indeed, many species, seemingly highly destructive under one set of conditions, are eminently useful under another set of conditions. What imparts a profoundly critical function to ecology is the fact that man's destructive activities raise the question: What are the conditions that have turned man into a destructive parasite? What produces a form of human parasitism that results not only in vast natural imbalances, but also threatens the very existence of humanity itself?

The truth is that man has produced imbalances not only in nature, but more fundamentally, in his relations with his fellow man—in the very structure of his society. To state this thought more precisely: The imbalances man has produced in the natural world are caused by the imbalances he has produced in the social world. A century ago it would have been possible to regard air pollution and water contamination as the result of greed, profit-seeking, and competition—in short, as the result of the activities of industrial barons and self-seeking bureaucrats. Today, this explanation would be a gross over-simplification. It is doubtless true that most bourgeois enterprises are still guided by a public-be-damned attitude, as witness the reactions of power utilities, automobile concerns, and steel corporations to pollution problems. But a more deep-rooted problem than the attitude of the owners is the size of the firms themselves—their enormous physical proportions, their location in a region, their density with respect to a community or a waterway, their requirements for raw materials and water, and their role in the national division of labour.

What we are seeing, today, is a crisis not only in natural ecology but, above all, in social ecology. Modern society, especially as we know it in the United States and Europe, is being organized around immense urban belts at one extreme, a highly industrialized agriculture at the other extreme, and capping both, a swollen, bureaucratized, anonymous state apparatus. If we leave all values aside, for the moment, and examine the physical structure of this society, what must necessarily impress us is the incredible logistical problems it must try to solve—problems of transportation, of density, of supply (raw materials, manufactured commodities, and foodstuffs), of economic and political organization, of industrial location, and so forth. The burden this type of urbanized and centralized society places on any continental area is enormous. If the process of urbanizing man and industrializing agriculture were to continue unabated, it would make much of the earth inhospitable for viable, healthy human beings and render vast areas utterly uninhabitable.

Ecologists are often asked, rather tauntingly, to locate with scientific exactness the ecological breaking point of nature—presumably, the point at which the natural world will cave in on man. This is equivalent to asking a psychiatrist for the precise moment when a neurotic will become a nonfunctional psychotic. No such answer is ever likely to be available. But the ecologist can supply a strategic insight into the directions man seems to be following as a result of his split with the natural world.

From the standpoint of ecology, man is dangerously simplifying his environment. The modern city represents a regressive encroachment of the synthetic on the natural, of the inorganic (concrete, metals, and glass) on the organic, of crude, elemental stimuli on variegated, wide-ranging ones. The vast urban belts now developing in industrialized areas of the world are not only grossly offensive to eye and ear, but they are becoming chronically smog-ridden, noisy, and virtually immobilized by congestion. This process of simplifying man's environment and rendering it increasingly elemental and crude has a cultural as well as a physical dimension. The need to manipulate immense urban populations—to transport, feed, employ, educate, and some-how entertain millions of densely concentrated people daily—leads to a crucial decline in civic and social standards. A mass concept of human relations—totalitarian, centralistic, and regimented in orientation—tends to dominate the more individuated concepts of the past. Bureaucratic techniques of social management tend to replace humanistic approaches. All that is spontaneous, creative, and individuated is circumscribed by the standardized, the regulated, and the massified. The space of the individual is steadily narrowed by restrictions imposed upon him by a faceless, impersonal social apparatus. Any recognition of unique personal qualities is increasingly sur-rendered to the needs—more precisely, the manipulation—of the group, indeed, of the lowest common denominator of the mass. A quantitative, statistical approach, a beehive manner of dealing with man, tends to triumph over that precious, individualized-qualities approach which places its strongest emphasis on personal uniqueness, free expression, and cultural complexity.

The same regressive simplification of the environment occurs in modern agriculture.[1] The manipulated people in modern cities must be fed, and to feed them involves an extension of industrial farming. Food plants must be cultivated in a manner that allows for a high degree of mechanization—not to reduce human toil but to increase productivity, efficiency, maximize invest-ments, exploit the biosphere. Accordingly the terrain must be reduced to a flat plain—to a factory floor if you will—and natural variations in topography must be diminished as much as possible. Plant growth must be closely regu-lated to meet the tight schedules of food-processing plants. Ploughing, soil fertilization, sowing, and harvesting must be handled on a mass scale, often in total disregard of the natural ecology of the area. Large areas of the land must be used to cultivate a single crop, a form of plantation agriculture that not only lends itself to mechanization but also to pest infestation—a single

[1] For an insight into this problem, I wish to urge the reader to consult The Ecology of Invasions by Charles S. Elton (John Wiley & Sons, New York, 1958), Soil and Civilization by Edward Hyams (Thames & Hudson, London, 1952), Our Synthetic Environment by Lewis Herber (Knopf, New York, 1962), and a re-reading of Silent Spring by Rachel Carson—the last to be read not so much as a diatribe against pesticides but as a plea for ecological diversification.

crop being the ideal environment for the proliferation of individual pest species. Finally, chemical agents must be used lavishly to deal with the problems created by insects, weeds, plant diseases; to regulate crop production and maximize soil exploitation. The real symbol of agriculture is not the sickle or, for that matter, the tractor, but the aeroplane. The modern food cultivator is represented not by the peasant, yeoman, or even the agronomist —men who could be expected to have an intimate relationship with the unique qualities of the land on which they grow crops—but the pilot and chemist, for whom soil is a mere resource, an inorganic raw material.

The simplification process is carried still further by an exaggerated regional, indeed a national division of labour. Immense areas of the planet are increasingly reserved for specific industrial tasks or reduced to depots of raw materials. Others are turned into centres of urban population, largely occupied with commerce and trade. Cities and regions, in fact countries and continents, are specifically identified with special products—Pittsburgh, Cleveland, and Youngstown with steel, New York with finance, Bolivia with tin, Arabia with oil, Europe and America with industrial goods, and the rest of world with raw materials of one kind or another. The complex ecosystems which make up the regions of a continent are submerged, in effect, by an organization of entire nations into economically rationalized entities, each a way-station in a vast industrial belt system, global in its dimensions. By the same token, it is only a matter of time before the most attractive areas of the countryside will succumb to the concrete mixer, just as most of the Eastern seashore areas of the United States have already succumbed to subdividers and bungalows. What will remain in the way of natural beauty will be debased by trailer lots, canvas slums, "scenic" highways, motels, food stalls, and the oil slicks of motor boats.

The point is that man is literally undoing the work of organic evolution. By creating vast urban agglomerations of concrete, metal, and glass, by overriding and undermining the complex, often subtly organized ecosystems that constitute local differences in the natural world—in short, by replacing a highly complex, organic environment by a simplified, inorganic one—man is disassembling the biotic pyramid that supported humanity for countless millenia. In the course of replacing the complex ecological relationships on which all advanced living things depend for more elementary relationships, man is steadily restoring the biosphere to a stage which will be able to support only simpler forms of life. If this great reversal of the evolutionary process continues, it is by no means fanciful to suppose that the preconditions for higher forms of life will be irreparably destroyed and the earth will be incapable of supporting man himself.

Ecology derives its critical edge not only from the fact that it alone, among all the sciences, presents this awesome message to humanity, but because it also presents this message in a new social dimension. From an ecological viewpoint, the reversal of organic evolution is the result of appalling contradictions between town and country, state and community, industry and husbandry, mass manufacture and craftsmanship, centralism and regionalism, the bureaucratic scale and the human scale.

## The Reconstructive Nature of Ecology

Until recently, attempts to resolve the contradictions created by urbanization, centralization, bureaucratic growth, and statification were viewed as a vain counterdrift to "progress"—a counterdrift that, at best, could be dismissed as chimerical and, at worst, reactionary. The anarchist was regarded as a forlorn visionary, a social outcast, filled with nostalgia for the peasant village or the medieval commune. His yearnings for a decentralized society, for a humanistic community at one with nature and the needs of the individual —spontaneous and unfettered by authority—were viewed as the reactions of a romantic, of a declassed craftsman or an intellectual "misfit." His protest against centralization and statification seemed all the less persuasive because it was supported primarily by ethical considerations, by utopian, ostensibly "unrealistic" notions of what man could be, not what he was. To this protest, opponents of anarchist thought—liberals, rightists, and authoritarian "leftists" —argued that they were the voices of historic reality, that their statist, centralist, and political notions were rooted in the objective, practical world.

Time is not very kind to the conflict of ideas. Whatever may have been the validity of libertarian and nonlibertarian views a few generations ago, historical development has rendered virtually all objections to anarchist thought meaningless today. The modern city and state, the massive coal-steel technology of the Industrial Revolution, the later, more rationalized systems of mass production and assembly-line systems of labour organization, the centralized nation, the state and its bureaucratic apparatus—all, have reached their limits. Whatever progressive or liberatory role they may have possessed has clearly become entirely regressive and oppressive. They are regressive not only because they erode the human spirit and drain the community of all its cohesive, solidarity, and ethico-cultural standards; they are regressive from an objective standpoint, from an ecological standpoint. For they undermine not only the human spirit and the human community but also the viability of the planet and all living things on it.

What I am trying to say—and it cannot be emphasized too strongly— is that the anarchist concept of a balanced community, a face-to-face democracy, a humanistic technology, and a decentralized society—these rich libertarian concepts are not only desirable but they are also necessary. They belong not only to the great visions of man's future but they now constitute the preconditions for human survival. The process of social development has carried them from an ethical, subjective dimension into a practical, objective dimension. What was once regarded as impractical and visionary has now become eminently practical. And what was once regarded as practical and objective has become eminently impractical and irrelevant in terms of man's development toward a fuller, unfettered existence. If community, face-to-face democracy, a humanistic, liberatory technology, and decentralization are conceived of merely as reactions to the prevailing state of affairs—a vigorous "nay" to the "yea" of what exists today—a compelling, objective case can now be made for the practicality of an anarchist society.

This reflex-like reaction, this rejection of the prevailing state of affairs accounts, I think, for the explosive growth of intuitive anarchism among young people today. Their love of nature is a reaction against the highly synthetic qualities of our urban environment and its shabby products. Their informality of dress and manners is a reaction against the formalized, standardized nature of modern institutionalized living. Their predisposition for direct action is a reaction against the bureaucratization and centralization of society. Their tendency to drop out, to avoid toil and the rat-race reflects a growing anger toward the mindless industrial routine bred by modern mass manufacture, be it in the factory, office, or university. Their intense individualism is, in its own elemental way, a *de facto* decentralization of social life—a personal abdication from the demands of a mass society.

What is most significant about ecology is its ability to convert this rejection of the *status quo,* often nihilistic in character, into an emphatic affirmation of life—indeed, into a reconstructive credo for a humanistic society. The essence of ecology's reconstructive message can be summed up in the word "diversity." From an ecological viewpoint, balance and harmony in nature, in society, and by inference, in behaviour, is achieved not by mechanical standardization, but precisely by its opposite, organic differentiation. This message can be understood clearly only by examining its practical meaning on several levels of experience.

Let us consider the ecological principle of diversity—what Charles Elton calls the "conservation of variety"—as it applies to biology, specifically to agriculture. A number of studies—Lotka's and Volterra's mathematical models, Gause's experiments with protozoa and mites in controlled environments, and extensive field research—clearly demonstrate that fluctuations in populations, ranging from mild to pest-like proportions, depend heavily upon the number of species in an ecosystem and the degree of variety in the environment. The greater the variety of prey and predators, the more stable the population; the more diversified the environment in terms of flora and fauna, the less likely is there to be ecological instability. Complexity, variety, and diversity—choose whatever term you will—are a function of stability. If the environment is simplified and the variety of animal and plant species is reduced, fluctuations in population become marked and tend to get out of control. They tend to reach pest proportions.

In the case of pest control, many ecologists now conclude that we can avoid the repetitive use of toxic chemicals such as insecticides and herbicides by allowing for a greater interplay between living things. We must accord more room for natural spontaneity, for the diverse biological forces that make up an ecological situation. "European entomologists now speak of managing the entire plant-insect community," observes Robert L. Rudd. "It is called manipulation of the biocenose.[2] The biocenetic environment is varied, complex and dynamic. Although numbers of individuals will constantly change, no one species will normally reach pest proportions. The special conditions which allow high populations of a single species in a complex ecosystem are rare events. Management of the biocenose or ecosystem should become our goal, challenging as it is."

To "manipulate" the biocenose in a meaningful way, however, presupposes a far-reaching decentralization of agriculture. Wherever feasible, industrial agriculture must give way to soil and agricultural husbandry; the factory floor must yield to gardening and horticulture. I do not wish to imply that we must surrender the gains acquired by large-scale agriculture and mechanization. What I *do* contend, however, is that the land must be cultivated as though it were a garden—its flora diversified and carefully tended, balanced by a fauna and tree shelter appropriate to the region. Decentralization is important, moreover, not only for the development of the agricultural situation, but also for the development of the agriculturist. Food cultivation, practised in a truly ecological sense, presupposes that the agriculturist is familiar with all the features and subtleties of the terrain on which the crops are grown. By this I mean that he must have a thorough knowledge of the physiography of the land, its variegated soils—crop land, forest land, pasture land; mineral and organic content—its microclimate, and he must be engaged in a continuing study of the effects produced by new flora and fauna. He must acquire a sensitivity to its possibilities and needs to a point where he becomes an organic part of the agricultural situation. We can hardly hope to achieve this high degree of sensitivity and integration in the food cultivator without reducing agriculture to a human scale, without bringing agriculture within the scope of the individual. To meet the demands of an ecological approach to food cultivation, agriculture must be rescaled from huge industrial farms to moderate-sized units.

The same reasoning applies to a rational development of energy resources. The Industrial Revolution increased the *quantity* of energy available

[2] Rudd's use of the word "manipulation" is likely to create the erroneous impression that an ecological situation can be reduced to simple mechanical terms. Lest this impression arise, I would like to emphasize that our knowledge of an ecological situation and the practical use of this knowledge is a matter of insight and understanding rather than power. Elton, I think, states the case for the management of an ecological situation when he writes: "The world's future has to be managed, but this management would not be just like a game of chess—[but] more like steering a boat."

to industry, but it diminished the *variety* of energy resources used by man. Although it is certainly true that pre-industrial societies relied primarily on animal power and human muscles, complex energy patterns developed in many regions of Europe, involving a subtle integration of resources such as wind and water power, and a variety of fuels (wood, peat, coal, vegetable starches, and animal fats).

The Industrial Revolution overwhelmed and largely destroyed these regional energy patterns, initially replacing them by a single energy system (coal) and later by a dual system (coal and petroleum). Regions disappeared as models of integrated energy patterns—indeed, the very concept of *integration through diversity* was obliterated. As I indicated earlier, many regions became predominantly mining areas, devoted to the extraction of a single resource, while others were turned into immense industrial areas, often devoted to the production of a few commodities. We need not review the role this breakdown in true regionalism has played in producing air and water pollution, the damage it has inflicted on large areas of the countryside, and the prospect we face in the depletion of our precious hydrocarbon fuels.

We can, of course, turn to nuclear fuels. Conceived as a single-energy-resource, it is chilling to think of the lethal radioactive wastes that would require disposal as power reactors replace conventional fuel systems. Eventually, an energy system based on radioactive materials would lead to the widespread contamination of the environment—at first, in a subtle form, but later on a massive and palpably destructive scale.

Or we could apply ecological principles to the solution of our energy problems. We could try to reestablish earlier regional energy patterns—a combined system of energy provided by wind, water, and solar power. But today we would be aided by more sophisticated devices than any known in the past. We have now designed wind turbines that could supply electricity in a number of mountainous areas to meet the electric-power needs of a community of 50,000 people. We have perfected solar-energy devices that yield temperatures high enough in our warmer latitudes to deal with most metallurgical problems. Used in conjunction with heat pumps, many solar devices could provide as much as three-quarters—if not all—of the heat required to comfortably maintain a small family house. And at this writing the French are completing a tidal dam at the mouth of the Rance River in Brittany that is expected to produce more than 500 million kilowatt-hours of electricity a year. In time, the Rance River project will meet most of the electrical needs of northern France.[3]

[3] These examples are merely glimpses of the liberatory potential of little-known areas of modern technology. In a later article on the subject of modern technology and decentralization, I plan to explore the problem in much greater detail, and I propose to show that it is possible to *humanize* technology in such a way that machines will no longer appear to be the masters of men, but quite to the contrary, will enter fully into the service of man's fulfillment, both spiritually and materially.

Solar devices, wind turbines, and hydroelectric resources—each, taken singly, does not provide a solution for our energy problems and the ecological disruption created by conventional fuels. Pieced together as a mosaic, more precisely, as an organic energy pattern *developed from the potentialities of a region,* they could amply meet the needs of a decentralized society. In warm, sunny latitudes, we could rely more heavily on solar energy than on combustible fuels. In areas marked by atmospheric turbulence, we could rely more heavily on wind devices, and in suitable coastal areas or inland regions with a good network of rivers, the greater part of our energy would come from hydroelectric installations. In all cases, we would use a *mosaic* of noncombustible energy resources, filling whatever gaps develop by combustible and nuclear fuels. The point I wish to make is that by diversifying our use of energy resources, by organizing them into an ecologically balanced pattern, we could combine wind, solar, and water power in a given region to meet all the industrial and domestic needs of a community with only a minimal use of hazardous fuels. And eventually, we would sophisticate all our noncombustion energy devices to a point where all harmful sources of energy could be eliminated from the pattern.

As in the case of agriculture, however, the application of ecological principles to energy resources presupposes a far-reaching decentralization of society and a truly regional concept of social organization. To maintain a large city requires immense packages of fuel—

> mountains of coal and veritable oceans of petroleum. By contrast, solar, wind, and tidal energy can reach us mainly in small packets; except for spectacular tidal dams, the new devices seldom provide more than a few thousand kilowatt-hours of electricity. It is difficult to believe that we will ever be able to design solar collectors that can furnish us with immense blocks of electric power produced by a giant steam plant; it is equally difficult to conceive of a battery of wind turbines that will provide us with enough electricity to illuminate Manhattan Island. If homes and factories are heavily concentrated, devices for using clean sources of energy will probably remain mere playthings, but if urban communities are reduced in size and widely dispersed over the land, there is no reason why these devices cannot be combined to provide us with all the amenities of an industrialized civilization. To use solar, wind and tidal power effectively, the megalopolis must be decentralized. A new type of community, carefully tailored to the characteristics and resources of a region, must replace the sprawling urban belts that are emerging today.[4]

An objective case for decentralization, to be sure, does not end with a discussion of agriculture and the problems created by combustible energy resources. The validity of the decentralist case can be demonstrated for nearly all the "logistical" problems of our time. At the risk of being cursory,

[4] Lewis Herber, *Crisis in Our Cities* (Prentice-Hall Inc., Englewood Cliffs, N.J.; 1965, p. 194).

let me cite an example from a problematical area such as transportation. A great deal has been written quite recently about the harmful effects of petrol-driven motor vehicles—their wastefulness, their role in urban air pollution, the noise they contribute to the city environment, the enormous death toll they claim annually in the large cities of the world and on highways. In a highly urbanized civilization, it would be meaningless to replace these noxious vehicles by clean, efficient, virtually noiseless, and certainly safer battery-powered vehicles. The best of our electric cars must be recharged about every hundred miles—a feature which limits their usefulness for transportation in large cities. In a small, decentralized community, however, it becomes eminently feasible to use these electric vehicles for intra-urban or regional transportation and establish monorail networks for long-distance transportation.

It is fairly well known, today, that petrol-powered vehicles contribute enormously to urban air pollution, and there is a strong sentiment to "engineer" the more noxious features of the automobile into oblivion. Our age characteristically tries to solve all its irrationalities with a gimmick—blow-by devices and after-burners for toxic petrol fumes, antibiotics for ill-health, tranquillizers for psychic disturbances. The problem of urban air pollution is more intractable than we care to believe. Basically, air pollution is caused by high population densities, by an excessive concentration of people in a small area. The fact is that millions of people, densely concentrated in a large city, necessarily produce serious *local* air pollution merely by their day-to-day activities. They must burn fuels for domestic and industrial reasons; they must construct or tear down buildings (the aerial debris produced by these activities is a major source of urban air pollution); they must dispose of immense quantities of rubbish; they must travel on roads with rubber tires (again, the particles produced by the erosion of tires and roadway materials add significantly to air pollution). Quite aside from the pollution-control devices we add to automobiles and power plants, it should be fairly clear that whatever improvements these devices will produce in the quality of urban air will be more than cancelled out by future megalopolitan growth.

The social possibilities opened by decentralization could be discussed indefinitely and, in any case, there is more to anarchism than decentralized communities. If I have examined these possibilities in some detail, it has been to demonstrate that an anarchist society, far from being a remote ideal, has become a pre-condition for the practice of ecological principles. To sum up the critical message of ecology: If we diminish variety in the natural world, we debase its unity and wholeness. We destroy the forces making for natural harmony and stability, for a lasting equilibrium, and what is even more significant, we introduce an absolute retrogression in the development of the natural world, eventually rendering the environment unfit for advanced forms of life. To sum up the reconstructive message of ecology: If we wish to advance the unity and stability of the natural world, if we wish to harmonize it on ever higher levels of development, we must conserve and promote variety. To be sure, mere variety for its own sake is a vacuous goal. In nature, variety emerges spontaneously. The capacities of a new species are tested by the rigours of climate, by its ability to deal with predators, by its capacity to establish and enlarge its niche. *Yet the species that succeeds in enlarging its niche in the environment also enlarges the ecological situation as a whole.* To borrow E. A. Gutkind's phrase, it "expands the environment," both for itself and for the species with which it enters into a balanced relationship.[5]

How do these concepts apply to social theory? To many, I suppose, it should suffice to say that, inasmuch as man is part of nature, an expanding natural environment enlarges the basis for social development. But the answer to the question, I think, goes much deeper than many ecologists and libertarians suspect. Again, allow me to return to the ecological principle of wholeness and balance as a product of diversity. Keeping this principle in mind, the first step toward an answer is provided by a passage in Herbert Read's *The Philosophy of Anarchism.* In presenting his "measure of progress," Read observes: "Progress is measured by the degree of differentiation within a society. If the individual is a unit in a corporate mass, his life will be limited, dull, and mechanical. If the individual is a unit on his own, with space and potentiality for separate action, then he may be more subject to accident or chance, but at least he can expand and express himself. He can develop—develop in the only real meaning of the word—develop in consciousness of strength, vitality, and joy."

Read's thought, unfortunately, is not fully developed, but it provides an interesting point of departure for our discussion. Leaving the quotation aside, for the moment, what first strikes us is that both the ecologist and the anarchist place a strong emphasis on spontaneity. The ecologist, in so far as he is more than a technician, tends to reject the notion of "power" over nature. He speaks instead of "steering" his way through an ecological situation, of *managing* rather than *recreating* an ecosystem. The anarchist, in turn, speaks in terms of social spontaneity, of releasing the potentialities of society and humanity, of giving free and unfettered reign to the creativity of people. Both, in their own ways, regard authority as inhibitory, as a weight limiting the creative potential of a natural and social situation. Their object is not to *rule* a domain, but to *release* it. They regard insight, reason, and knowledge as means for fulfilling the potentialities of a situation, as facilitating the working out of the logic of a situation, not of replacing these potentialities with preconceived notions or distorting their development with dogmas.

Turning, now, to Read's words, the next thing that strikes us is that both the ecologist and anarchist view differentiation as a measure of progress. The

[5] I do not wish to saddle Gutkind with the notions I have advanced above, but I believe the reader would benefit enormously by reading Gutkind's little book, a masterful discussion of communities, *The Expanding Environment* (Freedom Press).

ecologist uses the term "biotic pyramid" in speaking of biological advances; the anarchist, the word "individuation" to denote social advances. If we go beyond Read, we will observe that, to both the ecologist and anarchist, an ever-enlarging unity is achieved by growing differentiation. *An expanding whole is created by the diversification and enrichment of the parts.*

Just as the ecologist seeks to elaborate the range of an ecosystem and promote a freer interplay between species, so the anarchist seeks to elaborate the range of social experience and remove all fetters to its development. To state my point more concretely: Anarchism is not only a stateless society but also a harmonized society which exposes man to the stimuli provided by both agrarian and urban life, physical activity and mental activity, unrepressed sensuality and self-directed spirituality, communal solidarity and individual development, regional uniqueness and world-wide brotherhood, spontaneity and self-discipline, the elimination of toil and the promotion of craftsmanship. In our schizoid society, these goals are regarded as mutually exclusive dualities, sharply opposed to each other. To a large extent, they appear as dualities because of the very logistics of present-day society—the separation of town and country, the specialization of labour, the atomization of man— and it would be preposterous, I think, to believe that these dualities could be resolved without a general idea of the *physical* structure of an anarchist society. We can gain some idea of what such a society would be like by reading William Morris's *News From Nowhere* and the writings of Peter Kropotkin. But these are mere glimpses. They do not take into account the post-war developments of technology and the contributions made by the development of ecology. This is not the place to embark on "utopian writing," but certain guide lines can be presented even in a general discussion. And in presenting these guide lines, I am eager to emphasize not only the more obvious ecological premises that support them, but also the humanistic ones.

An anarchist society should be a decentralized society not only to establish a lasting basis for the harmonization of man and nature, *but also to add new dimensions to the harmonization of man and man.* The Greeks, we are often reminded, would have been horrified by a city whose size and population precluded a personal, often familiar, relationship between citizens. However true this precept may have been in practice two thousand years ago it is singularly applicable today. There is plainly a need to reduce the dimensions of the human community—partly to solve our pollution and transportation problems, partly also to create *real* communities. In a sense, we must *humanize* humanity. There should be a minimum of electronic devices—telephones, telegraphs, radios, television receivers and computers—to mediate the relations between people. In making collective decisions—and the ancient Athenian ecclesia was, in some ways, a model for making social decisions during the classical period—all members of the community should have an opportunity to acquire in full the measure of anyone who addresses the assembly. They should be in a position to absorb his attitudes, study his expressions, weigh his motives as well as his ideas in a direct personal encounter and through full debate, face-to-face discussion and inquiry.

Our small communities should be economically balanced and well rounded, partly so that they can make full use of local raw materials and energy resources, partly also to enlarge the agricultural and industrial stimuli to which individuals are exposed. The member of a community who has a predilection for engineering, for instance, should be encouraged to steep his hands in humus; the man of ideas should be encouraged to employ his musculature; the "inborn" farmer should gain a familiarity with the workings of a rolling mill. To separate the engineer from the soil, the thinker from the spade, and the farmer from the industrial plant may well promote a degree of vocational overspecialization that would lead to a dangerous measure of social control by specialists. What is equally important, professional and vocational specialization would prevent society from achieving a vital goal: the humanization of nature by the technician and the naturalization of society by the biologist.

I submit that an anarchist community, in effect, would approximate a clearly definable ecosystem—diversified, balanced, and harmonious. It is arguable whether such an ecosystem would acquire the configuration of an urban entity with a distinct centre, such as we find in the Greek *polis* or the medieval commune, or whether, as Gutkind proposes, society would consist of widely dispersed communities without a distinct centre. In either case, the ecological scale for any of these communities would be the smallest biome capable of supporting a moderate-sized population.

A relatively self-sufficient community, visibly dependent on its environment for the means of life, would gain a new respect for the organic interrelationships that sustain it. In the long run, the attempt to approximate self-sufficiency would, I think, prove more efficient than the prevailing system of a national division of labour. Although there would doubtless be many duplications of small industrial facilities from community to community, the familiarity of each group with its local environment and its rootedness in the area would make for a more intelligent and more loving use of its environment. I submit that far from producing provincialism, relative self-sufficiency would create a new matrix for individual and communal development—a oneness with the surroundings that would vitalize the community.

The rotation of civic, vocational, and professional responsibilities would awaken all the senses in the being of the individual, stimulating and rounding out new dimensions in self-development. In a complete society we could hope again to create complete men; in a rounded community, rounded men. In the Western world, the Athenians, for all their shortcomings and limitations, were the first to give us a notion of this completeness. "The *polis* was made for the amateur," Kitto tells us. "Its ideal was that every citizen (more or less, according as the *polis* was democratic or oligarchic) should play his part in all of its many activities—an ideal that is recognizably descended from the generous Homeric conception of *arete* as an all-round excellence and an all-round activity. It implies a respect for the wholeness or the oneness of life, and a consequent dislike of specialization. It implies a contempt for efficiency—or rather a much higher ideal of efficiency; an efficiency which exists not in one department of life, but in life itself."[6] An anarchist society, although it would surely aspire for more, could hardly hope to achieve less than this state of mind.

If the foregoing attempt to mesh ecological with anarchist principles is ever achieved in practice, social life would yield a sensitive development of human and natural diversity, falling together into a well-balanced, harmonious unity. Ranging from community, through region, to entire continents, we would see a colourful differentiation of human groups and ecosystems, each developing its unique potentialities and exposing members of the community to a wide spectrum of economic, cultural, and behavioural stimuli. Falling within our purview would be an exciting, often dramatic, variety of communal forms —here, marked by architectural and industrial adaptations to semi-arid biomes, there to grasslands, elsewhere to forest lands. We would witness a dynamic interplay between individual and group, community and environment, man and nature. Freed from an oppressive routine, from paralysing repressions and insecurities, from the burdens of toil and false needs, from the trammels of authority and irrational compulsion, the individual would finally be in a position, for the first time in history, to fully realize his potentialities as a member of the human community and the natural world.

[6] H. D. F. Kitto, *The Greeks* (Aldine Publishing Co., Chicago; 1951, p. 161).

# THE COMING SOLAR AGE

*Peter van Dresser*

*Man has persisted in thinking of nature as something to be con-
quered when the major technical advances have been the result of
cooperating with nature. Now, of course, we have developed our society
according to this aggressive spirit, and there is some doubt that our
industrial society can be nonviolent and less destructive. Those with interests in oil or nuclear power would, in all likelihood, not react kindly to
the proposal offered by Peter van Dresser in the following article. Perhaps as the present fuel sources are depleted, they will have no choice.*

Nuclear technology, now so indisputably in the limelight of world
attention, may be thought of as the ultimate development of that drive toward
the conquest of nature and the mastery of demiurgic forces, which began in
the time of Roger Bacon and led through the steam and iron age to the
emerging era of electronics and automation. The kind of civilization which
we can foresee as fission-produced power comes more into play is an
intensification of that which is already at work about us: one in which precision
and intensity of rational control are the factors of achievement and survival;
one in which the machine of necessity is the pattern of life, and centralization
and coordination the watchwords. For how long this drive, which Spengler
aptly called Faustian, will continue, we cannot, of course, tell, but it is interesting to be made aware of the birth-stirrings of a new orientation which in time
may lead us along a route quite different from that which seems so inevitable
at the moment.

Solar energy, in the last analysis, has always been the basis not only of
civilization, but of life; from the primeval sun-basking plankton to modern man

SOURCE: Peter van Dresser, "The Coming Solar Age," *Landscape*, vol. 5, no. 3, Spring 1956.
Reprinted by permission of the publisher.

harvesting his fields and burning coal and oil beneath his boilers, solar energy has provided the ultimate moving force. But its direct utilization at a higher level of technology is a new phenomenon, and rich with new potentialities at this stage of human affairs. Should it become feasible and usual within the next generation or two for man to heat a large proportion of his structures in the colder climates, and cool those in the hotter ones, by solar absorption and storage techniques; should he find it possible to energize industrial and metallurgical processes through an intensification of these techniques; should the generation of electrical or mechanical power from the same source become practicable—then sweeping changes in the economic and social future would be inevitable, changes as powerful in their effect on the landscape as were those which accompanied the industrial revolution.

Most obviously, the pattern of distribution of utilizable solar radiation is very different from that of deposits of coal, oil and, latterly, uranium ores. Zones of clear atmosphere and steady insolation would be the favored ones in a solar-oriented economy; the vast arid and semiarid belts and uplands, the hitherto energy-poor tropical lands, would come into their own. And over these great regions, the distribution of this new wealth of energy would be uniform and utterly impartial. No strategic locations of site or geologic accident would occur; the entire pattern of urban industrial concentration at coal-rich centers, around which our present national economies crystallized, would be nullified, and a new pattern of dispersion and decentralization would emerge in its place.

The second significant characteristic of solar radiant energy at the earth's surface is its relatively low intensity. Here again is implicit a fresh technological *esprit*. Controlled violence is the essence of contemporary mechanism; the violence of fire under forced draft, of exploding gases, of superheated steam, of high-voltage electricity, and bombarded atomic nuclei. This is the technology of Vulcan closely allied to, and often arising directly out of, the demands of war.

A future solar technology would appear to be largely free of this core of violence. Utilizing fluxes of low-gradient energy controlled by diurnal and seasonal cycles, its processes would of necessity be akin to the processes of plant growth. A society built around them would tend to reflect, on a higher level of science and organization, the esthetic values and the virtues traditionally associated with sylvan and agrarian cultures. A greater intimacy with, and respect for, the subtle phenomena of the natural world would be enforced by the geographic dispersal of the homes and work places utilizing solar energy, and their necessary affinity to climatic and celestial rhythms. A logical and close relationship to the husbandry of field and forest would develop, as biologic raw materials displace mineral raw materials with the further evolution of the plastic and kindred industrial techniques. The intensive organic cultivation and on-the-spot consumption or preservation of diversified food crops would integrate naturally into the living pattern of this earth-and-sunlight-oriented society. A diminution of the role of massive transportation, which dominates our present economy, would tend to result, and this in turn would serve to further demechanize the landscape.

Whether or not such a vision will displace—or at least modify—the faintly nightmarish atomic age that now looms before us, it does seem a fact that at long last a good deal of scientific and engineering attention is being focused on the problem of effective solar-energy utilization—a proposition which has engaged the attention of occasional visionary individuals for centuries. Last fall's World Symposium on Applied Solar Energy, held in Tucson and Phoenix, Arizona, seems to have brought together an array of talent, and to have concentrated a degree of public attention that is quite unprecedented in connection with this subject.

Sponsored by the Stanford Research Institute and the newly organized Association for Applied Solar Energy, this international symposium occupied in November 1955 an intensive week of lectures, papers, and panel discussions presented by researchers and engineers mainly from government services of a chain of low-latitude countries that virtually encircle the globe. The material ranged from highly mathematical discussion of optical absorption and emissivity phenomena to the techniques of boiling rice in solar cook-stoves and the harvesting of mass-cultured algae for fuel and food. The physics and economics of distillation of sea water, of space heating and refrigeration, of mechanical and electrical power generation, of irrigation water pumping, of photosynthesis, of high-temperature metallurgy—all by solar activation—were treated at considerable length. An accompanying exposition of working solar machines and devices at the Phoenix Civic Center, probably the first of its kind in the world, served as an interesting peephole into a possibly realizable future.

One general reflection engendered by the wealth of material so presented was on the tremendous structural change necessary in our living and social-organizational techniques to make effective use of the principle discussed. Our economy is, above all, a mass-transportation complex, evolved to transmute concentrated sources of mineral energy into an ever-increasing industrial dynamism. Our rail, highway, and air networks, our pipelines and transmission grids, interconnecting and energizing vast urban, manufacturing, and mineral-processing centers, are the very essence of our civilization. Our economists, businessmen, and engineers are conditioned to think of such complexes as synonymous with civilization. Our financial institutions are developed almost entirely around the function of channeling investment into their operation, while the citizen invests in them emotionally through ownership of his automobile and daily participation in the psychodrama of fast highway traffic.

By contrast, the utilization of solar energy usually requires extensive static installations, handling energy-fluxes of low intensity, and yielding useful work or products primarily for localized consumption. Athough the ultimate in efficient and nondestructive utilization of natural resources, such installations are inefficient from the point of view of capital investment, since their relative throughput is small and they are incapable of the high returns which intensive operations can yield. They are in fact the antithesis of the "capital-saving" technique, which contemporary economists see as characteristic of a progressive industrialism.

It is perhaps for such reasons that the most marked interest in solar technology seems to be associated with the underdeveloped areas of the world, where at present largely agrarian peasant or shepherd populations prevail. In such regions, the extensive but relatively simple structures needful for many sun-powered operations may be built by local labor, with the minimum importation of high-cost machinery. Once constructed, the yield of such installations would be free of fuel costs and the attendant complexity of fuel or power transmission, and would contribute to the local economy without burdening it with credit and import costs. Thus one finds engineers and administrators from India, North Africa, Israel, and kindred lands much more aware of the potentialities of solar energy than their counterparts in the highly industrialized countries, although owing to the technical inferiority of their economies they have been unable so far to carry out extensive pilot installations.

It is also interesting to note that leadership in solar metallurgy and power generation has appeared in France, a nation of high culture and prestige, which seems nevertheless in process of withdrawing from the ambitions of a "major power" and perhaps for this very reason is turning toward the development of nonexploitive resources. The Laboratoire de L'Energie Solaire in Montlouis in the French Pyrenees, with that of Algiers in North Africa, comprises by far the most complete installation for high-temperature solar metallurgy in the world. Similarly, the Claude-system power plant now under construction near Dakar, and designed to serve as a municipal plant operating from the energy of sun-heated water, is unique.

One also notes in Israel, a tiny newly formed country intent on developing an advanced economy in a resources-poor land, a striking concentration on the potentialities of solar energy. At the Physics Laboratory in Jerusalem, some of the most interesting work described at the Arizona symposium has been carried on in the field of solar-energy absorption, to the end of developing more efficient surfaces for the construction of power-collectors.

It seems still too early to guess intelligently whether the economic revolution implicit in these slight beginnings will maintain itself. Certainly in terms of total energy available, the promise of solar technology far exceeds that of nuclear technology; in terms of total benefits to mankind, it ranks at least equal. But whether we can accomplish the profound changes in our technical, economic, and psychological structure orientation to realize these potentials it is at present impossible to know.

# TOWARD A PLANNER'S GUIDE TO ECOLOGY

*Eugene Anderson, Jr.*

*The following address is the result of Professor Anderson's teaching of courses on the Far East and folklore at the University of California, Riverside. His training in anthropology provides him with a unique vantage point for discussing the problems of our environment. In this selection he makes the connection between the ancient science of planning known as feng-shui or geomancy and the contemporary practice of overplanning our urban sprawl.*

All planners are now aware of the environment. Pollution, loss of open space, lack of recreational areas and disappearance of green belts have made planners aware of the fact that the overall American environment is very rapidly deteriorating, and must be reconstituted. Yet most planners remain rather ignorant of the science of ecology: the relationships of living things to each other and to their environmental setting. Because of this ignorance, planning has in many cases brought terrible damage to an area, or threatened destruction of a major resource.

Not only do the usual planners' reasons, aesthetic and practical, make attention to ecology mandatory, but far more serious concerns enter into the picture. Most planners seem to feel, for instance, that agricultural land is pleasant as an amenity but is not necessary, and that urban sprawl is reprehensible but could be wonderful (as in Frank Lloyd Wright's *Broadacres*). Apparently the point is seldom made that the human organism needs food. When all the agricultural land is taken, more than aesthetics will suffer. The planner, of course, assumes that since America is an "economy of abundance"

SOURCE: Eugene Anderson, Jr., "Toward a Planner's Guide to Ecology," *Ecology, a Journal of Cultural Transformation*, vol. 1, no. 1. Reprinted by permission of the author.

such waste of land can be afforded. The hard fact [is] that America is not isolated in the world. Our surplus is now keeping alive several tens of millions of people in India alone, and many more in other countries. At the present rate of attrition of agricultural land and (more deadly) population increase, we will soon have no more surplus. Estimates of the fatal year vary, but all emphasize that we will be a food-importing nation within a generation, unless present trends are very markedly changed.

As surplus food goes, so goes the world. And where will we obtain further food? Who will be exporting in another generation? It is time we realized that America and the rest of the world have much more crucial reasons than aesthetic ones for preserving agricultural land. This rather crude and obvious fact is hardly sophisticated ecology, but at least it provides an example of the need for planners and biological thinkers to work together more closely.

Basic to ecology is the concept of an ecosystem. An ecosystem is a system of organisms linked together by mutual dependence. They live in the same area, eat each other, interact with each other, grow from soil made by each other's decay. The members of the famous self-sufficient village, where everyone lived by taking in one another's washing, were in the same situation as are animals and plants in an ecosystem. Most ecosystems are tremendously resilient as long as certain factors are maintained—the soil, the water, the acid balance—but exceedingly fragile; the alteration of one factor in the environment causes a progressive degradation. Fewer and fewer species and individuals of organisms can survive. A forest is cut over; in a few years it grows back. But if the topsoil is scraped off meanwhile, only the toughest of "ruderals" (weeds) will survive. In the forest there were hundreds of plant species, but in the weedlot there are a dozen. Timber, medicinal plants, edible plants, all are gone; ragweed and Russian thistle inherit the earth, whether bulldozers, atomic bombs (as at Yucca Flat) or air pollution have done the initial killing. Another and less obvious fact of ecology is the "edge effect." Where several kinds of habitat join, as where forest intergrades through scrub into grassland, the variety and number of plants and especially of animals is much greater than elsewhere. Game animals in particular seem wedded to edges. Deer, quail, pheasants and grouse have multiplied exceedingly in old lumbering slashes, abandoned farms and other brushy transitional country. Yet another ecological fact is the limited nature of certain goods. Water in deserts, treed areas in open country, brush patches in otherwise monotonous land, and of course edges are examples of limiting conditions. A small patch of trees in an open plain may have as many animals as a large continuous forest; on the plains, trees are a refuge for animals that can feed widely over the open land, but can find safe shelter only in the woodlot. An estuary or shallow marshy bay is always a critical place; it is an edge, where water meets land; it has currents from the sea and streams from the land to bring in great quantities of food. Marshes and swamps, wherever found, are critical to all surrounding life. These habitats where land meets water invariably show the highest counts of species and individuals of any habitat within an area; cold, monotonous, uniform, dryish country shows the lowest.

Planners consistently display appalling indifference to, or ignorance of, these facts. In my city, Riverside, an excellent and justly famous planning firm was called to produce a regional plan. They set aside a large area of scrubby, human-wrecked, barren and uninteresting land for a park; but the best they could do with the river that gives the city its name was slash a (proposed) freeway straight up the bottoms. Now, besides the fact that this freeway would be flooded every so often, there are ecological considerations. The river-bottoms are still fairly wild and extremely beautiful; rich in plant and animal life; a joy and delight to the town's citizens; and the only large, extensive wet area for miles. They are the last surviving large piece of river country in southern California. The freeway would utterly destroy the wildness and . . . I need not go on. (Citizen protest forced the firm to plan a park, not a freeway, in the river bottoms.) Similarly, in the East Bay region of California —the Oakland-Berkeley area—a really fine regional park plan was set up and implemented, with the result that the East Bay hills are almost one continuous park. Yet the San Francisco Bay marshes were not included. The cities would not set them aside and the planners didn't care. In both these cases, dedicated local groups exist to save the wetlands; the planners could have easily found out that their plans took into account relatively uninteresting and biologically unproductive country, at the expense of unique and biologically irreplaceable land. I ask a planning student: Do you know that marshes are the most productive of wildlife of any kind of terrain? He says: "No, they teach us that marshes are only something to fill." ("They" refers to a CRP department which shall remain nameless.)

Now I am assuming that planners, if given the choice, will opt for more animals, more birds, more rare and unusual plants, more trees. With the American heritage, the Japanese influence and the rest of it, we have learned to love wilderness—and even the scrubby, anarchist little fragments of the wild that are represented by weedy lots and bits of second-growth forest on former farms. A few apostles of Development, left over from an earlier aesthetic, still apostrophize the cleared landscape or say that because there is no more true wilderness we should not try to preserve anything. But the public—unless monetarily motivated—is not with them. Even a lumber merchant seeks the forests on his day off, and even a land-reclaimer hunts ducks, though his job is to massacre them by filling in their breeding pools. But beyond the obvious aesthetic and recreational uses of ecologically critical spots, there is the far more serious consideration of future food and resources. America's wetlands constitute our greatest undeveloped resource. Cropping ducks—by hunting— is extensively practised, but even so we allow or encourage the destruction of the habitat; a marsh that produced tons of fine protein—fish and bird—is filled to produce a few pounds of wheat, or to make a base for a few wretched houses. And when the inevitable floods come, the local authorities get a few million from the government to repair the damage. Who will repair the damage to our protein reserves? Who will repair the aesthetic chaos? And the city planner goes home from his urban office to his remodeled farmhouse in the hills . . . .

The losses of the Jersey marshes near New York and the Bay marshes of San Francisco are national tragedies; the loss of Yukon Flats to the proposed Rampart Dam would be a world tragedy, destroying so many thousand salmon and game animals that the total world stocks of edible protein would be quite perceptibly depleted. It is the duty of all planners to oppose these, both on traditional planning grounds and on ecological ones. Pollution is also a problem for both planners and biologists. Besides being irritations to eye and lung, pollutants kill. Pesticides are the most dangerous; one part of DDT in a *billion* will kill some water animals. Much of the lower Mississippi, once the finest game and fish and wildlife country in the continent, has been virtually sterilized by pesticides washed from upstream because of users' carelessness. Carelessness, again, has caused the die off of millions of fish in California. On a more human scale, I remember the yearly spraying of elms, in the New England spring, and the yearly die off of robins that accompanied it: nearly all the robins killed in hopes of saving certain moribund and overage elms from the Dutch elm disease, against which DDT has no absolutely proven effect.

Ecology has begun to concern itself with human aggregations. Biologists, such as Edward Deevey, have become concerned with a new fear. Among many social animals, too many social contacts are almost as dangerous as too few. As isolation-reared rats and monkeys are abnormal and pathological in gross anatomical features as well as in behavior, so those that are in too close and constant contact with others of their species develop glandular and heart troubles and in some cases die of something very like shock. It has been found that too many stimuli from too many social mates can trigger this—as it triggers the suicide marches of lemmings to the sea. (The lemmings do not necessarily go to sea; they may go in any direction—as long as it is away from the crowd. If lemmings had suburbs . . .) The application of these studies to humans is controversial, but there are indications. The Meban, an African herding people, were studied in their homes and in the city of Khartoum. Starving in their native bush, they had no ulcers, no high blood pressure, no heart trouble—or at least very few cases. But Khartoum-dwelling Meban developed all of these and more—all the nervous ailments. It was hypothesized that once again overstimulation was the key: too much noise, too much confusion, too much strain and dirt and tension. Planners are well aware of these ills and of the pathology of cities, but few realize that actual physical destruction is the result of a badly planned city. Biologists have not found what makes an ideal city, but planners will no doubt be able to tell by introspection what cities are successful and what cities merely raise random, nerve-wracking stimuli to an intolerable power.

From all the preceding, you will have realized that planners have their own ecological philosophy—as M. Jourdain had his prose. It is interesting to study not only the biologists' principles of resource management, but also those of traditional farmers, city-dwellers, villagers. Such people have of necessity reached a harmony with their world; ecologically and aesthetically they have adapted. Rudofsky, in *Architecture without Architects,* has taught us that "primitive" architecture is often better than our own. The same may be said of planning. The traditional villages of south China were sited and planned according to an ancient and theoretically exact science known as *feng-shui,* "wind and water," and by us called "geomancy." The goal of this body of theory was siting buildings and graves in such a way that they would be protected from storms and floods, open to warmth, closely related to farmland and other useful terrain, and situated so as to have a beautiful view. Geomancy also protected against ghosts and evil influences while maximizing good influences and making the spirits and ancestors happy, but this makes it no less scientific in the eyes of the practitioners, for whom such invisibles as good and evil fortune are as real as the equally invisible yet tangible winds of the typhoon. A perfect geomanic site is on a hill, near the foot, looking out over a broad valley with tranquil waters. It is protected by a thick grove of trees upslope, small streams flowing down to meet in the valley before it, and partially encircling ridges on each side. The flood-prone valley floor is rice paddy. The steep hills behind the site yield firewood and game. The view is varied and harmonious, with a pattern of wild land, farmland, towns and water. In rural Hong Kong I have had occasion to observe villages under stress: cold waves, heat waves, floods and typhoon conditions. In every case, villages with bad *feng-shui* suffered terribly, while those that had paid attention to the geomanters were far less seriously affected. Ecologically, the well-sited village was also well-cropped; rice, vegetables, pigs, chickens, tree crops were grown together, and often fish also. The fish lived in the rice paddies eating insects, the pigs and chickens ate pests and provided fertilizer, the rice straw fed the plow animals, and the fruit trees provided shade and windbreaks. Everything had several uses, and everything was part of a complete system that allowed each village to be almost self-sufficient. Here a native traditional philosophy of planning was proving itself so successful that the British government of Hong Kong was working to encourage traditional siting practices among the all too modernized villagers.

Southern California presents a sort of complementary case. Usually described as underplanned, it is in fact overplanned—every slum and slurb is carefully laid out, but by planners whose sole concern is for quick profit, rather than for long-term satisfaction. It is in the planning that the ugliness arises— all the houses alike, all the streets identical, all the land used the same way.

When southern Californians plan for themselves, the result is quite different. I have studied, with the help of others of similar persuasion, some parts of the large rural slum that extends in a crazy pattern over much of the hilly land near Riverside. The houses here are built by individuals whose architectural knowledge is zero—persons of wildly different incomes, social backgrounds, and ideas. Planning is a matter of anarchistic opportunism—but also of mutual accommodation and neighborhood consultation. The result is a remarkably even, fairly wide spacing of houses, with a small amount of clustering around road junctions. Disturbance of the landscape is minimal; wildlife comes to the doors, rare wild plants grow in the yards. Each person has his own garden, and the individuality of the houses has encouraged a spectacular proliferation of imaginative landscapings. Here is diffuse settlement without urban sprawl, striking and original architecture and planning without the help of professionals —people, allowed to do as they want, have produced a settlement that is infinitely superior to the deadly monotonous, stereotyped, totally unwild, biologically ruined tract homes that the more "fortunate" inhabitants of the city can buy. The rural slum-dwellers operated on their own planning principles: diffuse settlement, very small-scale disturbance of the natural environment, fitting of houses into individual landscape niches, use of natural materials in building, and maximum individuality of house and garden design. The result is far from perfect—if only because of a lack of resources—but is equally far above the level of most planned communities.

In these small communities, the line between ecology and planning does not exist. The rural Chinese and the rural Riversiders have their specific ways of looking at the landscape. They have come to terms with it. Biology and aesthetics are merely two aspects of man's relationship with the land. This is the most important lesson we can learn from biology, from the Chinese peasants and from the rural citizens of America. There is no possible divorce between the two aspects; ignoring the one does not make it go away. Willy-nilly we are all changing America. The planners must take the full range of effects into account. What will a new regional plan do to America's food future? What will a new pesticide do to the aesthetic resources of the continent? Conservation is no longer a matter for farm advisors and lady birdwatchers. It is the most desperately important issue in the world—no less in well-fed America than in the ravaged Near East. It is an issue on which all experts must come together.

# THE ECONOMICS OF THE COMING
# SPACESHIP EARTH

*Kenneth E. Boulding*

*Kenneth E. Boulding has taught at universities in Scotland, Canada, and the United States and has written numerous books, pamphlets, and articles that demonstrate his interdisciplinary approach to human problems. Perhaps his best-known work is* The Meaning of the Twentieth Century *(1964). In this article he works from those images by which modern man understands himself and his world to specific suggestions on how our economy can be altered to face the environmental crisis. The essay is an excellent example of how a truly learned scholar can communicate some of the most difficult ideas to be found in his own special area of scholarship.*

We are now in the middle of a long process of transition in the nature of the image which man has of himself and his environment. Primitive men, and to a large extent also men of the early civilizations, imagined themselves to be living on a virtually illimitable plane. There was almost always somewhere beyond the known limits of human habitation, and over a very large part of the time that man has been on earth, there has been something like a frontier. That is, there was always some place else to go when things got too difficult, either by reason of the deterioration of the natural environment or a deterioration of the social structure in places where people happened to live. The image of the frontier is probably one of the oldest images of mankind, and it is not surprising that we find it hard to get rid of.

SOURCE: Kenneth E. Boulding, "The Economics of the Coming Spaceship Earth," *Environmental Quality in a Growing Economy*, ed. Henry Jarrett (Baltimore: The Johns Hopkins Press, 1966). Published by The Johns Hopkins Press for Resources For The Future, Inc. Reprinted by permission of the publisher.

Gradually, however, man has been accustoming himself to the notion of the spherical earth and a closed sphere of human activity. A few unusual spirits among the ancient Greeks perceived that the earth was a sphere. It was only with the circumnavigations and the geographical explorations of the fifteenth and sixteenth centuries, however, that the fact that the earth was a sphere became at all widely known and accepted. Even in the nineteenth century, the commonest map was Mercator's projection, which visualizes the earth as an illimitable cylinder, essentially a plane wrapped around the globe, and it was not until the Second World War and the development of the air age that the global nature of the planet really entered the popular imagination. Even now we are very far from having made the moral, political, and psychological adjustments which are implied in this transition from the illimitable plane to the closed sphere.

Economists in particular, for the most part, have failed to come to grips with the ultimate consequences of the transition from the open to the closed earth. One hesitates to use the terms "open" and "closed" in this connection, as they have been used with so many different shades of meaning. Nevertheless, it is hard to find equivalents. The open system, indeed, has some similarities to the open system of von Bertalanffy,[1] in that it implies that some kind of a structure is maintained in the midst of a throughput from inputs to outputs. In a closed system, the outputs of all parts of the system are linked to the inputs of other parts. There are no inputs from outside and no outputs to the outside; indeed, there is no outside at all. Closed systems, in fact, are very rare in human experience, in fact almost by definition unknowable, for if there are genuinely closed systems around us, we have no way of getting information into them or out of them; and hence if they are really closed, we would be quite unaware of their existence. We can only find out about a closed system if we participate in it. Some isolated primitive societies may have approximated to this, but even these had to take inputs from the environment and give outputs to it. All living organisms, including man himself, are open systems. They have to receive inputs in the shape of air, food, water, and give off outputs in the form of effluvia and excrement. Deprivation of input of air, even for a few minutes, is fatal. Deprivation of the ability to obtain any input or to dispose of any output is fatal in a relatively short time. All human societies have likewise been open systems. They receive inputs from the earth, the atmosphere, and the waters, and they give outputs into these reservoirs; they also produce inputs internally in the shape of babies and outputs in the shape of corpses. Given a capacity to draw upon inputs and to get rid of outputs, an open system of this kind can persist indefinitely.

There are some systems—such as the biological phenotype, for instance the human body—which cannot maintain themselves indefinitely by inputs

[1] Ludwig von Bertalanffy, *Problems of Life* (New York: John Wiley and Sons, 1952).

and outputs because of the phenomenon of aging. This process is very little understood. It occurs, evidently, because there are some outputs which cannot be replaced by any known input. There is not the same necessity for aging in organizations and in societies, although an analogous phenomenon may take place. The structure and composition of an organization or society, however, can be maintained by inputs of fresh personnel from birth and education as the existing personnel ages and eventually dies. Here we have an interesting example of a system which seems to maintain itself by the self-generation of inputs, and in this sense is moving toward closure. The input of people (that is, babies) is also an output of people (that is, parents).

Systems may be open or closed in respect to a number of classes of inputs and outputs. Three important classes are matter, energy, and information. The present world economy is open in regard to all three. We can think of the world economy or "econosphere" as a subset of the "world set," which is the set of all objects of possible discourse in the world. We then think of the state of the econosphere at any one moment as being the total capital stock, that is, the set of all objects, people, organizations, and so on, which are interesting from the point of view of the system of exchange. This total stock of capital is clearly an open system in the sense that it has inputs and outputs, inputs being production which adds to the capital stock, outputs being consumption which subtracts from it. From a material point of view, we see objects passing from the noneconomic into the economic set in the process of production, and we similarly see products passing out of the economic set as their value becomes zero. Thus we see the econosphere as a material process involving the discovery and mining of fossil fuels, ores, etc., and at the other end a process by which the effluents of the system are passed out into noneconomic reservoirs—for instance, the atmosphere and the oceans—which are not appropriated and do not enter into the exchange system.

From the point of view of the energy system, the econosphere involves inputs of available energy in the form, say, of water power, fossil fuels, or sunlight, which are necessary in order to create the material throughput and to move matter from the noneconomic set into the economic set or even out of it again; and energy itself is given off by the system in a less available form, mostly in the form of heat. These inputs of available energy must come either from the sun (the energy supplied by other stars being assumed to be negligible) or it may come from the earth itself, either through its internal heat or through its energy of rotation or other motions, which generate, for instance, the energy of the tides. Agriculture, a few solar machines, and water power use the current available energy income. In advanced societies this is supplemented very extensively by the use of fossil fuels, which represent as it were a capital stock of stored-up sunshine. Because of this capital stock of energy, we have been able to maintain an energy input into the system, particularly over the last two centuries, much larger than we would have been able to do with existing techniques if we had had to rely on the current input of available energy from the sun or the earth itself. This supplementary input, however, is by its very nature exhaustible.

The inputs and outputs of information are more subtle and harder to trace, but also represent an open system, related to, but not wholly dependent on, the transformations of matter and energy. By far the larger amount of information and knowledge is self-generated by the human society, though a certain amount of information comes into the sociosphere in the form of light from the universe outside. The information that comes from the universe has certainly affected man's image of himself and of his environment, as we can easily visualize if we suppose that we lived on a planet with a total cloud-cover that kept out all information from the exterior universe. It is only in very recent times, of course, that the information coming in from the universe has been captured and coded into the form of a complex image of what the universe is like outside the earth; but even in primitive times, man's perception of the heavenly bodies has always profoundly affected his image of earth and of himself. It is the information generated within the planet, however, and particularly that generated by man himself, which forms by far the larger part of the information system. We can think of the stock of knowledge, or as Teilhard de Chardin called it, the "noosphere," and consider this as an open system, losing knowledge through aging and death and gaining it through birth and education and the ordinary experience of life.

From the human point of view, knowledge or information is by far the most important of the three systems. Matter only acquires significance and only enters the sociosphere or the econosphere insofar as it becomes an object of human knowledge. We can think of capital, indeed, as frozen knowledge or knowledge imposed on the material world in the form of im-probable arrangements. A machine, for instance, originated in the mind of man, and both its construction and its use involve information processes im-posed on the material world by man himself. The cumulation of knowledge, that is, the excess of its production over its consumption, is the key to human development of all kinds, especially to economic development. We can see this preeminence of knowledge very clearly in the experiences of countries where the material capital has been destroyed by a war, as in Japan and Germany. The knowledge of the people was not destroyed, and it did not take long, therefore, certainly not more than ten years, for most of the material capital to be reestablished again. In a country such as Indonesia, however, where the knowledge did not exist, the material capital did not come into being either. By "knowledge" here I mean, of course, the whole cognitive structure, which includes valuations and motivations as well as images of the factual world.

The concept of entropy, used in a somewhat loose sense, can be applied to all three of these open systems. In the case of material systems, we can distinguish between entropic processes, which take concentrated materials and diffuse them through the oceans or over the earth's surface or into the atmosphere, and anti-entropic processes, which take diffuse materials and concentrate them. Material entropy can be taken as a measure of the uniformity of the distribution of elements and, more uncertainly, compounds and other structures on the earth's surface. There is, fortunately, no law of increasing material entropy, as there is in the corresponding case of energy, as it is quite possible to concentrate diffused materials if energy inputs are allowed. Thus the processes for fixation of nitrogen from the air, processes for the extraction of magnesium or other elements from the sea, and processes for the desalinization of sea water are anti-entropic in the material sense, though the reduction of material entropy has to be paid for by inputs of energy and also inputs of information, or at least a stock of information in the system. In regard to matter, therefore, a closed system is conceivable, that is, a system in which there is neither increase nor decrease in material entropy. In such a system all outputs from consumption would constantly be recycled to become inputs for production, as for instance, nitrogen in the nitrogen cycle of the natural ecosystem.

In regard to the energy system there is, unfortunately, no escape from the grim Second Law of Thermodynamics; and if there were no energy inputs into the earth, any evolutionary or developmental process would be impossible. The large energy inputs which we have obtained from fossil fuels are strictly temporary. Even the most optimistic predictions would expect the easily available supply of fossil fuels to be exhausted in a mere matter of centuries at present rates of use. If the rest of the world were to rise to American standards of power consumption, and still more if world population continues to increase, the exhaustion of fossil fuels would be even more rapid. The development of nuclear energy has improved this picture, but has not fundamentally altered it, at least in present technologies, for fissionable material is still relatively scarce. If we should achieve the economic use of energy through fusion, of course, a much larger source of energy materials would be available, which would expand the time horizons of supplementary energy input into an open social system by perhaps tens to hundreds of thousands of years. Failing this, however, the time is not very far distant, historically speaking, when man will once more have to retreat to his current energy input from the sun, even though this could be used much more effectively than in the past with increased knowledge. Up to now, certainly, we have not got very far with the technology of using current solar energy, but the possibility of substantial improvements in the future is certainly high. It may be, indeed, that the biological revolution which is just beginning will produce a solution to this problem, as we develop artificial organisms which are capable of much more efficient transformation of solar energy into easily available forms than any that we now have. As Richard Meier has suggested, we may run our machines in the future with methane-producing algae.[2]

The question of whether there is anything corresponding to entropy in the information system is a puzzling one, though of great interest. There are certainly many examples of social systems and cultures which have lost knowledge, especially in transition from one generation to the next, and in which the culture has therefore degenerated. One only has to look at the folk culture of Appalachian migrants to American cities to see a culture which started out as a fairly rich European folk culture in Elizabethan times and which seems to have lost both skills, adaptability, folk tales, songs, and almost everything that goes up to make richness and complexity in a culture, in the course of about ten generations. The American Indians on reservations provide another example of such degradation of the information and knowledge system. On the other hand, over a great part of human history, the growth of knowledge in the earth as a whole seems to have been almost continuous, even though there have been times of relatively slow growth and times of rapid growth. As it is knowledge of certain kinds that produces the growth of knowledge in general, we have here a very subtle and complicated system, and it is hard to put one's finger on the particular elements in a culture which make knowledge grow more or less rapidly, or even which make it decline. One of the great puzzles in this connection, for instance, is why the take-off into science, which represents an "acceleration," or an increase in the rate of growth of knowledge in European society in the sixteenth century, did not take place in China, which at that time (about 1600) was unquestionably ahead of Europe, and one would think even more ready for the breakthrough. This is perhaps the most crucial question in the theory of social development, yet we must confess that it is very little understood. Perhaps the most significant factor in this connection is the existence of "slack" in the culture, which permits a divergence from established patterns and activity which is not merely devoted to reproducing the existing society but is devoted to changing it. China was perhaps too well-organized and had too little slack in its society to produce the kind of acceleration which we find in the somewhat poorer and less well-organized but more diverse societies of Europe.

The closed earth of the future requires economic principles which are somewhat different from those of the open earth of the past. For the sake of picturesqueness, I am tempted to call the open economy the "cowboy economy," the cowboy being symbolic of the illimitable plains and also associated with reckless, exploitative, romantic, and violent behavior, which is characteristic of open societies. The closed economy of the future might similarly be

---

[2] Richard L. Meier, *Science and Economic Development* (New York: John Wiley and Sons, 1956).

called the "spaceman" economy, in which the earth has become a single spaceship, without unlimited reservoirs of anything, either for extraction or for pollution, and in which, therefore, man must find his place in a cyclical ecological system which is capable of continuous reproduction of material form even though it cannot escape having inputs of energy. The difference between the two types of economy becomes most apparent in the attitude toward consumption. In the cowboy economy, consumption is regarded as a good thing and production likewise; and the success of the economy is measured by the amount of the throughput from the "factors of production," a part of which, at any rate, is extracted from the reservoirs of raw materials and noneconomic objects, and another part of which is output into the reservoirs of pollution. If there are infinite reservoirs from which material can be obtained and into which effluvia can be deposited, then the throughput is at least a plausible measure of the success of the economy. The gross national product is a rough measure of this total throughput. It should be possible, however, to distinguish that part of the GNP which is derived from exhaustible and that which is derived from reproducible resources, as well as that part of consumption which represents effluvia and that which represents input into the productive system again. Nobody, as far as I know, has ever attempted to break down the GNP in this way, although it would be an interesting and extremely important exercise, which is unfortunately beyond the scope of this paper.

By contrast, in the spaceman economy, throughput is by no means a desideratum, and is indeed to be regarded as something to be minimized rather than maximized. The essential measure of the success of the economy is not production and consumption at all, but the nature, extent, quality, and complexity of the total capital stock, including in this the state of the human bodies and minds included in the system. In the spaceman economy, what we are primarily concerned with is stock maintenance, and any technological change which results in the maintenance of a given total stock with a lessened throughput (that is, less production and consumption) is clearly a gain. This idea that both production and consumption are bad things rather than good things is very strange to economists, who have been obsessed with the income-flow concepts to the exclusion, almost, of capital-stock concepts.

There are actually some very tricky and unsolved problems involved in the questions as to whether human welfare or well-being is to be regarded as a stock or a flow. Something of both these elements seems actually to be involved in it, and as far as I know there have been practically no studies directed toward identifying these two dimensions of human satisfaction. Is it, for instance, eating that is a good thing, or is it being well fed? Does economic welfare involve having nice clothes, fine houses, good equipment, and so on, or is it to be measured by the depreciation and the wearing out of these things? I am inclined myself to regard the stock concept as most fundamental, that is, to think of being well fed as more important than eating, and to think even of so-called services as essentially involving the restoration of a depleting psychic capital. Thus I have argued that we go to a concert in order to restore a psychic condition which might be called "just having gone to a concert," which, once established, tends to depreciate. When it depreciates beyond a certain point, we go to another concert in order to restore it. If it depreciates rapidly, we go to a lot of concerts; if it depreciates slowly, we go to few. On this view, similarly, we eat primarily to restore bodily homeostasis, that is, to maintain a condition of being well fed, and so on. On this view, there is nothing desirable in consumption at all. The less consumption we can maintain a given state with, the better off we are. If we had clothes that did not wear out, houses that did not depreciate, and even if we could maintain our bodily condition without eating, we would clearly be much better off.

It is this last consideration, perhaps, which makes one pause. Would we, for instance, really want an operation that would enable us to restore all our bodily tissues by intravenous feeding while we slept? Is there not, that is to say, a certain virtue in throughput itself, in activity itself, in production and consumption itself, in raising food and in eating it? It would certainly be rash to exclude this possibility. Further interesting problems are raised by the demand for variety. We certainly do not want a constant state to be maintained; we want fluctuations in the state. Otherwise there would be no demand for variety in food, for variety in scene, as in travel, for variety in social contact, and so on. The demand for variety can, of course, be costly, and sometimes it seems to be too costly to be tolerated or at least legitimated, as in the case of marital partners, where the maintenance of a homeostatic state in the family is usually regarded as much more desirable than the variety and excessive throughput of the libertine. There are problems here which the economics profession has neglected with astonishing singlemindedness. My own attempts to call attention to some of them, for instance, in two articles,[3] as far as I can judge, produced no response whatever; and economists continue to think and act as if production, consumption, throughput, and the GNP were the sufficient and adequate measure of economic success.

It may be said, of course, why worry about all this when the spaceman economy is still a good way off (at least beyond the lifetimes of any now living), so let us eat, drink, spend, extract and pollute, and be as merry as we can, and let posterity worry about the spaceship earth. It is always a little hard to find a convincing answer to the man who says, "What has posterity ever done for me?" and the conservationist has always had to fall back on rather vague ethical principles postulating identity of the individual with

[3] K. E. Boulding, "The Consumption Concept in Economic Theory," *American Economic Review*, 35:2 (May 1945), pp. 1–14; and "Income or Welfare?," *Review of Economic Studies*, 17 (1949–50), pp. 77–86.

some human community or society which extends not only back into the past but forward into the future. Unless the individual identifies with some community of this kind, conservation is obviously "irrational." Why should we not maximize the welfare of this generation at the cost of posterity? "Après nous, le déluge" has been the motto of not insignificant numbers of human societies. The only answer to this, as far as I can see, is to point out that the welfare of the individual depends on the extent to which he can identify himself with others, and that the most satisfactory individual identity is that which identifies not only with a community in space but also with a community extending over time from the past into the future. If this kind of identity is recognized as desirable, then posterity has a voice, even if it does not have a vote; and in a sense, if its voice can influence votes, it has votes too. This whole problem is linked up with the much larger one of the determinants of the morale, legitimacy, and "nerve" of a society, and there is a great deal of historical evidence to suggest that a society which loses its identity with posterity and which loses its positive image of the future loses also its capacity to deal with present problems, and soon falls apart.[4]

Even if we concede that posterity is relevant to our present problems, we still face the question of time-discounting and the closely related question of uncertainty-discounting. It is a well-known phenomenon that individuals discount the future, even in their own lives. The very existence of a positive rate of interest may be taken as at least strong supporting evidence of this hypothesis. If we discount our own future, it is certainly not unreasonable to discount posterity's future even more, even if we do give posterity a vote. If we discount this at 5 percent per annum, posterity's vote or dollar halves every fourteen years as we look into the future, and after even a mere hundred years it is pretty small—only about 1 ½ cents on the dollar. If we add another 5 percent for uncertainty, even the vote of our grandchildren reduces almost to insignificance. We can argue, of course, that the ethical thing to do is not to discount the future at all, that time-discounting is mainly the result of myopia and perspective, and hence is an illusion which the moral man should not tolerate. It is a very popular illusion, however, and one that must certainly be taken into consideration in the formulation of policies. It explains, perhaps, why conservationist policies almost have to be sold under some other excuse which seems more urgent, and why, indeed, necessities which are visualized as urgent, such as defense, always seem to hold priority over those which involve the future.

All these considerations add some credence to the point of view which says that we should not worry about the spaceman economy at all, and that we should just go on increasing the GNP and indeed the gross world product,

[4] Fred L. Polak, *The Image of the Future*, vols. I and II, translated by Elise Boulding (New York: Sythoff, Leyden and Oceana, 1961).

or GWP, in the expectation that the problems of the future can be left to the future, that when scarcities arise, whether this is of raw materials or of pollutable reservoirs, the needs of the then present will determine the solutions of the then present, and there is no use giving ourselves ulcers by worrying about problems that we really do not have to solve. There is even high ethical authority for this point of view in the New Testament, which advocates that we should take no thought for tomorrow and let the dead bury their dead. There has always been something rather refreshing in the view that we should live like the birds, and perhaps posterity is for the birds in more senses than one; so perhaps we should all call it a day and go out and pollute something cheerfully. As an old taker of thought for the morrow, however, I cannot quite accept this solution; and I would argue, furthermore, that tomorrow is not only very close, but in many respects it is already here. The shadow of the future spaceship, indeed, is already falling over our spendthrift merriment. Oddly enough, it seems to be in pollution rather than in exhaustion that the problem is first becoming salient. Los Angeles has run out of air, Lake Erie has become a cesspool, the oceans are getting full of lead and DDT, and the atmosphere may become man's major problem in another generation, at the rate at which we are filling it up with gunk. It is, of course, true that at least on a microscale, things have been worse at times in the past. The cities of today, with all their foul air and polluted waterways, are probably not as bad as the filthy cities of the pretechnical age. Nevertheless, that fouling of the nest which has been typical of man's activity in the past on a local scale now seems to be extending to the whole world society; and one certainly cannot view with equanimity the present rate of pollution of any of the natural reservoirs, whether the atmosphere, the lakes, or even the oceans.

I would argue strongly also that our obsession with production and consumption to the exclusion of the "state" aspects of human welfare distorts the process of technological change in a most undesirable way. We are all familiar, of course, with the wastes involved in planned obsolescence, in competitive advertising, and in poor quality of consumer goods. These problems may not be so important as the "view with alarm" school indicates, and indeed the evidence at many points is conflicting. New materials especially seem to edge toward the side of improved durability, such as, for instance, neolite soles for footwear, nylon socks, wash and wear shirts, and so on. The case of household equipment and automobiles is a little less clear. Housing and building construction generally almost certainly has declined in durability since the Middle Ages, but this decline also reflects a change in tastes toward flexibility and fashion and a need for novelty, so that it is not easy to assess. What is clear is that no serious attempt has been made to assess the impact over the whole of economic life of changes in durability, that is, in the ratio of capital in the widest possible sense to income. I suspect that we have underestimated, even in our spendthrift society, the gains from increased durability, and that this might very well be one of the places where the price system needs correction through government-sponsored research and development. The problems which the spaceship earth is going to present, therefore, are not all in the future by any means, and a strong case can be made for paying much more attention to them in the present than we now do.

It may be complained that the considerations I have been putting forth relate only to the very long run, and they do not much concern our immediate problems. There may be some justice in this criticism, and my main excuse is that other writers have dealt adequately with the more immediate problems of deterioration in the quality of the environment. It is true, for instance, that many of the immediate problems of pollution of the atmosphere or of bodies of water arise because of the failure of the price system, and many of them could be solved by corrective taxation. If people had to pay the losses due to the nuisances which they create, a good deal more resources would go into the prevention of nuisances. These arguments involving external economies and diseconomies are familiar to economists, and there is no need to recapitulate them. The law of torts is quite inadequate to provide for the correction of the price system which is required, simply because where damages are widespread and their incidence on any particular person is small, the ordinary remedies of the civil law are quite inadequate and inappropriate. There needs, therefore, to be special legislation to cover these cases, and though such legislation seems hard to get in practice, mainly because of the widespread and small personal incidence of the injuries, the technical problems involved are not insuperable. If we were to adopt in principle a law for tax penalties for social damages, with an apparatus for making assessments under it, a very large proportion of current pollution and deterioration of the environment would be prevented. There are tricky problems of equity involved, particularly where old established nuisances create a kind of "right by purchase" to perpetuate themselves, but these are problems again which a few rather arbitrary decisions can bring to some kind of solution.

The problems which I have been raising in this paper are of larger scale and perhaps much harder to solve than the more practical and immediate problems of the above paragraph. Our success in dealing with the larger problems, however, is not unrelated to the development of skill in the solution of the more immediate and perhaps less difficult problems. One can hope, therefore, that as a succession of mounting crises, especially in pollution, arouse public opinion and mobilize support for the solution of the immediate problems, a learning process will be set in motion which will eventually lead to an appreciation of and perhaps solutions for the larger ones. My neglect of the immediate problems, therefore, is in no way intended to deny their importance, for unless we at least make a beginning on a process for solving the immediate problems we will not have much chance of solving the larger ones. On the other hand, it may also be true that a long-run vision, as it were, of the deep crisis which faces mankind may predispose people to taking more interest in the immediate problems and to devote more effort for their solution. This may sound like a rather modest optimism, but perhaps a modest optimism is better than no optimism at all.

# AN AMERICAN LAND ETHIC

## N. Scott Momaday

*For most people, conservation is what is meant by a land ethic in America. When the condition becomes too serious for conservationists, we then argue politically on grounds of self-enlightenment or we encourage the major violators to see the environmental problems from an economic point of view. None of these solutions—conservation, self-enlightened care of our own property, or economic inducement—is in itself enough to repair what we have already destroyed. What is needed now is a land ethic, a code of behavior toward the land as a living biotic mechanism. We have to acknowledge that we have obligations to the land and that certain ethical rules must be initiated in order that we all adhere to those obligations. N. Scott Momaday, the author of* The Way to Rainy Mountain *(1969) and* House Made of Dawn *(1968), presently teaches English at the University of California, Berkeley.*

## I.

One night a strange thing happened. I had written the greater part of *The Way to Rainy Mountain*—all of it, in fact, except the epilogue. I had set down the last of the old Kiowa tales, and I had composed both the historical and the autobiographical commentaries for it. I had the sense of being out of breath, of having said what it was in me to say on that subject. The manuscript lay before me in the bright light, small, to be sure, but complete; or nearly so. I had written the second of the two poems in which that book is framed. I had uttered the last word, as it were. And yet a whole, penultimate piece was missing. I began once again to write.

SOURCE: N. Scott Momaday, "An American Land Ethic," *Ecotactics*. Copyright © 1970, by the Sierra Club. Reprinted by permission of Pocket Books/division of Simon & Schuster, Inc.

During the first hours after midnight on the morning of November 13, 1833, it seemed that the world was coming to an end. Suddenly the stillness of the night was broken; there were brilliant flashes of light in the sky, light of such intensity that people were awakened by it. With the speed and density of a driving rain, stars were falling in the universe. Some were brighter than Venus; one was said to be as large as the moon.

I went on to say that that event, the falling of the stars on North America, that explosion of Leonid meteors which occurred 137 years ago, is among the earliest entries in the Kiowa calendars. So deeply impressed upon the imagination of the Kiowas is that old phenomenon that it is remembered still; it has become a part of the racial memory.

"The living memory," I wrote, "and the verbal tradition which transcends it, were brought together for me once and for all in the person of Ko-sahn." It seemed eminently right for me to deal, after all, with that old woman. Ko-sahn is among the most venerable people I have ever known. She spoke and sang to me one summer afternoon in Oklahoma. It was like a dream. When I was born she was already old; she was a grown woman when my grandparents came into the world. She sat perfectly still, folded over on herself. It did not seem possible that so many years—a century of years—could be so compacted and distilled. Her voice shuddered, but it did not fail. Her songs were sad. An old whimsy, a delight in language and in remembrance, shone in her one good eye. She conjured up the past, imagining perfectly the long continuity of her being. She imagined the lovely young girl, wild and vital, she had been. She imagined the Sun Dance:

> There was an old, old woman. She had something on her back. The boys went out to see. The old woman had a bag full of earth on her back. It was a certain kind of sandy earth. That is what they must have in the lodge. The dancers must dance upon the sandy earth. The old woman held a digging tool in her hand. She turned toward the south and pointed with her lips. It was like a kiss, and she began to sing:
> We have brought the earth.
> Now it is time to play;
> As old as I am, I still have the feeling of play.
> That was the beginning of the Sun Dance.

By this time I was back into the book, caught up completely in the act of writing. I had projected myself—imagined myself—out of the room and out of time. I was there with Ko-sahn in the Oklahoma July. We laughed easily together; I felt that I had known her all of my life—all of hers. I did not want to let her go. But I had to come to the end. I set down, almost grudgingly, the last sentences:

It was—all of this and more—a quest, a going forth upon the way to Rainy Mountain. Probably Ko-sahn too is dead now. At times, in the quiet of evening, I think she must have wondered, dreaming, who she was. Was she become in her sleep that old purveyor of the sacred earth, perhaps, that ancient one who, old as she was, still had the feeling of play? And in her mind, at times, did she see the falling stars?

For some time I sat looking down at these words on the page, trying to deal with the emptiness that had come about inside of me. The words did not seem real. The longer I looked at them, the more unfamiliar they became. At last I could scarcely believe that they made sense, that they had anything whatsoever to do with meaning. In desperation almost, I went back over the final paragraphs, backwards and forwards, hurriedly. My eyes fell upon the name Ko-sahn. And all at once everything seemed suddenly to refer to that name. The name seemed to humanize the whole complexity of language. All at once, absolutely, I had the sense of the magic of words and of names. Ko-sahn, I said. And I said again KO-SAHN.

Then it was that that ancient, one-eyed woman Ko-sahn stepped out of the language and stood before me on the page. I was amazed, of course, and yet it seemed to me entirely appropriate that this should happen.

"Yes, grandson," she said. "What is it? What do you want?"

"I was just now writing about you," I replied, stammering. "I thought —forgive me—I thought that perhaps you were . . . that you had . . ."

"No," she said. And she cackled, I thought. And she went on. "You have imagined me well, and so I am. You have imagined that I dream, and so I do. I have seen the falling stars."

"But all of this, this *imagining*," I protested, "this has taken place—is taking place in my mind. You are not actually here, not here in this room." It occurred to me that I was being extremely rude, but I could not help myself. She seemed to understand.

"Be careful of your pronouncements, grandson," she answered. "You imagine that I am here in this room, do you not? That is worth something. You see, I have existence, whole being, in your imagination. It is but one kind of being, to be sure, but it is perhaps the best of all kinds. If I am not here in this room, grandson, then surely neither are you."

"I think I see what you mean," I said meekly. I felt justly rebuked. "Tell me, grandmother, how old are you?"

"I do not know," she replied. "There are times when I think that I am the oldest woman on earth. You know, the Kiowas came into the world through a hollow log. In my mind's eye I have seen them emerge, one by one, from the mouth of the log. I have seen them so clearly, how they were dressed, how delighted they were to see the world around them. I *must* have been there. And I must have taken part in that old migration of the Kiowas from the Yellowstone to the Southern Plains, for I have seen antelope bounding in the tall grass near the Big Horn River, and I have seen the ghost forests in the Black Hills. Once I saw the red cliffs of Palo Duro Canyon. I was with those who were camped in the Wichita Mountains when the stars fell."

"You are indeed very old," I said, "and you have seen many things."

"Yes, I imagine that I have," she replied. Then she turned slowly around, nodding once, and receded into the language I had made. And then I imagined I was alone in the room.

## II.

Once in his life a man ought to concentrate his mind upon the remembered earth, I believe. He ought to give himself up to a particular landscape in his experience, to look at it from as many angles as he can, to wonder about it, to dwell upon it. He ought to imagine that he touches it with his hands at every season and listens to the sounds that are made upon it. He ought to imagine the creatures there and all the faintest motions of the wind. He ought to recollect the glare of noon and all the colors of the dawn and dusk.

The Wichita Mountains rise out of the Southern Plains in a long crooked line that runs from east to west. The mountains are made of red earth, and of rock that is neither red nor blue but some very rare admixture of the two like the feathers of certain birds. The yellow, grassy knoll that is called Rainy Mountain lies a short distance to the north and west. There, on the west side, is the ruin of an old school where my grandmother went as a wild young girl in blanket and braids to learn of numbers and of names in English. And there she is buried.

## III.

I am interested in the way that a man looks at a given landscape and takes possession of it in his blood and brain. For this happens, I am certain, in the ordinary motion of life. None of us lives apart from the land entirely; such an isolation is unimaginable. We have sooner or later to come to terms with the world around us—and I mean especially the physical world, not only as it is revealed to us immediately through our senses, but also as it is perceived more truly in the long turn of seasons and of years. And we must come to moral terms. There is no alternative, I believe, if we are to realize and maintain our humanity, for our humanity must consist in part in the ethical as well as the practical ideal of preservation. And particularly here and now is that true. We Americans need now more than ever before—and indeed more than we know—to imagine who and what we are with respect to the earth and sky. I am talking about an act of the imagination essentially, and the concept of an American land ethic.

It is no doubt more difficult to imagine in 1970 the landscape of America than it was in, say, 1900. Our whole experience as a nation in this century has been a repudiation of the pastoral idea which informs so much of the art and literature of the nineteenth century. One effect of the Technological Revolution has been to uproot us from the soil. We have become disoriented, I believe; we have suffered a kind of psychic dislocation of ourselves in time and space. We may be perfectly sure of where we are in relation to the supermarket and the next coffee break, but I doubt that any of us knows where he is in relation to the stars and to the solstices. Our sense of the natural order has become dull and unreliable. Like the wilderness itself, our sphere of instinct has diminshed in proportion as we have failed to imagine truly what it is. And yet I believe that it is possible to formulate an ethical idea of the land—a notion of what it is and must be in our daily lives—and I believe moreover that it is absolutely necessary to do so.

It would seem on the surface of things that a land ethic is something that is alien to, or at least dormant in, most Americans. Most of us in general have developed an attitude of indifference toward the land. In terms of my own experience, it is difficult to see how such an attitude could ever have come about.

## IV.

Ko-sahn could remember where my grandmother was born. "It was just there," she said, pointing to a tree, and the tree was like a hundred others that grew up in the broad depression of the Washita River. I could see nothing to indicate that anyone had ever been there, spoken so much as a word, or touched the tips of his fingers to the tree. But in her memory Ko-sahn could see the child. I think she must have remembered my grandmother's voice, for she seemed for a long moment to listen and to hear. There was a still, heavy heat upon that place; I had the sense that ghosts were gathering there.

And in the racial memory, Ko-sahn had seen the falling stars. For her there was no distinction between the individual and the racial experience, even as there was none between the mythical and the historical. Both were realized for her in the one memory, and that was of the land. This landscape, in which she had lived for a hundred years, was the common denominator of everything that she knew and would ever know—and her knowledge was profound. Her roots ran deep into the earth, and from those depths she drew strength enough to hold still against all the forces of chance and disorder. And she drew therefrom the sustenance of meaning and of mystery as well. The falling stars were not for Ko-sahn an isolated or accidental phenomenon. She had a great personal investment in that awful commotion of light in the night sky. For it remained to be imagined. She must at last deal with it in words; she must appropriate it to her understanding of the whole universe. And, again, when she spoke of the Sun Dance, it was an essential expression of her relationship to the life of the earth and to the sun and moon.

In Ko-sahn and in her people we have always had the example of a deep, ethical regard for the land. We had better learn from it. Surely that ethic is merely latent in ourselves. It must now be activated, I believe. We Americans must come again to a moral comprehension of the earth and air. We must live according to the principle of a land ethic. The alternative is that we shall not live at all.

# from THE MYTH OF OBJECTIVE CONSCIOUSNESS

*Theodore Roszak*

*So much of our thinking is evaluated on the strength of its ability to deliver certain objectively derived truths, but, as the artists of our age have said repeatedly, truth if anything is relative. Theodore Roszak is fully conscious of this contradiction and suggests that we substitute the problem of living for the problem of knowing. His essay, "The Myth of Objective Consciousness," from which the following excerpt is taken, points out the necessity for us to respond to the "homely magic" of our environment and to consider every action one of creation. By now the notion of living creatively and finding the reward in the endeavor itself is an old notion. The Romantic poets and philosophers of the early nine-teenth century were well aware of it. Mr. Roszak's genius lies in his ability to say it once more with feeling. Perhaps, this time, we will listen.*

We have C. P. Snow to thank for the notion of the "two cultures." But Snow, the scientific propagandist, scarcely grasps the terrible pathos that divides these two cultures; nor for that matter do most of our social scientists and scientistic humanists. While the art and literature of our time tell us with ever more desperation that the disease from which our age is dying is that of alienation, the sciences, in their relentless pursuit of objectivity, raise aliena-tion to its apotheosis as our *only* means of achieving a valid relationship to reality. Objective consciousness *is* alienated life promoted to its most honorific status as the scientific method. Under its auspices we subordinate nature to our command only by estranging ourselves from more and more of what we experience, until the reality about which objectivity tells us so much finally becomes a universe of congealed alienation. It is totally within our intellectual and technical power . . . and it is a worthless possession. For "what does it profit a man that he should gain the whole world, but lose his soul?"

When, therefore, those of us who challenge the objective mode of consciousness are faced with the question "but is there any *other* way in which we can know the world?" I believe it is a mistake to seek an answer on a narrowly epistemological basis. Too often we will then find ourselves struggling to discover some alternative method to produce the same sort of knowledge we now derive from science. There is little else the word "knowledge" any longer means besides an accumulation of verifiable propositions. The only way we shall ever recapture the sort of knowledge Lao-tzu referred to in his dictum "those who know do not speak," is by subordinating the question "how shall we know?" to the more existentially vital question "how shall we live?"

To ask this question is to insist that the primary purpose of human existence is not to devise ways of piling up ever greater heaps of knowledge, but to discover ways to live from day to day that integrate the whole of our nature by way of yielding nobility of conduct, honest fellowship, and joy. And to achieve those ends, a man need perhaps "know" very little in the conventional, intellectual sense of the word. But what he does know and may only be able to express by eloquent silence, by the grace of his most commonplace daily gestures, will approach more closely to whatever reality is than the most dogged and disciplined intellectual endeavor. For if that elusive concept "reality" has any meaning, it must be that toward which the entire human being reaches out for satisfaction, and not simply some fact-and-theory-mongering fraction of the personality. What is important, therefore, is that our lives should be as *big* as possible, capable of embracing the vastness of those experiences which, though yielding no articulate, demonstrable propositions, nevertheless awake in us a sense of the world's majesty.

The existence of such experiences can hardly be denied without casting out of our lives the witness of those who have been in touch with such things as only music, drama, dance, the plastic arts, and rhapsodic utterance can express. How dare we set aside as a "nothing but," or a "merely," or a "just" the work of one artist, one poet, one visionary seer, without diminishing our nature? For these, as much as any scientist or technician, are our fellow human beings. And they cry out to us in song and story, in the demanding beauty of line, color, shape, and movement. We have their lives before us as testimony that men and women have lived—and lived magnificently—in communion with such things as the intellective consciousness can do no justice to. If their work could, after some fashion, be explained, or explained away, if it could be computerized—and there are those who see this as a sensible project —it would overlook the elemental fact that in the making of these glorious things, these images, these utterances, these gestures, there was a supreme joy, and that the achievement of that joy was the purpose of their work. In the making, the makers breathed an ecstatic air. The technical mind that by-passes the making in favor of the made has already missed the entire meaning of this thing we call "creativity."

When we challenge the finality of objective consciousness as a basis for culture, what is at issue is the size of man's life. We must insist that a culture which negates or subordinates or degrades visionary experience commits the sin of diminishing our existence. Which is precisely what happens when we insist that reality is limited to what objective consciousness can turn into the stuff of science and of technical manipulation. The fact and the dire cost of this diminishing is nothing that can be adequately proved by what I write here, for it is an experience which every man must find in his own life. He finds it as soon as he refuses to block, to screen out, to set aside, to discount the needs his own personality thrusts upon him in its fullness, often in its terrifying fullness. Then he sees that the task of life is to take this raw material of his total experience—its need for knowledge, for passion, for imaginative exuberance, for moral purity, for fellowship—and to shape it *all,* as laboriously and as cunningly as a sculptor shapes his stone, into a comprehensive style of life. It is not of supreme importance that a human being should be a good scientist, a good scholar, a good administrator, a good expert; it is not of supreme importance that he should be right, rational, knowledgeable, or even creatively productive of brilliantly finished objects as often as possible. Life is not what we are in our various professional capacities or in the practice of some special skill. What *is* of supreme importance is that each of us should become a person, a whole and integrated person in whom there is manifested a sense of the human variety genuinely experienced, a sense of having come to terms with a reality that is awesomely vast.

It is my own conviction that those who open themselves in this way and who allow what is Out-There to enter them and to shake them to their very foundations are not apt to finish by placing a particularly high value on scientific or technical progress. I believe they will finish by subordinating such pursuits to a distinctly marginal place in their lives, because they will realize that the objective mode of consciousness, useful as it is on occasion, cuts them off from too much that is valuable. They will therefore come to see the myth of objective consciousness as a poor mythology, one which diminishes life rather than expands it; and they will want to spend little of their time with it. That is only my hunch; I could be wrong.

But of this there can be no doubt: that in dealing with the reality our non-intellective powers grasp, *there are no experts.* The expansion of the personality is nothing that is achieved by special training, but by a naive openness to experience. Where and when the lightning will strike that unaccountably sets one's life on fire with imaginative aspirations is beyond prediction. Jakob Boehme found his moment when a stray beam of sunlight set a metal dinner dish flashing. Supposedly the Zen master Kensu achieved illumination upon biting into a shrimp he had just caught. Tolstoy was convinced that the moment came in the experience of self-sacrifice to one's fellows, no matter how inconsequential and obscure the act. The homely magic of such turning points waits for all of us and will find us if we let it. What befalls us then is an experience of the personality suddenly swelling beyond all that we had once thought to be "real," swelling to become a greater and nobler identity than we had previously believed possible. It is precisely this sense of the person we should look for in all those who purport to have something to teach us. We should ask: "Show us this person you have made of yourself. Let us see its full size. For how can we judge what you know, what you say, what you do, what you make, unless in the context of the whole person?" It is a matter of saying, perhaps, that truth ought not to be seen as the property of a proposition, but of the person.

This would mean that our appraisal of any course of personal or social action would not be determined simply by the degree to which the proposal before us squares with objectively demonstrable knowledge, but by the degree to which it enlarges our capacity to experience: to know ourselves and others more deeply, to feel more fully the awesomeness of our environment. This, in turn, means that we must be prepared to trust that the expanded personality becomes more beautiful, more creative, more humane than the search for objective correctness can make it. To take this attitude is, I think, far from eccentric. Is it not the attitude we feel spontaneously compelled to assume whenever we find ourselves in the presence of an authentically great soul? I, who do not share any of Tolstoy's religion or that of the prophets of Israel, and who do not believe that a single jot of Dante's or Blake's world view is "true" in any scientific sense, nevertheless realize that any carping I might do about the correctness of their convictions would be preposterously petty. Their words are the conduit of a power that one longs to share. One reads their words only with humility and remorse for having lived on a lesser scale than they, for having at any point foregone the opportunity to achieve the dimensions of their vision.

When a man has *seen* and has *spoken* as such men did, the criticisms of the objective consciousness fade into insignificance. What men of this kind invite us to do is to grow as great with experience as they have, and in so doing to find the nobility they have known. Compared with the visionary powers that moved in these souls, what is the value of all the minor exactitudes of all the experts on earth?

Were we prepared to accept the beauty of the fully illuminated personality as our standard of truth—or (if the word "truth" is too sacrosanctly the property of science) of ultimate meaningfulness—then we should have done with this idiocy of making fractional evaluations of men and of ourselves. We should stop hiding behind our various small-minded specializations and pretending that we have done all that is expected of us when we have flourished a tiny banner of expertise. We should be able to ask every man who desires to lead us that he step forward and show us what his talents have made of him as a whole person. And we should reject the small souls who know only how to be correct, and cleave to the great who know how to be wise.

# PASSAGE TO MORE THAN INDIA

## Gary Snyder

> It will be a revival, in higher form, of the liberty, equality, and fraternity of the ancient gentes.
>
> —Lewis Henry Morgan

*Gary Snyder is a poet-ecologist who realizes that man must regain the old associations with nature not only to combat the present environmental difficulties but also to live life as it should be lived. In this sense, what he has to say is not especially new. However, the practical procedures leading to the renaissance he imagines possible are obviously at this time against the American grain. In the following article, for example, he argues for the widespread use of drugs and a radical restructuring of the American family.*

## The Tribe

The celebrated human Be-In in San Francisco, January of 1967, was called "A Gathering of the Tribes." The two posters: one based on a photograph of a Shaivite sadhu with his long matted hair, ashes and beard; the other based on an old etching of a Plains Indian approaching a powwow on his horse—the carbine that had been cradled in his left arm replaced by a guitar. The Indians, and the Indian. The tribes were Berkeley, North Beach, Big Sur, Marin County, Los Angeles, and the host, Haight-Ashbury. Outriders were present from New York, London and Amsterdam. Out on the polo field

SOURCE: Gary Snyder, *Earth House Hold.* Copyright © 1968, 1969 by Gary Snyder. Reprinted by permission of New Directions Publishing Corporation. "Passage to More Than India" first published in *Evergreen Review.*

that day the splendidly clad ab/originals often fell into clusters, with children, a few even under banners. These were the clans.

Large old houses are rented communally by a group, occupied by couples and singles (or whatever combinations) and their children. In some cases, especially in the rock-and-roll business and with light-show groups, they are all working together on the same creative job. They might even be a legal corporation. Some are subsistence farmers out in the country, some are contractors and carpenters in small coast towns. One girl can stay home and look after all the children while the other girls hold jobs. They will all be cooking and eating together and they may well be brown-rice vegetarians. There might not be much alcohol or tobacco around the house, but there will certainly be a stash of marijuana and probably some LSD. If the group has been together for some time it may be known by some informal name, magical and natural. These households provide centers in the city and also out in the country for loners and rangers; gathering places for the scattered smaller hip families and havens for the questing adolescent children of the neighborhood. The clan sachems will sometimes gather to talk about larger issues—police or sheriff department harassments, busts, anti-Vietnam projects, dances and gatherings.

All this is known fact. The number of committed total tribesmen is not so great, but there is a large population of crypto-members who move through many walks of life undetected and only put on their beads and feathers for special occasions. Some are in the academies, others in the legal or psychiatric professions—very useful friends indeed. The number of people who use marijuana regularly and have experienced LSD is (considering it's all illegal) staggering. The impact of all this on the cultural and imaginative life of the nation—even the politics—is enormous.

And yet, there's nothing very new about it, in spite of young hippies just in from the suburbs for whom the "beat generation" is a kalpa away. For several centuries now Western Man has been ponderously preparing himself for a new look at the inner world and the spiritual realms. Even in the centers of nineteenth-century materialism there were dedicated seekers—some within Christianity, some in the arts, some within the occult circles. Witness William Butler Yeats. My own opinion is that we are now experiencing a surfacing (in a specifically "American" incarnation) of the Great Subculture which goes back as far perhaps as the late Paleolithic.

This subculture of illuminati has been a powerful undercurrent in all higher civilizations. In China it manifested as Taoism, not only Lao-tzu but the later Yellow Turban revolt and medieval Taoist secret societies; and the Zen Buddhists up till early Sung. Within Islam the Sufis; in India the various threads converged to produce Tantrism. In the West it has been represented largely by a string of heresies starting with the Gnostics, and on the folk level by "witchcraft."

Buddhist Tantrism, or Vajrayana as it's also known, is probably the finest and most modern statement of this ancient shamanistic-yogic-gnostic-socio-economic view: that mankind's mother is Nature and Nature should be tenderly respected; that man's life and destiny is growth and enlightenment in self-disciplined freedom; that the divine has been made flesh and that flesh is divine; that we not only should but do love one another. This view has been harshly suppressed in the past as threatening to both Church and State. Today, on the contrary, these values seem almost biologically essential to the survival of humanity.

## The Family

Lewis Henry Morgan (d. 1881) was a New York lawyer. He was asked by his club to reorganize it "after the pattern of the Iroquois confederacy." His research converted him into a defender of tribal rights and started him on his career as an amateur anthropologist. His major contribution was a broad theory of social evolution which is still useful. Morgan's *Ancient Society* inspired Engels to write *Origins of the Family, Private Property and the State* (1884, and still in print in both Russia and China), in which the relations between the rights of women, sexuality and the family, and attitudes toward property and power are tentatively explored. The pivot is the revolutionary implications of the custom of matrilineal descent, which Engels learned from Morgan; the Iroquois are matrilineal.

A schematic history of the family:

Hunters and gatherers—a loose monogamy within communal clans usually reckoning descent in the female line, i.e., matrilineal.

Early agriculturalists—a tendency toward group and polyandrous marriage, continued matrilineal descent and smaller-sized clans.

Pastoral nomads—a tendency toward stricter monogamy and patrilineal descent; but much premarital sexual freedom.

Iron-Age agriculturalists—property begins to accumulate and the family system changes to monogamy or polygyny with patrilineal descent. Concern with the legitimacy of heirs.

Civilization so far has implied a patriarchal, patrilineal family. Any other system allows too much creative sexual energy to be released into channels which are "unproductive." In the West, the clan, or gens, disappeared gradually, and social organization was ultimately replaced by political organization, within which separate male-oriented families compete: the modern state.

Engels' Marxian classic implies that the revolution cannot be completely achieved in merely political terms. Monogamy and patrilineal descent may well be great obstructions to the inner changes required for a people to truly live by "communism." Marxists after Engels let these questions lie. Russia and China today are among the world's staunchest supporters of monogamous, sexually turned-off families. Yet Engels' insights were not entirely ignored. The Anarcho-Syndicalists showed a sense for experimental social reorganization. American anarchists and the I.W.W. lived a kind of communalism, with some lovely stories handed down of free love—their slogan was more than just words: "Forming the new society within the shell of the old." San Francisco poets and gurus were attending meetings of the "Anarchist Circle"—old Italians and Finns—in the 1940's.

## The Redskins

In many American Indian cultures it is obligatory for every member to get out of the society, out of the human nexus, and "out of his head," at least once in his life. He returns from his solitary vision quest with a secret name, a protective animal spirit, a secret song. It is his "power." The culture honors the man who has visited other realms.

Peyote, the mushroom, morning-glory seeds and Jimson-weed are some of the best-known herbal aids used by Indian cultures to assist in the quest. Most tribes apparently achieved these results simply through yogic-type disciplines: including sweat-baths, hours of dancing, fasting and total isolation. After the decline of the apocalyptic fervor of Wovoka's Ghost Dance religion (a pan-Indian movement of the 1880's and 1890's which believed that if all the Indians would dance the Ghost Dance with their Ghost shirts on, the Buffalo would rise from the ground, trample the white men to death in their dreams, and all the dead game would return; America would be restored to the Indians), the peyote cult spread and established itself in most of the western American tribes. Although the peyote religion conflicts with preexisting tribal religions in a few cases (notably with the Pueblo), there is no doubt that the cult has been a positive force, helping the Indians maintain a reverence for their traditions and land through their period of greatest weakness—which is now over. European scholars were investigating peyote in the twenties. It is even rumored that Dr. Carl Jung was experimenting with peyote then. A small band of white peyote users emerged, and peyote was easily available in San Francisco by the late 1940's. In Europe some researchers on these alkaloid compounds were beginning to synthesize them. There is a karmic connection between the peyote cult of the Indians and the discovery of lysergic acid in Switzerland.

Peyote and acid have a curious way of tuning some people in to the local soil. The strains and stresses deep beneath one in the rock, the flow and fabric of wildlife around, the human history of Indians on this continent. Older powers become evident: west of the Rockies, the ancient creator-trickster, Coyote. Jaime de Angulo, a now-legendary departed Spanish shaman and anthropologist, was an authentic Coyote-medium. One of the most relevant poetry magazines is called *Coyote's Journal*. For many, the invisible presence of the Indian, and the heartbreaking beauty of America work without fasting or herbs. We make these contacts simply by walking the Sierra or Mohave, learning the old edibles, singing and watching.

## The Jewel in the Lotus

At the Congress of World Religions in Chicago in the 1890's, two of the most striking figures were Swami Vivekananda (Shri Ramakrishna's disciple) and Shaku Soyen, the Zen Master and Abbot of Engaku-ji, representing Japanese Rinzai Zen. Shaku Soyen's interpreter was a college student named Teitaro Suzuki. The Ramakrishna-Vivekananda line produced scores of books and established Vedanta centers all through the Western world. A small band of Zen monks under Shaku Sokatsu (disciple of Shaku Soyen) was raising strawberries in Hayward, California, in 1907. Shigetsu Sasaki, later to be known as the Zen Master Sokei-an, was roaming the timberlands of the Pacific Northwest just before World War I, and living on a Puget Sound Island with Indians for neighbors. D. T. Suzuki's books are to be found today in the libraries of biochemists and on stone ledges under laurel trees in the open-air camps of Big Sur gypsies.

A Californian named Walter Y. Evans-Wentz, who sensed that the mountains of his family's vast grazing lands really did have spirits in them, went to Oxford to study the Celtic belief in fairies and then to Sikkim to study Vajrayana under a lama. His best-known book is *The Tibetan Book of the Dead*.

Those who do not have the money or time to go to India or Japan, but who think a great deal about the wisdom traditions, have remarkable results when they take LSD. The *Bhagavad-Gita,* the Hindu mythologies, *The Serpent Power,* the *Lankavatara-sūtra,* the *Upanishads,* the *Hevajra-tantra,* the *Mahanirvana-tantra*—to name a few texts—become, they say, finally clear to them. They often feel they must radically reorganize their lives to harmonize with such insights.

In several American cities traditional meditation halls of both Rinzai and Soto Zen are flourishing. Many of the newcomers turned to traditional meditation after initial acid experience. The two types of experience seem to inform each other.

## The Heretics

> When Adam delved and Eve span,
> Who was then a gentleman?

The memories of a Golden Age—the Garden of Eden—the Age of the Yellow Ancestor—were genuine expressions of civilization and its discontents. Harking back to societies where women and men were more free with each other; where there was more singing and dancing; where there were no serfs and priests and kings.

Projected into future time in Christian culture, this dream of the Millennium became the soil of many heresies. It is a dream handed down right to our own time—of ecological balance, classless society, social and economic freedom. It is actually one of the possible futures open to us. To those who stubbornly argue "it's against human nature," we can only patiently reply that you must know your own nature before you can say this. Those who have gone into their own natures deeply have, for several thousand years now, been reporting that we have nothing to fear if we are willing to train ourselves, to open up, explore and grow.

One of the most significant medieval heresies was the Brotherhood of the Free Spirit, of which Hieronymus Bosch was probably a member. The Brotherhood believed that God was immanent in everything, and that once one had experienced this God-presence in himself he became a Free Spirit; he was again living in the Garden of Eden. The brothers and sisters held their meetings naked, and practiced much sharing. They "confounded clerics with the subtlety of their arguments." It was complained that "they have no uniform . . . sometimes they dress in a costly and dissolute fashion, sometimes most miserably, all according to time and place." The Free Spirits had communal houses in secret all through Germany and the Lowlands, and wandered freely among them. Their main supporters were the well-organized and affluent weavers.

When brought before the Inquisition they were not charged with witchcraft, but with believing that man was divine, and with making love too freely, with orgies. Thousands were burned. There are some who have as much hostility to the adepts of the subculture today. This may be caused not so much by the outlandish clothes and dope, as by the nutty insistence on "love." The West and Christian culture on one level deeply wants love to win—and having decided (after several sad tries) that love can't, people who still say it will are like ghosts from an old dream.

Love begins with the family and its network of erotic and responsible relationships. A slight alteration of family structure will project a different love-and-property outlook through a whole culture . . . thus the communism and free love of the Christian heresies. This is a real razor's edge. Shall the lion lie down with the lamb? And make love even? The Garden of Eden.

## White Indians

The modern American family is the smallest and most barren family that has ever existed. Each newly married couple moves to a new house or apartment—no uncles or grandmothers come to live with them. There are seldom more than two or three children. The children live with their peers and leave home early. Many have never had the least sense of family.

I remember sitting down to Christmas dinner eighteen years ago in a communal house in Portland, Oregon, with about twelve others my own age, all of whom had no place they wished to go home to. That house was my first discovery of harmony and community with fellow beings. This has been the experience of hundreds of thousands of men and women all over America since the end of World War II. Hence the talk about the growth of a "new society." But more; these gatherings have been people spending time with each other—talking, delving, making love. Because of the sheer amount of time "wasted" together (without TV) they know each other better than most Americans know their own family. Add to this the mind-opening and personality-revealing effects of grass and acid, and it becomes possible to predict the emergence of groups who live by mutual illumination—have seen themselves as of one mind and one flesh—the "single eye" of the heretical English Ranters; the meaning of sahajiya, "born together"—the name of the latest flower of the Tantric community tradition in Bengal.

Industrial society indeed appears to be finished. Many of us are, again, hunters and gatherers. Poets, musicians, nomadic engineers and scholars; fact-diggers, searchers and re-searchers scoring in rich foundation territory. Horse-traders in lore and magic. The super hunting-bands of mercenaries like Rand or CIA may in some ways belong to the future, if they can be transformed by the ecological conscience, or acid, to which they are very vulnerable. A few of us are literally hunters and gatherers, playfully studying the old techniques of acorn flour, seaweed-gathering, yucca-fiber, rabbit snaring and bow hunting. The densest Indian population in pre-Columbian America north of Mexico was in Marin, Sonoma and Napa Counties, California.

And finally, to go back to Morgan and Engels, sexual mores and the family are changing in the same direction. Rather than the "breakdown of the family" we should see this as the transition to a new form of family. In the near future, I think it likely that the freedom of women and the tribal spirit will make it possible for us to formalize our marriage relationships in any way we please—as groups, or polygynously or polyandrously, as well as monogamously. I use the word "formalize" only in the sense of make public and open the relationships, and to sacramentalize them; to see family as part of the divine ecology. Because it is simpler, more natural, and breaks up tendencies toward property accumulation by individual families, matrilineal descent seems ulitmately indicated. Such families already exist. Their children are different in personality structure and outlook from anybody in the history of Western culture since the destruction of Knossos.

The American Indian is the vengeful ghost lurking in the back of the troubled American mind. Which is why we lash out with such ferocity and passion, so muddied a heart, at the black-haired young peasants and soldiers who are the "Viet Cong." That ghost will claim the next generation as its own. When this has happened, citizens of the USA will at last begin to be Americans, truly at home on the continent, in love with their land. The chorus of a Cheyenne Indian Ghost dance song—"hi-niswa' vita'ki'ni"—"We shall live again."

> Passage to more than India!
> Are thy wings plumed indeed for such far flights?
> O soul, voyagest thou indeed on voyages like those?

# THE NATURALISTS

## Ian McHarg

*Ian McHarg is the chairman of the Department of Landscape Architecture and Planning at the University of Pennsylvania. His recently published* Design with Nature *(1969), from which the following chapter is taken, is a book that treats large land areas very much in the tradition of the eighteenth-century English landscape designer. It has at its core a vision of the world as a garden alive with the splendor of creative regeneration in which man, the gardener-steward, shares with nature the cooperative function of imaginative planning. What he suggests is a new aesthetic based on sound ecological principles, which will in time provide us with a design for living. McHarg's writing is extremely enthusiastic, but his prose never leads him into the realms of the unrealistic. Of* Design with Nature *Lewis Mumford wrote, "here are the foundations for a civilization that will replace the polluted, bulldozed, machine-dominated, dehumanized, explosion-threatened world that is even now disintegrating before our eyes."*

In an enterprise such as this quest [to find a solution to the environmental crisis], there is the ever-present temptation to resolve the problem by creating a Utopia wherein live all those admirable people whose views correspond with one's own. Yet, this must be avoided, because if Utopias vary greatly for one man, how much more for many men. There are occasions when a sprig of cherry blossom is utopian and other times when survival itself is the single yearning. But a more modest objective may be achieved—not the philosophy for a Utopia, but only the simplest, most basic views, which can ensure survival and life and which may produce a rational basis for human affairs. This will

SOURCE: From *Design with Nature* by Ian McHarg, copyright © 1969 by Ian McHarg. Reprinted by permission of Doubleday & Company, Inc.

not inhibit the great flights of courage or love, the unpredictable perceptions or creations. These can be left to their own devices, unaffected by the rut of men and simple rational laws. We can collect the evidence that has been presented fragmentarily and assemble it into some coherent sequence. But, rather than presenting this as a narrative description of natural law, it might be more palatable if invested in a people. They would look much like ourselves but would differ in their attitudes to nature and to man, their ethics and ethos, planning, management and art. These would be entirely based upon the natural sciences, ecology and the ecological view.

These Naturalists, for this is an appropriate name for them, have concluded that evolution has proceeded as much from cooperation as competition; conquest has no primacy in their lexicon, while the quest to understand nature, which is also to say man, dominates their preoccupations. This view, being the basis for the successful evolution of the species, pervades the entire population instead of reposing, as it does with us, in a small number of rather recent and retiring scientists and a few poets. It is, of course, accepted by the Naturalists that the earth and its denizens are involved in a totally creative process and that there is a unique and important role for man. It is agreed that evolution is directional, that it has recognizable attributes and that man is involved in its orderings.

Their cosmography is much different from ours—less encompassing, less certain, less romantic, more modest, and not at all man-centered. They disclaim all knowledge of the origins of the universe, although they seek to learn all that they can of this great genesis. Their knowledge begins with that time after the beginning when there was hydrogen. From this followed the evolution of the elements—helium, lithium, and the rest forged in cosmic cauldrons. This line of evolution terminated when the heaviest elements proved insubstantial and impermanent. The evolution of compounds followed, permitting combination after combination, increasing in complexity until, with the amino acids, evolution stood on the threshold of a new type of organization—life. Their understanding of the evolution of life forms corresponds closely with ours, although their sense of this journey is more immediate and vivid.

Every cosmography contains a creation story and the Naturalists are no different in this respect, save in the nature of the evidence employed. The unknown is the threshold of their minds; wonder is a common companion, but mysticism is not conspicuous. In support of their concept of creation they employ, not mysticism, but replicable experiment—indeed, an experiment which can be conducted by the least of them. This involves simply a glass cubicle enclosing a sterile environment. It is observed that the sunlight falls upon this and that the heat lost equals the heat falling upon the surface. In the companion experiment, a plant, some nutrients, and decomposers in a water medium are introduced into the cubicle. It is observed that the heat lost is less than that gained. Some of the sunlight is utilized by the plant which grows and proliferates. It is observed that the sunlight has been transmuted, with matter, from a lesser to a higher order. Some of the sunlight that otherwise would have been lost is now an ingredient of the plant. Some of the sun's energy had been entrapped on its path to entropy. This is defined as creation —the raising of matter from lower to higher order, negentropy.

Now this is perhaps the most modest creation myth ever advanced, but, as you consider it, it is seen to accommodate all physical and biological evolution. Is it as satisfactory in accounting for cultural evolution and for art? Is the symphony more ordered than random noise, the painting more ordered than the pigments in tubes and the waiting canvas, the poem a higher order than static? One must answer yes, but while this distinction is accurate, it is clearly not sufficient. Yet, this is a modest cosmography; it is enough that its claims are correct even though they are incomplete.

The conception of creation as movement from lower to higher order has its antithesis in destruction, the reduction from higher to lower levels. Evolution is then seen as a creative process, retrogression as reductive.

Creation and reduction, evolution and retrogression, are thought to have attributes. The replicable experiments demonstrating this involve two environments, both equal in area: the first, a sand dune, and the second, a primeval forest covering an ancient sand dune. In the first case, only a few decades have elapsed since the emergence of the dune from the sea; it is sparsely populated by some grasses and herbs; it supports some bacteria and insects, but no mammals. In contrast, the forest has existed undisturbed for millennia, so one could expect it to represent the highest evolutionary expression that the long time period and available denizens could support. The young dune is on the same path, but has not yet attained the creativity of the older example.

What are the attributes of these two systems, the first primitive in an evolutionary scale of which the other is the climax? The dune is simple, dominated by a few physical processes; it consists of a few physical constituents, mainly sand; it contains a few inhabitants and the relations between these can also be described as simple. When the forest is examined in these terms, it is seen to be inordinately complex. The physical processes that occurred, the numbers of species, the variety of habitats and niches (which is to say the roles which were performed) could only be encompassed within the term complex.

If you multiply simplicities, the result is uniformity; the product of complexities is diversity, and so it is found in examining the respective environments. The dune is the result of the uniform behavior of sand particles, their angle of repose and the action of wind; the conspicuous organisms are the grasses, bent to the wind, reflecting the sunlight, a constancy of uniformity. The forest is completely otherwise—uniformities are nowhere to be found. Although there is a structure of creatures occupying different trophic layers and different levels of stratification, the variation present is a permutation of the large numbers of species, environments, roles and pathways which are, indeed, multiplications of complexities.

The next attribute to be examined is relative instability and stability. The dune is, of course, unstable, subject to the vicissitudes of wind and ocean, tempered only by the anchoring vegetation. The forest has transformed the dune that was its origin; its own internal climate, microclimate and water regimen are all products of the evolution of the forest. The processes themselves are the basis of stability and the measure of this is not only the implacable, unmoving aspect which it portrays, but the age of its creatures.

For each of those environments, equal in area, the incident energy is the same. In the case of the dune, most of the sunlight that falls is reflected by the sand and only a small proportion is utilized by the few grasses. In the forest, the incident sunlight powers the entire ecosystem; the light reflected is from the leaves of the canopy; the variation in light down to the shadowed floor is utilized by existing creatures. Clearly, in the dune entropy is high, in the forest it is low. If we consider entropy as a measure of greater randomness, disorder and uniformity, then it is apparent that the dune better qualifies for this description than the forest. Indeed, the forest can be described by Lawrence K. Frank's term "organized complexity" while the dune is, in comparison, a less organized simplicity. If high entropy reveals low order, then the dune is low, the forest an expression of high order, of negentropy.

A further measure of creation is the number of species. It is a proposition that species survive only insofar as they can perform a role. Where two species perform identical roles in the same place and time, one will surely succumb. Therefore, the number of species present is an indication of the number of roles being performed. In the dune there are obviously few species; in the forest these are legion. In the dune, with few species, but relatively large populations, interactions will be preponderantly intraspecies. The forest with many species would exhibit interspecies, as well as intraspecies interaction. These relationships might be described, from the point of view of species interaction, as exhibiting independence in the case of the dune, and interdependence in the forest.

**Evolution** ———————————————————→

| primitive state | advanced state |
|---|---|
| simplicity | complexity |
| uniformity | diversity |
| instability | stability (steady state) |
| low number of species | high number of species |
| low number of symbioses | high number of symbioses |
| high entropy | low entropy |

**Retrogression** ←———————————————————

The cosmography has now linked creation with the increase in order in a system, and demonstrated that this is the path of evolution, that the antithesis is destruction, the path of retrogression consisting in the reduction of order from higher to lower levels. Both creation and destruction are seen to have distinctive, descriptive attributes.

Does this hold true outside, in the world at large? Apparently evolution has proceeded from simple to complex, whether we consider elements, compounds, life forms or communities. It seems clear that if you multiply simplicities, then uniformities will result; complexities similarly treated will produce diversity. Observe the difference between an algal bloom and a forest. It then follows that simple, uniform systems will tend to be unstable as a function of these characteristics. They are inordinately vulnerable to epidemic disease in that they provide large, uniform populations for any parasite. In contrast, complex and diverse systems are unlikely to provide large populations of single organisms which are so vulnerable. Moreover, the larger the number of species, the larger the genetic pool capable of adapting to any exigency. On all counts, the complex environment will be more stable. If it is true that simple and uniform systems by definition cannot occupy all available niches, then energy available to the system will not be as fully utilized as in the complex diverse system. Thus, entropy will be high, order low in the simple uniform system; high order and low entropy will characterize the complex diverse ecosystem. Complexity and diversity are describable in terms of numbers of species—therefore, the higher the order, the more the species; and finally, where the environment consists of a community of many species, the interactions are likely to be interspecies, whereas the alternative—large populations of few species—will emphasize intraspecies interaction.

The evidence is available to us whether we examine the regeneration of an abandoned field on its way to becoming a forest, or if we look at the healing scab of ailanthus, sumac and ragweed clothing the railroad embankment. It appears that creation, viewed in thermodynamic terms, does have attributes. This offers a considerable utility both for diagnosis and prescription. The Naturalists could conclude on the state of any system on an evolutionary scale and, moreover, could decide whether it was evolving or retrogressing.

The Naturalists employ both conceptions of fitness, that propounded by Henderson and that by Darwin. Thus the environment is fit for life, for the forms which had preexisted, those which do now exist and those of the future. In addition, the surviving organism or ecosystem is fit for the environment. The process of achieving a fitting between the organism and the environment is a continuous and dynamic one—physical processes are dynamic, but even more the presence of organisms composing environments, themselves changing, is the major component of change. Where fitness is reflected in equilibrium, this is a dynamic equilibrium. Evolution then consists of a tendency toward increasing fitness whereby the organism adapts the environment to make it more fitting and, through mutation and natural selection, adapts itself toward the same end. As the process of fitting exhibits the direction from simplicity to complexity, uniformity to complexity, instability to stability, low to high number of species, low to high number of symbioses and thus high to low entropy, it corresponds to the most basic creative processes in the earth. Fitting and the movement toward fitness were thus creative. The failure to accomplish a fitting, the misfit, is not creative. Processes whereby the system reverts from complexity to simplicity and so on are therefore entropic and destructive. There are two polar conditions, the first creative fitting and the other a destructive unfitting. The measure of fitness and fitting is evolutionary survival, success of the species or ecosystem, and, in the short run, health.

This conception is not modified in any essential way when man is considered nor even when sociocultural factors are introduced. There would be an environment fit for a man, and a man fit for the environment; the creative process requires that the environment be made more fit, that the man adapt the environment and himself. Tools of culture are fundamentally no different than mutation and natural selection although they can accomplish change at a much greater rate. The creative test is to accomplish a creative fitting. This involves identifying those environments intrinsically fit for an organism or process, identifying the organism, species or institution fit for the environment and inaugurating the process whereby the organism and the environment is adapted to accomplish a better fitting.

As the Naturalists deny themselves the luxury of mysticism and assume that all meaning and purpose can be inferred from the operation of the biophysical world, it is here that they have searched for an ethic. Being natural scientists, their mode of examination takes them into studies of the relations between creatures, in the expectation that those that are operative before the emergence of man might equally hold for the relations between men and nature, between men and men.

They have noted that no organism can exist independently. As each organism has adapted to certain foods, it will expel certain wastes. The product of such a situation would be the exploitation of all available foodstuffs and the creation of a sea of wastes. Remember that the amoeba recoils from its excrement. So there must be at least two organisms; one of these must be photosynthetic, the other would be, in such a minimum situation, a decomposer. Here, in this example of a theoretical situation, the plant, utilizing sunlight, would produce wastes—leaves and detritus—which would be consumed by the decomposers. Clearly, these two creatures are interdependent: they are related as to numbers; they are cooperating for survival. The decomposer has adapted to utilize the wastes of the plant; the plant to utilize the wastes of decomposers. This is described by the Naturalists as altruism— the concession of some autonomy toward the ends of mutual benefit for the creatures involved.

Now the principles affecting the organisms do not change when the numbers of species increase or energy pyramids enlarge or when the pathways become inordinately complicated. In every case, in these astonishingly complex relations in an elaborate ecosystem, all of the organisms must concede some part of their autonomy, which is to say their freedom, toward the end of sustaining the system and the other co-tenants of it. This corresponds very closely with the proposition of intercellular altruism which has been advanced for us by Dr. Hans Selye. He had noted that while a man consists of some thirty billion billion cells, the original cells are unspecialized and evolve to occupy specialist niches, as tissue, organs and blood. The organism only exists because these cells assume interdependent roles within the totality of a single integrated organism. When first formed, each unspecialized cell is similar to independent unicellular creatures, with an origin, a metabolism, and the capacity to replicate. Yet, each cell concedes some part of this freedom, inherent in nonspecialization, and assumes a cooperative role in the maintenance of the single organism.

Selye extrapolated from intercellular to interpersonal altruism, but for the Naturalists, the entire biosphere exhibits altruism. Every organism occupies a niche in an ecosystem and engages in cooperative arrangements with the other organisms sustaining the biosphere. In every case this involves a concession of some part of the individual freedom toward the survival and evolution of the biosphere.

Now in the consideration of altruism, it is important to reject sentimentality. While the wolf culls the old and weak caribou or the lion the antelope, there is no doubt about the fear of the prey and ferocity of the predator. Numbers are thus regulated and the fittest survive and reproduce, but it is not a picture of lions sleeping with lambs. Yet, if it does not fulfill dreams of idyllic nature, a world without competition, it nonetheless does provide an understanding of the relationships that demonstrably do exist in the living world. One might wish them different, but it is important to find out what they are.

Now energy in a system can just as well be considered as information. The heat that falls upon a creature can inform that creature of the heat falling upon it. But, the information provided has meaning only if the matter or organism can perceive and respond to it. The direction of evolution, or at least the Naturalist's conception of this, is toward higher order, more negentropy, but it is seen that if energy is reconsidered as information, then the capacity to attribute meaning to this energy is also a measure of evolution. If this is so, then apperception is that capacity by which meaning is perceived.

I am sure you have observed that there seem to be several concurrent value systems operating in this cosmography. The first of these is based upon negentropy, and can be measured in entropy units. Thus, creatures can be seen as the makers of negentropy. In this value scale it is clear that the plants are supreme, that behind them fall the indispensable decomposers and that all other life forms have relatively much lower values. It is also clear that the major work is being performed by the smallest creatures; the marine plants lead by a great margin, the terrestrial plants a poor second, and so, in the animal world, the major function of putting plants into more elaborate orders is accomplished by small marine organisms, the small and pervasive herbivores.

If we consider energy as information and use apperception as a value, then quite different creatures assume ascendency. The evolution of more complex perceiving creatures reflects this value, and here man ranks very high indeed.

If we examine the second criterion, that of cooperative mechanisms ensuring survival and directing the arrow of evolution, we confront a more difficult task. We can see in the lichen an early testimony to symbiosis, the alga and the fungus interfused into a single organism; we can identify the indispensable roles of the aminofying, nitrate and nitrite bacteria, of the pollinating insects and flowering plants, termites and cellulose bacteria and many other examples but in man, symbioses are more highly developed at the involuntary level—as in intercellular altruism—than in social organization. But, apperception is the key to symbiosis, and man is the most perceptive of creatures. This then is his potential: by perceiving and understanding nature, he can contribute to its operation, manage the biosphere, and in so doing, enhance his apperception, which with symbiosis, appears to be the arrow of evolution.

Now, the Naturalists believe less than we do in the divisibility of man from the rest of the biosphere; they think of man in nature rather than against nature. They have a vivid sense of the other creatures in the earth as being of themselves. They know that the beginnings were accomplished by the simplest creatures and that they had not been superseded by subsequent life forms, but only augmented. They know well that most of the world's work is still performed by these early forms. They were his ancestors, they were history, he had been there in times past, his past is still here. They know that their lineage is still in the sea and upon the land. They know of evolutionary successes in distant time that took their kin into the shallow bays and marshes, to the dry earth, elaborating as they colonized the land, reaching into more and more hostile environments, simplifying again as they reached these extreme environments, until only the most simple pioneers existed at the fringes of life in the arctic and antarctic, the summits of mountains and the oceanic depths. Every man could extend himself through this lineage out to its hostile limits and his. This is no metaphor; it is true and known to be true. So, the value system was not demeaning to ancient and simple forms; they were no more simple than his own unspecialized cells and as indispensable to the biosphere as the emergent simple cells of his marrow were to himself.

But the search for man's role was not to be found in a thermodynamic role—this other creatures could do much better; it was essential that he was not destructive in these terms. Apperception was surely the key to man's role, he was the uniquely perceptive and conscious animal, he who had developed language and symbols, and this was clearly his opportunity. What of his role? Surely it was as a cooperative mechanism sustaining the biosphere, and this was the great value of apperception, the key to man's role as steward, the agent of symbioses.

Believing, as natural scientists, that meaning could be found in earth, and its processes, both physical and biological, they continuously examined the phenomenal world for that evidence necessary to permit them to conduct that intelligent stewardship which they assume to be their responsibility. In examining all things over long periods of time, they have reached a startling conclusion. They observe that while creatures exhibited many similarities, sufficient to place them in discrete groups, minute examination discloses that neither two sand grains nor any two creatures are in fact identical. Some small reflection confirms that this might, indeed, be anticipated. Similarities increase as the species recedes into simple forms although identical pairs can never be found. As the creatures examined become more complex, the degree of their distinction increases, not only as a function of antecedents and over time (births not simultaneous), but also in the subsequent life experience. This study has led the Naturalists to conclude that every thing is its unique self, never having pre-existed, never to be succeeded by an identical form. That is, the matter of the creature is absolutely, not metaphorically, unique. It is the single pathway that is itself and will exist only once.

Thus, uniqueness is the basis of their attitude to all things and all life forms; it is upon this that deference and consideration is based. However, uniqueness has the unusual attribute of being singular, but also of being ubiquitous. It, therefore, concedes neither superiority nor inferiority, but simply uniqueness. How much better a claim this is than equality, which is insupportable in fact and a mere claim in comparison.

In this preoccupation with the development of an ethic, no subject has received more attention than that of freedom. The attribution of uniqueness is the basis for the individual's claim for consideration and deference; it is also the basis for his freedom. Clearly each individual has a responsibility for the entire biosphere and is required to engage in creative, cooperative activities. Freedom is thought to be inherent in uniqueness and in the infinite opportunities afforded by the environment, that is, modes of existence and expression are unlimited and the unique individual has these inherent opportunities. Anarchy is rejected because it replaces creation with randomness. Tyranny is rejected because it suppresses the uniqueness of the individual and his freedom. Poised between these two extremes is the concept of creation, linked to uniqueness, freedom, and the responsibility wherein the organism might perform any role that is creative and enhances the biosphere and the evolution of apperception and symbioses.

We know very well that the same evidence can be differently interpreted, and while the Naturalists are familiar with Darwin, they have chosen to emphasize the importance of cooperation, or rather altruism, in the evolution of the biosphere, rather than competition. That is, they see the elaboration of creatures as evidence of increase in creativity, as an increase in apperception, and, most important of all, as an increase in altruism.

The relations between predator and prey cause them no trouble. The creatures so related are mutually beneficial. The wolf culls the aged, infirm and unfit caribou, and, thus, serves their evolutionary development; the caribou feed the wolf—both regulate the numbers of both. This offers no difficulty in the cosmography; nor do parasite-host relationships. This was surely only an early point in the evolution of a mutually beneficial arrangement. It behooved the host to learn to derive benefit from the parasite which the latter had so clearly accomplished. The relationship would in time become mutually beneficial or when the hosts succumbed it would no longer persist.

A great importance is given to roles. As you might expect, their language reflects this. We once called men Weaver and Carpenter, Smith and Wheelwright, Thatcher and Farmer, Potter and Tailor, but the language of the Naturalists encompasses mountains and mosses as well as men. The sun is known as the first giver; mountains have many attributes, among them those which brought this or that from the ancient seas, the bringers of rain. The snowcaps and icesheets are known as those that hold water in reserve, the source of the cool winds. Rivers and streams are mainly known as those that bring water to us. The oceans are the second givers, home of ancient life; the chloroplast and the plant are the third givers, while the essential decomposers are the fourth-order givers, those that return all things.

All creatures are seen in terms of succession. The simplest creatures are known simply as the pioneers, those of the first wave who brought simple order to places of little order. The second wave followed the pioneers and raised the level of order. In this company are not only plants, but animals and the simplest men. Each successive group has assumed a role in the increase of order until the final group consists of the climaxes, those communities of creatures that represented the zenith capable of accomplishment by those beings that existed.

As we have seen, there is nothing pejorative in these descriptions; the conception of uniqueness and the sense of unity that embrace all of the biosphere allowed distinction to be made without allocating either superiority or inferiority. The general cell in the self is neither superior nor inferior to the specialized cell. Having said this, it is nevertheless true that the decomposers are especially regarded, for they alone can ensure the recycling in the system. Volcanoes and lightning are treasured, as are the sea birds that bring phosphorus back to the land, and the spawning fish that bring rich nutrients back from the sea to deposit them high in mountain streams before they die, bringing nourishment to the forests. The bacteria in the soil are seen as a great resource, and these were cultured in soils and considered to be among the highest accomplishments of all creation.

There are also the creatures of the special functions, the pollinating animals, those creatures that aerate the soils, the nitrogen bacteria, and then there are the indicators of successional stages or of retrogression. There are also the communities of the highest expression—those that express most vividly the glory of birth that is the spring, the glory of the working summer, the glory of death that is the autumn, and, of course, the glory of introspection and preparation that is the winter.

If one can view the biosphere as a single superorganism, then the Naturalist considers that man is an enzyme capable of its regulation, and conscious of it. He is of the system and entirely dependent upon it, but has the responsibility for management, derived from his apperception. This is his role—steward of the biosphere and its consciousness.

Now the Naturalists turn to the zoologists and physical anthropologists among them to reveal the nature of man. It is well understood that he has a common ancestry with the apes and that he is a raised ape rather than a fallen angel. They are convinced, as apparently we are not, that his evolutionary success results from exploitation of weapons and a capacity to kill: he was a successful predator. Observation of animals in the wild has convinced them that dominance is a reality, that rank orders are true for all creatures and thus for man. The defense of territory was observed in animals much simpler than man and holds true for him. Among all of the evidence used to discern the historic attitudes that are our traditional response to the fellow and the stranger, the most convincing is the response of the organism to a graft: no matter how beneficial such surgery might be, the body continuously rejects this foreign intrusion. This confirms the observation of the basic hostility of the organism to the unfamiliar. Altruism within the community is the rule; hostility to the stranger is as strongly instilled. As the community enlarges then so must the umbrella of altruism, but it is well to recognize the primitive origins of parochial hostility.

The Naturalists, of course, believe that man is natural, and therefore there are no divisions between the natural and the social sciences. Indeed, if there is a realm of knowledge concerned with the affairs of man it need be no less scientific than any other. They recognized that there are some special problems in dealing with man. While it is easy to relegate an alga or flatworm to species and abstraction, this detachment is more difficult with the evident personality of the individual human. Nonetheless some things are known. They believe that the foetus is influenced by anxiety experienced by the mother and so pay inordinate attention to the conditions attending her and the child she carries. They too have observed orphans and waifs revert to moronity without affection and have concluded that love and cherishing are indispensable to the growth of children. They know too that trauma from early experience is difficult, if not impossible, to eradicate and so they ensure that the conditions attending the life of the infant and young child are the most felicitous that can be arranged. They have observed the assuaging power of grief and seen in this a great capacity for healing. Their devotion to the conception of uniqueness among all things is employed in dealings with man. Thus the environment, both physical and social, must offer the maximum opportunity for the elaboration of each unique personality. Diversity is seen as an important component of this quest—the provision of the maximum number of opportunities and pathways. As reality consists only in the response to those stimuli impinging upon the individual, then the greater the number and diversity of these, the greater the choice. Sensory deprivation produces an impoverished environment and can induce hallucination. Diversity offers the maximum opportunity for the emergence of the unique individual. But it is important to distinguish noise from information.

Their hierarchy of requirements for man runs the gamut from survival to fulfillment. Beyond survival is mere existence, found simply in the satisfaction of physiological needs. The next level is identified by the presence of dignity; here existence is transcended. The last stage is fulfillment and is known to be unrealizable although it is the omnipresent quest and involves healthy men who not only solve problems but who seek them. In this evolution from survival to fulfillment is a corresponding hierarchy of symbioses. The least of these are the cooperative mechanisms necessary for survival, which ascend in number and complexity and reach their highest state in these symbioses which are altruistic and can be better described as love.

Cooperative relationships are as essential for survival as for fulfillment, but their nature has changed in this evolution from a mutuality of interest essential to survival to the transcendent form of love.

Now the attitudes the Naturalists bring to the roles of men are no different than those they bring to the remainder of the biosphere. Men are assumed to be as natural as other creatures—neither apperception nor consciousness suspend natural laws, but only reflect them. Every man, just as every creature, is required to be creative; destructiveness is intolerable. As there were pioneers among plants, so are there among men. The simplest societies are hunting and gathering communities, another type of predator in the forest or the seas, surviving in numbers related to their prey, cautioned to be neither depletive nor destructive, serving in the maintenance of the system while the other creatures performed the major works of creation—as the forest developed in complexity, the soils deepened and the community elaborated, or in the seas where the biota evolved, filling more and more ordering niches.

The next are societies of itinerant farmers. By burning and cultivating, they act as another decomposer and recycler. They perform a slightly creative role, but the major work is still being performed by the forest and the creatures within it. The fixed farmers assume a potentially more creative role but are required to elaborate the biota in order to compensate for the simplifications that monocultures produce. Nonetheless, successful agriculture involves the farmer in a creative role not accomplished in the simpler societies. The terracers are a special group of fixed farmers who arrest nutrients and soils on their path to the sea and in so doing accomplish valuable conservation.

If, as do the Naturalists, you attribute uniqueness to all things and all creatures, and, further, you agree that that which is being considered is that single pathway, itself, which can never be replicated, and can thus never recur, then you have assumed a position vis-à-vis the phenomenal world. This Naturalist view is much more encompassing than the reverence for some life that Schweitzer proposed; it does not end with those creatures having a utility to man, but encompasses all matter and all creatures. That which is, is justified by being; it is unique, it needs no other justification.

The consequence of these views is the ensurance that the Naturalists will not change preexisting conditions unless they can demonstrate that such changes are creative. Of course, they recognize that change is inevitable—change is accomplished by simply being. They also rightly assume that their existence permits them to claim that which is necessary to sustain them, but these claims are always subject to the necessity of showing an increase of negentropy or an increase in apperception of the system, resulting from such change.

Now the observed fact that life persists because life eats life is not seen as any contradiction of their propositions, nor, indeed, is death any problem to their cosmography. The operation of the biological world requires that the substances of living creatures and their wastes be consumed by other creatures in the creative process of the world. Man too subscribes to this, knowing that his wastes in life, and his substance after death, will be consumed by other creatures in a creative process. Death is seen in a like way, an indispensable part of a creative process. It is only when death is examined out of context that it appears as a reduction from higher to lower levels of order. As the basis for evolution, itself moving to higher orders, it is creative.

The Naturalists have turned to the world at large in order to find laws and forms of government that might work satisfactorily. They have observed that the world is an ordered place and infer that the creatures respond to physical and biological laws that are intrinsic and self-enforcing. Survival is contingent upon operation of "the way of things." This is the basis for the laws: does this or that correspond to "the way of things?" This way has no central authority, although it does have overweening laws; it has relative hierarchies but no absolute scales; the individual is the basic unit of law and of government, the overwhelming presumption is "in favor of the natural."[1] But then, there is no unnatural; there are the unknowns and those actions, which, while natural, do not correspond with "the way of things."

When one attempts to create people in whom repose such wisdom and rationality, there is a real danger that there is engendered, not admiration, but annoyance. How sanctimonious they appear to be. Yet that is simply because we have not looked closely enough to see their warts and squints, to see that many of them are ill-formed, bald and fat, that they reveal pettiness, bitterness and jealousy, superciliousness, and even stupidity for, of course, they are thoroughly human. Indeed, as we do look more closely, they appear to be much too human and contentious to be the appropriate repositories for such vital knowledge. The microbiologists are the aristocrats, frightfully superior yet largely ignorant of the visible world; the geologists know too little of the living; botanists too little zoology; and it is not only the taxonomists who are guilty of an ignorance of ecology. Moreover, as a group, their great sin is not that they are human, but that they have a certain professional myopia, they tend to be rather disinterested in human problems, and bring perhaps too clinical a view to art. One of the most serious of criticisms is that they are thoroughly irresolute in the absence of impeccable evidence, and this is a profound weakness in a world which is finally unknowable.

Yet, two things return them to our concern: they are committed to the acquisition of knowledge and in them is encapsulated a great realm of human understanding. In addition, they have in their company not only scientists but humanists who have espoused the ecological view.

[1] Clarence Morris, "The Rights and Duties of Beasts and Trees: A Law Teacher's Essay for Landscape Architects," *Journal of Legal Education*, vol. 17, 1964, pp. 185–192.

# from THE HERO WITH A THOUSAND FACES

## Joseph Campbell

*Joseph Campbell's* The Hero with a Thousand Faces *(1949) is a classic study of mythology as it influences literature. In it Campbell traces the various mythological patterns in past cultures and explains how they are an essential part of modern man's culture and psychological inheritance. The following excerpt suggests the role of the hero at the present time. There is, as Campbell says, "an inexhaustible and multifariously wonderful divine existence that is the life in all of us." Our task is to create once again the symbolic connection with that mysterious existence.*

The democratic ideal of the self-determining individual, the invention of the power-driven machine, and the development of the scientific method of research, have so transformed human life that the long-inherited, timeless universe of symbols has collapsed. In the fateful, epoch-announcing words of Nietzsche's Zarathustra: "Dead are all the gods."[1] One knows the tale; it has been told a thousand ways. It is the hero-cycle of the modern age, the wonder-story of mankind's coming to maturity. The spell of the past, the bondage of tradition, was shattered with sure and mighty strokes. The dream-web of myth fell away; the mind opened to full waking consciousness; and modern man emerged from the ancient ignorance, like a butterfly from its cocoon, or like the sun at dawn from the womb of mother night.

SOURCE: Joseph Campbell, *The Hero with a Thousand Faces*, Bollingen Series XVII, rev. ed. (Princeton, N.J.: Princeton University Press, 1968). Copyright © 1949 by Bollingen Foundation. Reprinted by permission of Princeton University Press.

[1] Nietzsche, *Thus Spake Zarathustra*, 1. 22. 3.

It is not only that there is no hiding place for the gods from the searching telescope and microscope; there is no such society any more as the gods once supported. The social unit is not a carrier of religious content, but an economic-political organization. Its ideals are not those of the hieratic pantomime, making visible on earth the forms of heaven, but of the secular state, in hard and unremitting competition for material supremacy and resources. Isolated societies, dream-bounded within a mythologically charged horizon, no longer exist except as areas to be exploited. And within the progressive societies themselves, every last vestige of the ancient human heritage of ritual, morality, and art is in full decay.

The problem of mankind today, therefore, is precisely the opposite to that of men in the comparatively stable periods of those great co-ordinating mythologies which now are known as lies. Then all meaning was in the group, in the great anonymous forms, none in the self-expressive individual; today no meaning is in the group—none in the world: all is in the individual. But there the meaning is absolutely unconscious. One does not know toward what one moves. One does not know by what one is propelled. The lines of communication between the conscious and the unconscious zones of the human psyche have all been cut, and we have been split in two.

The hero-deed to be wrought is not today what it was in the century of Galileo. Where then there was darkness, now there is light; but also, where light was, there now is darkness. The modern hero-deed must be that of questing to bring to light again the lost Atlantis of the co-ordinated soul.

Obviously, this work cannot be wrought by turning back, or away, from what has been accomplished by the modern revolution; for the problem is nothing if not that of rendering the modern world spiritually significant—or rather (phrasing the same principle the other way round) nothing if not that of making it possible for men and women to come to full human maturity through the conditions of contemporary life. Indeed, these conditions themselves are what have rendered the ancient formulae ineffective, misleading, and even pernicious. The community today is the planet, not the bounded nation; hence the patterns of projected aggression which formerly served to co-ordinate the in-group now can only break it into factions. The national idea, with the flag as totem, is today an aggrandizer of the nursery ego, not the annihilator of an infantile situation. Its parody-rituals of the parade ground serve the ends of Holdfast, the tyrant dragon, not the God in whom self-interest is annihilate. And the numerous saints of this anticult—namely the patriots whose ubiquitous photographs, draped with flags, serve as official icons—are precisely the local threshold guardians (our demon Sticky-hair) whom it is the first problem of the hero to surpass.

Nor can the great world religions, as at present understood, meet the requirement. For they have become associated with the causes of the factions, as instruments of propaganda and self-congratulation. (Even Buddhism has lately suffered this degradation, in reaction to the lessons of the West.) The universal triumph of the secular state has thrown all religious organizations into such a definitely secondary, and finally ineffectual, position that religious pantomime is hardly more today than a sanctimonious exercise for Sunday morning, whereas business ethics and patriotism stand for the remainder of the week. Such a monkey-holiness is not what the functioning world requires; rather, a transmutation of the whole social order is necessary, so that through every detail and act of secular life the vitalizing image of the universal god-man who is actually immament and effective in all of us may be somehow made known to consciousness.

And this is not a work that consciousness itself can achieve. Consciousness can no more invent, or even predict, an effective symbol than foretell or control tonight's dream. The whole thing is being worked out on another level, through what is bound to be a long and very frightening process, not only in the depths of every living psyche in the modern world, but also on those titanic battlefields into which the whole planet has lately been converted. We are watching the terrible clash of the Symplegades, through which the soul must pass—identified with neither side.

But there is one thing we may know, namely, that as the new symbols become visible, they will not be identical in the various parts of the globe; the circumstances of local life, race, and tradition must all be compounded in the effective forms. Therefore, it is necessary for men to understand, and be able to see, that through various symbols the same redemption is revealed. "Truth is one," we read in the Vedas; "the sages call it by many names." A single song is being inflected through all the colorations of the human choir. General propaganda for one or another of the local solutions, therefore, is superfluous—or much rather, a menace. The way to become human is to learn to recognize the lineaments of God in all of the wonderful modulations of the face of man.

With this we come to the final hint of what the specific orientation of the modern hero-task must be, and discover the real cause for the disintegration of all of our inherited religious formulae. The center of gravity, that is to say, of the realm of mystery and danger has definitely shifted. For the primitive hunting peoples of those remotest human millenniums when the sabertooth tiger, the mammoth, and the lesser presences of the animal kingdom were the primary manifestations of what was alien—the source at once of danger, and of sustenance—the great human problem was to become linked psychologically to the task of sharing the wilderness with these beings. An unconscious identification took place, and this was finally rendered conscious in the half-human, half-animal figures of the mythological totem-ancestors. The animals became the tutors of humanity. Through acts of literal imitation—such as to-day appear only on the children's playground (or in the madhouse)—an effective annihilation of the human ego was accomplished and society achieved a cohesive organization. Similarly, the tribes supporting themselves on plant-food became cathected to the plant; the life-rituals of planting and reaping were identified with those of human procreation, birth, and progress to maturity. Both the plant and the animal worlds, however, were in the end brought under social control. Whereupon the great field of instructive wonder shifted —to the skies—and mankind enacted the great pantomime of the sacred moon-king, the sacred sun-king, the hieratic, planetary state, and the symbolic festivals of the world-regulating spheres.

Today all of these mysteries have lost their force; their symbols no longer interest our psyche. The notion of a cosmic law, which all existence serves and to which man himself must bend, has long since passed through the preliminary mystical stages represented in the old astrology, and is now simply accepted in mechanical terms as a matter of course. The descent of the Occidental sciences from the heavens to the earth (from seventeenth-century astronomy to nine-teenth-century biology), and their concentration today, at last, on man himself (in twentieth-century anthropology and psychology), mark the path of a pro-digious transfer of the focal point of human wonder. Not the animal world, not the plant world, not the miracle of the spheres, but man himself is now the crucial mystery. Man is that alien presence with whom the forces of egoism must come to terms, through whom the ego is to be crucified and resurrected, and in whose image society is to be reformed. Man, understood however not as "I" but as "Thou": for the ideals and temporal institutions of no tribe, race, continent, social class, or century, can be the measure of the inexhaustible and multifariously wonderful divine existence that is the life in all of us.

The modern hero, the modern individual who dares to heed the call and seek the mansion of that presence with whom it is our whole destiny to be atoned, cannot, indeed must not, wait for his community to cast off its slough of pride, fear, rationalized avarice, and sanctified misunderstanding. "Live," Nietzsche says, "as though the day were here." It is not society that is to guide and save the creative hero, but precisely the reverse. And so every one of us shares the supreme ordeal—carries the cross of the redeemer—not in the bright moments of his tribe's great victories, but in the silences of his personal despair.